P9-EKS-698

DETERGENCY

THEORY AND TEST METHODS

SURFACTANT SCIENCE SERIES

CONSULTING EDITORS

Volume 1: NONIONIC SURFACTANTS, edited by Martin J. Schick

Volume 2: SOLVENT PROPERTIES OF SURFACTANT SOLUTIONS, edited by Kozo Shinoda

Volume 3: SURFACTANT BIODEGRADATION, by Robert D. Swisher

Volume 4: CATIONIC SURFACTANTS, edited by Eric Jungermann

Volume 5: DETERGENCY: THEORY AND TEST METHODS, edited by W. G. Cutler and R. C. Davis

DETERGENCY

THEORY AND TEST METHODS

(in two parts)

PART I

edited by W. G. CUTLER and R. C. DAVIS

Whirlpool Corporation
Research and Engineering Center
Monte Road
Benton Harbor, Michigan

MARCEL DEKKER, INC. New York 1972

MARCEL DEKKER, INC.
95 Madison Avenue, New York, New York 10016

LIBRARY OF CONGRESS CATALOG CARD NUMBER 79-163921

ISBN: 0-8247-1113-0

PRINTED IN THE UNITED STATES OF AMERICA

PREFACE

The detergent process is very widely used today. Despite its acceptance and wide usage, it is a process not completely defined and understood. There is no general agreement as to test methods for measuring the effectiveness of the detergent process. The present book is intended to provide both industrial and academic research workers with a compilation of test methods in use today and background information as to the theoretical basis of these tests. Perhaps such a volume will stimulate new interest in further investigation in the field of detergency.

The timing of this book is considered appropriate when viewed in the light of changes in detergent formulations that appear inevitable at this point in time. The entire concept of detergency and the relation of detergents to our environment are being scrutinized in many laboratories now.

The author list has been chosen to provide representatives from detergent producers, those specialized in applications of detergents, and universities. There has been no attempt to conform to a group presentation or render a group opinion. Each author has been left free, within the constraint of a general outline, to present and interpret data from his viewpoint and experience.

For the convenience of the reader and to expedite publication schedules, this volume of the Surfactant Science Series is being presented in two parts. Part I treats the fundamentals of the soil removal process and methods for its evaluation. Part II consists of topics related to a more complete understanding of the detergency process, particularly treating several topics of current interest. The indexes to the entire volume appear at the end of Part II. These indexes permit the reader to examine contributions to the theory of detergency in the scientific literature and serve to identify those researchers who have made contributions to this theory.

A number of people have contributed to this book. The editors wish to thank the contributing authors and their Companies. The support of the Whirlpool Corporation in permitting the editors to undertake the task of compiling this volume is also acknowledged. Special thanks are also due to Dr. Martin Schick for his many suggestions and for critical review of the book outline; to Richard Matthias, Whirlpool's electron microscopist, for the cover illustration and other electron microscopy work for the book; to Mr. Harvey Leland for the computerized approach to the indexes, and to Mrs. Carol Hauch for the secretarial assistance so necessary to this undertaking.

Benton Harbor, Michigan

W. G. Cutler
R. C. Davis

CONTRIBUTORS TO PART I

J. J. CRAMER, BASF Wyandotte Corporation, Chemical Specialties Division, Wyandotte, Michigan

W. G. CUTLER, Whirlpool Corporation, Benton Harbor, Michigan

RICHARD C. DAVIS, Whirlpool Corporation, Benton Harbor, Michigan

H. LANGE, Henkel and Cie GmbH, Duesseldorf, West Germany

OSCAR W. NEIDITCH, Lever Brothers Company, Edgewater, New Jersey

WILLIAM C. POWE, Whirlpool Corporation, Benton Harbor, Michigan

HANS SCHOTT, School of Pharmacy, Temple University, Philadelphia, Pennsylvania

B. A. SHORT*, Whirlpool Corporation, Benton Harbor, Michigan

W. G. SPANGLER, Colgate-Palmolive Co., Piscataway, New Jersey

O. VIERTEL, Waeschereiforschung Krefeld, Krefeld, West Germany

*Present address: Whirlpool Corporation, St. Paul, Minnesota.

CONTENTS OF PART II

CONTENTS OF PART I

Chapter 1

INTRODUCTION

W. G. Cutler
Whirlpool Corporation
Benton Harbor, Michigan

I. DETERGENT PROCESS: DEFINITION AND SCOPE

A simple definition of the detergent process—or detergency—is: the removal of soil (matter out of place) from a substrate immersed in some medium, generally through the application of a mechanical force, in the presence of a chemical substance which may lower the adhesion of the soil to the substrate. The process is completed when the soil is maintained in suspension so it can be rinsed away.

The detergent process has been known in some form or other since ancient times (1). It might be assumed, therefore, that it has been studied well and is completely understood and defined. Such, however, is not the case. When it is considered that the medium in which cleaning can take place may be aqueous, nonaqueous, or mixed; that many types of forces can be applied; and that the surfaces and soils involved are very numerous, then the real complexity of the detergent process is apparent, and it is perhaps not so surprising that there is much knowledge which is fundamental to an understanding of the detergent process that still needs to be gathered.

Although the process of detergency may not be thoroughly and completely understood, it is a process that is very widely used today. If we consider only the domestic cleaning of wearing apparel, this application alone serves to indicate the magnitude of the use of the detergency process. It is estimated (2) that in the United States today some 50-52 million clothes washers are in use in homes; and that about 90% of American homemakers own washing machines. Daily, these machines apply the detergency process to millions of pounds of soiled clothing. When we also consider the commercial laundry, the dry cleaner, domestic and commercial cleaning of dishes, hard surface cleaning, the cleaning of metals, etc., then we realize that the detergency process is a very important process in our daily lives.

II. TEST METHODS

Test methods to determine the effectiveness of detergents have long been sought. Considering the complexity of the detergency process, it is not unusual that there is a lack of general accord on recognized testing methods. Many investigators have decided that the only acceptable test is a practical one conducted under the conditions encountered in actual end-use application. There are, however, controlled laboratory tests that have value if only for purposes of screening samples for later practical tests. Harris (3) has collected and published both kinds of tests.

A number of organizations have attempted to standardize the test methods used in evaluating the detergency process. These organizations include the American Society for Testing and Materials (4, 5), the American Association of Textile Chemists and Colorists (6), the American Home Laundry Manufacturers Association (7) (now a part of the Association of Home Appliance Manufacturers), the Association of Home Appliance Manufacturers (8) and many others. They have not met with complete success, but they have stimulated further research into the detergency process.

III. BRIEF DESCRIPTION OF THIS WORK

The original intent of this book was to provide an updated version of the Harris (3) volume. However, it is the opinion of the editors that another compilation of test methods without an examination of the theory on which they are based is relatively meaningless. Therefore, it has been our intent in compiling this work to provide general background information about detergents and soils and a considerable discussion of the current status of knowledge of the detergency (soil removal) process.

When one begins an attempt to describe the detergency process it is difficult to limit the subject matter brought into the description. This difficulty has been encountered in compiling this book. It is the feeling of the editors that an understanding of detergency necessitates examining such topics as fluorescent whitening, bleaching, stain removal, action of enzymes, fabric abrasion, and others. The toxicological and dermatological effects of detergents are also important. The importance of the above topics has broadened the scope of this book.

It is the intent of the editors to provide in this book, then, a compilation of frequently used tests related to detergency and a state-of-the-art report on the theory of detergency. It is the editors' hope that such a volume will inspire additional research efforts toward a complete understanding of the detergency process and the generation of a series of definitive tests for evaluation of the effectiveness of this process.

REFERENCES

1. F. Bertrich, Kulturgeschicte des Waschens, Econ-Verlag GmbH., Duesseldorf, 1966.
2. R. C. Davis, Textile Chemist and Colorist, 1, 524 (1969).
3. J. C. Harris, Detergency Evaluation and Testing, Wiley-Interscience, New York, 1954.
4. ASTM Standards, Part 22, American Society for Testing and Materials, Philadelphia, 1968.
5. ASTM Standards, Part 24, American Society for Testing and Materials, Philadelphia, 1968.
6. Technical Manual and Yearbook, American Association of Textile Chemists and Colorists, Howes Publ. Co., New York, 1968.
7. American Home Laundry Manufacturers Association, private communication, 1966.
8. Household Washer Performance Evaluation Procedure, No. HLW1, Association of Home Appliance Manufacturers, Chicago, 1970.

Chapter 2

DEFINITION OF TERMS

Oscar W. Neiditch
Lever Brothers Company
Edgewater, New Jersey

I. INTRODUCTION

Although this book is primarily concerned with the test methods for surfactants and detergents, and although the various test methods are discussed in detail by the authors of the following chapters, an introduction to the subject of surfactants and detergents may be helpful to some readers. The purpose of this chapter is to provide, in brief form, a definition of some of the principal terms for surfactant studies, detergency studies, and test methods and to provide typical detergent formulations and discuss the purpose of their components.

II. DEFINITION OF TERMS FOR SURFACTANT STUDIES

A. Wetting, Wetting Agents, Rewetting

The ease with which a surface can be wetted by water or by other liquids is an important property from many considerations, including detergency, waterproofing, oilproofing, dyeing, dispersion of pigments, and water adsorption. Wetting agents are classified practically according to their ability and speed in displacing air from solid surfaces. A requirement of a wetting agent is evidently to reduce both the surface tension of the liquid and its interfacial tension against the solid.

B. Surface Tension

Surface tension is the work required to extend a surface by unit area. It is the consequence of the unsymmetrical force field acting on molecules in the surface which results in a net inward attraction on them perpendicular to the surface. Surface tension, a free energy, is expressed in ergs per square centimeter or in the mathematically equivalent unit of dynes per centimeter.

C. Interfacial Tension

Interfacial tension is the equivalent of surface tension at the boundary between two phases. It is the work required to extend the interface by unit area and has the same units as surface tension.

D. Contact Angle

Contact angle in a solid-liquid-gas system is the angle between the liquid-gas surface and the liquid-solid interface at the point where the three phases come together. It is a measure of the degree to which a solid body is wetted by a liquid.

E. Micelles, Critical Micelle Concentration

Micelles are aggregates of surfactant molecules. They form in surfactant solutions at or above a rather narrow range of concentration called the critical micelle concentration (cmc). Since the concentration of singly dispersed surfactant molecules is virtually constant above the cmc, the surfactant is at essentially its optimum of activity for many applications: e.g., wetting.

F. Zeta Potential

Zeta potential, as determined by electrokinetic techniques, is the electrical potential at the plane of shear near a surface or around a particle. It is frequently employed as an indicator of the electrical force of attraction or repulsion between particles or particles and a surface.

G. Peptizing, Peptizing Agent

Peptizing or peptization involves the processes of dispersing and suspending colloidal material or the formation of a stable sol from an agglomerated colloidal sol.

H. Solubilization, Solubilizing Agent

Solubilization in an aqueous system is the spontaneous dissolving of a normally insoluble substance by a relatively dilute aqueous solution of a surfactant. Solubilization differs from hydrotropy (see below) in that much lower amounts of solubilizing agents will suffice for effectiveness. Solubilization differs from emulsification in that the solubilized material is incorporated directly into (or around) the surfactant micelles so that, unless the solubilizate is colored, etc., the system resembles the surfactant itself and is thermodynamically stable.

I. Cosolvency, Cosolvent

Cosolvency is a phenomenon exhibited by a mixture of two solvents which can dissolve more of a given solute than either solvent alone can dissolve. Alternatively, cosolvency is exhibited by a mixture of solutes which is more soluble in a given solvent than either of the solutes.

J. Hydrotropy, Hydrotropes, Hydrotropic Agents

Hydrotropy is a solubilization effect in water brought about by agents which are not generally surface active and do not have to form micelles to exert their action. The agent is generally employed at a relatively high concentration.

K. Emulsification, Emulsifying Agent

Emulsification is the suspension of liquid particles within a second immiscible liquid phase. The particle size of the dispersed phase may vary from about 0.5 μ in diameter to particles visible to the naked eye.

III. DEFINITION OF TERMS FOR DETERGENCY STUDIES

A. Surfactant, Surface Active Agent, Active

Surfactant is a coined word used to describe all classes of surface active agents including wetting agents, emulsifying agents, and detergents. It is the basic component or "active" of a formulated detergent and functions by lowering surface and interfacial tension, thus making wetting easier. Other roles, such as, increasing negative zeta potentials of soil and fabric, converting the soil surface from hydrophobic to hydrophylic character, and "rolling up" and dispersing of oil into stabilized, suspended droplets to prevent soil redeposition, have also been assigned to the "active." The surfactants used in laundry detergents are divided chemically into (a) anionics, (b) nonionics, (c) ampholytics, and (d) zwitterionics. Anionic surfactants are characterized by ionizing in solution into two parts. The larger ion, which is the negatively charged portion or anion, is the active cleaning agent. The most commonly used anionics are alkylarylsulfonates and fatty alcohol sulfates. Nonionic surfactants do not ionize in solution. The whole molecule acts as the active cleaning agent. The most commonly used nonionics are ethylene oxide condensates of alkylphenols and fatty alcohols. Ampholytic surfactants have both positive and negative centers and can assume either a positive or negative charge depending on the pH of the solution. Examples of this class are sodium-3-dodecylaminoprorionate and sodium-3-dodecylaminopropanesulfonate. Zwitterionic surfactants also have positive and negative centers which are internally balanced, and like nonionic surfactants, the whole molecule acts as the active cleaning agent. An example of this class is 3-(N, N-dimethyl-N-hexadecylammonio) propane-1-sulfonate.

B. Detergency

Detergency refers to the process of cleaning the surfaces of a solid material by means of a liquid bath involving a physicochemical action other than simple solution. Generally it is considered to be an unusually enhanced cleaning effect of a liquid bath caused by the presence of a special agent, the detergent (1).

C. Detergent, Built Detergent

Detergent is a term usually used to describe a formulated product and most often a built detergent containing in addition to a surfactant such materials as builders, suds-controlling agents, soil-suspending or anti-redeposition agents, anticorrosion agents, fabric whiteners, blueing agents, bleaches, fabric softeners, antibacterial agents, enzymes, perfumes, and others.

D. Builders

Builders are certain components in a formulated detergent which enhance the measured cleaning effectiveness of the surfactant. Mechanistically, they perform a number of functions, the principal one being the removal of metal ions from the solution where they interfere with the action of the surfactant. Calcium and magnesium—the hardness metals—are most common among the metal ions found in water. They are removed by builders either in a soluble form by sequestration or in an insoluble form by precipitation. Among other functions performed by builders is maintenance of alkalinity in the wash solution. They also neutralize fatty acid components in soil forming soaps *in situ*. In addition, they increase the negative zeta potentials of soil and fabric and thus aid in preventing soil from redepositing during the washing process (2). Included in the list of builders are pentasodium tripolyphosphate, tetrasodium and tetrapotassium pyrophosphate, trisodium orthophosphate, nitrilotriacetic acid (NTA) sodium salt, ethylendiaminetetraacetic acid (EDTA) sodium salt, sodium carbonate, borax, silicates, citrates, organic polymers and polyelectrolytes, oxidized starches, and sodium oxydiacetate (ODA).

E. Foam or Suds Controllers

Foam or suds controllers are agents added to either boost or stabilize the sudsing properties of a normal sudsing detergent or to depress the suds level in a controlled or low-sudsing detergent. Examples of stabilizers or boosters are condensed amines, alkylolamides, fatty alcohols, and amine oxides. Soap is the most commonly used suds depressant but hydrocarbons and ethylene oxide or propylene oxide condensates are also used for this purpose.

F. Soil-Suspending or Antiredeposition Agents

Soil-suspending or antiredeposition agents are added to synthetic detergent formulations to help the surfactant hold soil particles in suspension and to prevent them from settling back on fabrics. Sodium carboxymethyl cellulose and polyvinyl alcohol are most often used for this purpose.

G. Fluorescent Whitening Agents, Optical Brighteners

Fluorescent whitening agents are substances added to detergents to make white fabrics appear whiter and some colors appear brighter. These agents are adsorbed on fabrics where they act by converting invisible ultraviolet light to visible blue light. The radiated blue or blue-violet light is complementary to the natural yellowness of fabric which can develop over a period of time. The result is a neutral or bluish-white shade.

H. Anticorrosion Agents and Tarnish Inhibitors

Anticorrosion agents such as sodium silicate are added to detergent formulations to protect washing-machine parts against corrosion but they also can function as auxiliary builders. Tarnish inhibitors are added to prevent tarnishing of cooking and eating utensils.

I. Dirt, Soil

Dirt or soil is classically defined as matter out of place. On fabrics, soils generally consist of fat (or oil) and finely divided solids or one of these components alone. Oily soil comes from human sebum and contact with fats and oils in the environment and also from soap residues from the use of towels for personal washing purposes. Solid or particulate soil can include carbonaceous material (soot), clays, metal oxides, as well as water-soluble materials.

J. Stains

Stains, in housewives' terms, are usually caused by substances (often colored) which are distinguished from soiling derived from normal use of the particular household article or item of clothing, and are usually localized.

In general terms, a stain is any undesirable matter present on a substrate which has an adverse effect on the appearance of the substrate. Stains can be divided broadly into two types: (1) stains such as rust, ink, and paint, which are not removed by normal or even abnormal wash treatment and require special chemical or solvent treatment; and (2) stains largely removed by appropriate wash treatment.

IV. DEFINITION OF TERMS FOR TEST METHODS

A. Wetting Tests

The tests most commonly employed to determine the wetting power of surfactants are the Draves-Clarkson skein test (3, 4) and the canvas disk method (5).

B. Rewetting Tests

The property of rewetting of fabrics which may be of importance in processing or in general for absorbency may be measured by a test developed by Shapiro (6) or by the drop absorption test (7).

C. Sudsing, Foaming, or Lathering Tests

The measurement of the properties of ease of foam generation, foam volume, and foam stability is conducted by numerous methods. The most widely used method is that of Ross and Miles (8).

D. Surface Tension Measurement

The measurement of surface tension is conducted by numerous methods. The most generally used methods employ the ring (9) and the Wilhelmy plate. A comparative study of the reliability of different methods of surface tension determination was reported by Boucher et al. (10).

E. Interfacial Tension Measurement

The measurement of interfacial tension may be conducted by several methods but is often measured by the ring method (9) used for surface tension.

F. Contact Angle Measurement

The wetting of a solid by a liquid may be measured by several techniques involving measurement of the angle of contact (11).

G. Critical Micelle Concentration Measurement (cmc)

The critical micelle concentration has been determined by measurement of conductivity, density, refractive index, pH, X-ray diffraction, surface tension, interfacial tension, bubble pressure, polarographic changes, freezing point and other "colligative" methods, light scattering, dye solubilization, spectral dye changes, equilibrium dialysis, and others. These measurements reflect a change in solution properties, which occurs at a definite concentration (12).

H. Detergency and Soil Redeposition Evaluation

Test methods for the evaluation of detergency and soil redeposition on fabrics and hard surfaces are reviewed in a manual by Harris (13), the books of Schwartz et al. (14, 15), and an article by Wolfrom (16).

TABLE 2.1

Classification of Domestic-Type Cleaners[a]

Physical form	Type of cleaner	Foaming property
1. Bars for toilet use	Soap-Synthetic	—
2. Bars for toilet use	Synthetic	—
3. Tablets for laundry use	—	High foaming
4. Tablets for laundry use	—	Low foaming
5. Powders and beads	Light duty	High foam
6. Powders and beads	Light duty	Low foam
7. Powders and beads	Heavy duty	High foam
8. Powders and beads	Heavy duty	Low foam
9. Liquids	Light duty	High foaming, clear
10. Liquids	Light duty	High foaming, opaque or creamy
11. Liquids	Light duty	Low foam, clear
12. Liquids	Light duty	Low foam, opaque, creamy
13. Liquids	Heavy duty	High foaming, clear
14. Liquids	Heavy duty	High foaming, opaque or creamy
15. Liquids	Heavy duty	Low foaming, clear
16. Liquids	Heavy duty	Low foaming, opaque, or creamy
17. Paste	Heavy duty, detergent	High foam
18. Paste	Heavy duty, detergent	Low foam
19. Abrasive cleaners	Powders	—
20. Abrasive cleaners	Liquids	—
21. Misc. type cleaners	Domestic, industrial	—
22. Cosmetic detergent material	—	—

[a]From (20) by courtesy of John W. McCutcheon, Inc.

I. Fluorescent Whitener Evaluation

Discussions of the factors involved in the evaluation of fluorescent whiteners were presented by Villaume (17), Zussman (18), and Stensby (19).

V. TYPICAL DETERGENT FORMULATIONS

A classification of domestic-type cleaners was presented by McCutcheon (20). This classification, shown in Table 2.1, divides products by physical form, as light and heavy duty, and by high and low foaming properties.

Examples of typical detergent formulations are abundant in the patent literature. Many of the examples are representative of formulations which were landmarks in the detergent industry. Examples of several types of detergents from the patent literature are tabulated below.

Light-Duty Liquid Detergent[a]

Components	Weight percent
Sodium dodecylbenzene sulfonate	17.4
Ammonium dodecylphenoxy-hexaoxyethylene sulfate	11.9
Sodium xylenesulfonate	7.0
Lauric diethanolamide	6.0
Polystyrene latex	0.8
Ethanol	2.8
Water	54.1
	100.0

[a]From (21).

Heavy-Duty Liquid Detergent[a]

Components	Weight percent
Potassium dodecylbenzene sulfonate	10.0
Tetrapotassium pyrophosphate	19.1

Sodium xylenesulfonate (commercial)	8.15
Lauric diethanolamide	3.8
Lauric isoproponolamide	3.2
Sodium silicate (37%) (Na_2O:SiO_2 of 1:2.5)	7.0
Optical brighteners	0.079
Water plus KOH to pH 12.1	48.171
Sodium carboxymethyl cellulose	0.04
Methyl cellulose	0.46
	100.00

[a]From (22).

Heavy-Duty Liquid Detergent[a]

Components	Weight percent
Dodecylphenol·10 mole ethylene oxide condensate	7.0
Polyoxyalkylene alkanol	2.0
Potassium pyrophosphate	25.0
Vinyl methyl ether and maleic anhydride interpolymer	2.0
Sodium xylenesulfonate	2.0
Oleic acid	2.0
Tallow fatty acids	1.0
Potassium hydroxide	2.28
Water and miscellaneous	56.72
	100.00

[a]From (23).

Heavy-Duty Powder Detergent[a]

Components	Weight percent
Sodium linear dodecylbenzene sulfonate	13.3
Sodium nitrilotriacetate	9.5
Sodium tripolyphosphate	41.2
Sodium silicate (SiO_2:Na_2O of 1.6:1)	10.0
Sodium sulfate	11.0
Sodium soap of hardened fish oil fatty acids	0.5
Sodium soap of hardened tallow fatty acids	1.5
Sodium carboxymethyl cellulose	0.33
Phosphonated random octadecene	0.1
Optical brightener	0.5
Moisture	Balance

[a]From (24).

Heavy-Duty Detergent Powder[a]

Components		Weight percent
Condensation product of one part of tall oil with 1.6 parts of ethylene oxide		15.0
Sodium tripolyphosphate		20.0
Tetrasodium pyrophosphate		20.0
Soda ash		20.0
Sodium silicate solution (38% solids) Na_2O:SiO_2 = 1:3.2		24.0
Sodium carboxymethyl cellulose		1.0
		100.00
Plus potassium tallate	up to	1.0

[a]From (25).

Heavy-Duty Detergent Tablet[a]

Components	Weight percent
Sodium tripolyphosphate	33.0
Sodium alkylbenzene sulfonate	22.0
Sodium sulfate	19.1
Sodium silicate (SiO_2:Na_2O of 1.6:1)	11.0
Sodium carboxymethyl cellulose	0.65
Optical brightener	0.2
Monoethanolamide of coconut oil fatty acids	3.0
Sodium chloride	7.25
Moisture	4.0

[a]From (26).

McCutcheon (20) described a number of formulas which are used as industrial specialty cleaning products. These products, used for general cleaning of hotels, office buildings, and institutions, are generally hard surface cleaners for floors, walls, tables, glass, toilet bowls, machine or manual dishwashing equipment, etc. Also included is a rug shampoo.

The specialty formulas which follow are reprinted from (20, pp. 28-31, 105) by courtesy of John W. McCutcheon, Inc.

Some Specialty Formulas

Some typical formulas covering industrial specialty cleaning products are given below to illustrate the variety and use of various inorganic salts in their preparation.

1. Electrolytic Metal Cleaner

Sodium metasilicate, anhyd.	42.7%
Soda ash	11.7%
LAS sodium salt	5.0%
NaOH	40.6%
	100.0%

2. Bottle Soak

NaOH	68%
Soda ash	14%
Trisodium phosphate	4%
Tetrasodium pyrophosphate	8%
Sodium metasilicate	6%
	100%

3. Car Wash Detergent

Soda ash	31.8%
Sodium metasilicate	34.4%
Tetrasodium pyrophosphate	27.6%
Dodecylbenzene sodium sulfonate	6.2%
	100.0%

4. Floor Cleanser

Sodium metasilicate	59.2%
Trisodium phosphate	10.5%
Tetrasodium pyrophosphate	27.9%
Sodium alkylaryl sulfonate	2.4%
	100.0%

5. Dairy Cleaner

Soda ash	7.2%
Sodium metasilicate, anhyd.	8.0%
Trisodium phosphate	45.5%
Sodium tripolyphosphate	15.4%
NaCl	18.4%
Surfactant	5.5%
	100.0%

6. Metal Cleaner

Trisodium polyphosphate	10%
Dodecylbenzene sodium sulfonate	2%
Hydroxyacetic acid	88%
	100%

7. Hard Surface Cleaner

Trisodium phosphate	80%
Trisodium polyphosphate	15%
Dodecylbenzene sodium sulfonate	5%
	100%

8. Hard Surface Cleaner for Floors

Sodium metasilicate	73.3%
Sodium tripolyphosphate	26.2%
Wetting agent	0.5%
	100.0%

9. Heavy-Duty Hard Surface Cleaner

Soda ash	26.5%
Sodium metasilicate, anhyd.	13.0%
Trisodium phosphate	34.0%
Tetrasodium pyrophosphate	14.5%
Sodium tripolyphosphate	9.5%
Dodecylbenzene sodium sulfonate	2.5%
	100.0%

10. Heavy-Duty Cleaner

Trisodium phosphate	3.2%
Disodium phosphate	38.1%
Soap: anhydrous tallow	58.7%
	100.0%

11. Vat Cleaner

Soda ash	6%
Trisodium phosphate, cryst.	30%
Metasilicate, cryst.	38%
Sodium tripolyphosphate, anhyd.	20%
Sodium bicarbonate	6%
	100%

12. Penetrant

Soda ash	32%
Sodium metasilicate, cryst.	52%
Pine oil	16%
	100%

13. Floor Cleaner

Soda ash	36%
Sodium metasilicate, anhyd.	52%
Tall oil soap	12%
	100%

14. Stripper

NaOH	68%
Soda ash	15%
Sodium metasilicate, cryst.	10%
Tall oil soap	7%
	100%

15. Hard Surface Cleaner-Sanitizer

Sodium tripolyphosphate	70%
Quaternary ammonium salt	1%
Sodium sulfate	19%
Carboxymethyl cellulose	3%
Water	7%
	100%

16. Abrasive-Sanitizer Cleaner

Trisodium phosphate, cryst.	20%
Tetrasodium pyrophosphate, cryst.	25%
Quaternary ammonium chloride	3%
Borax	15%
Bentonite	37%
	100%

17. Acid Cleaner (Bowl)

Phosphoric acid	31.25%
Nonionic surfactant	0.30%
Water	68.45%
	100.00%

18. Porcelain Cleaner

Sodium acid phosphate	91%
Pine oil	5%
Sodium sulfate	4%
	100%

19. Abrasive Acid Cleaner

Phosphoric acid	20%
Monosodium dihydrogen phosphate	4%
Nonionic surfactant	1%
Volcanic ash	65%
Water	10%
	100%

20. Acid Cleaner

Phosphoric acid	26%
Wetting agent (LAS)	2%
Water	72%
	100%

21. Hard Surface Bleach Cleaner

Chlorinated trisodium phosphate	25%
Sodium tripolyphosphate, anhyd.	75%
	100%

22. Alkali Stripper

Caustic soda	85%
Trisodium phosphate, anhyd.	5%
Tetrasodium pyrophosphate, anhyd.	10%
	100%

23. Bottle Washer

NaOH	95%
Sodium gluconate	5%
	100%

24. Machine Parts Cleaner

Soda ash	15%
Tetrasodium pyrophosphate, anhyd.	51%
Sodium metasilicate, cryst.	26%
Surfactant (LAS)	1%
Water	7%
	100%

25. Milk Can Washing Compound (Machine)

Sodium tripolyphosphate	40%
Sodium metasilicate	12%
Surfactant: Triton X-114	2%
Soda ash	20%
Sodium sulfate	26%
	100%

26. Acid Cleaner—Dairy for Milk Stone Removal

Phosphoric acid 85%	50%
Surfactant: Triton X-114	10%
Water	40%
	100%

27. Milk Can Washing

Dodecylbenzene sodium sulfonate, 40% active	10%
Triton X-100	4%
Sodium tripolyphosphate	25%
Sodium metasilicate	10%
Sodium sulfate	51%
	100%

28. Pipe Line Cleaner

Surfactant: Pluronic, Triton, etc.	3%
Sodium tripolyphosphate	25%
Sodium metasilicate	10%
Soda ash	30%
Sodium sulfate	32%
	100%

29. Steam Cleaner (for use with Steam Jenny)

Sodium hydroxide	25%
Sodium metasilicate	32%
Soda ash	30%
Sodium tripolyphosphate	10%
Dodecylbenzene sodium sulfonate	3%
	100%

30. Dairy Equipment Cleaner (Manual)

Dodecylbenzene sodium sulfonate (LAS), 40% active	10%
Nonionic: Triton X-100	4%

Sodium tripolyphosphate	25%
Sodium metasilicate	10%
Filler: sodium sulfate and/or borax	51%
	100%

31. Powdered Bleach Cleaner

Halane	12%
LAS sodium salt	2%
Disodium phosphate	6%
Monosodium phosphate	48%
Sodium sulfate	32%
	100%

32. Soak Cleaner for Machine Parts

Kerosene	5 parts
Triton X-100	3 parts
Oleic acid	3 parts

Mix kerosene and Triton, then add coupling agent oleic acid. Above concentrate may be diluted with 2-3 parts of additional kerosene.

33. Liquid Cleaner for Vehicle Use

Triton X-100	10%
Dodecylbenzene sulfonate, 85% active	10%
Water	80%
	100%

34. Dishwashing Detergent, Machine Use

Triton X-114	2%
Sodium tripolyphosphate	30%
Sodium metasilicate, cryst.	45%
Borax	10%
Soda ash	13%
	100%

35. Dishwashing Detergent, Hand Use

Triton X-100	4%
Dodecylbenzene sodium sulfonate, 40% active	10%
Sodium tripolyphosphate	20%
Sodium metasilicate, cryst.	5%
Sodium sesquicarbonate	30%
Sodium sulfate	31%
	100%

36. Concrete Floor Cleaner

Nonionic (Triton X-100)	3%
Sodium metasilicate, cryst.	97%
	100%

37. Light-Duty Concrete Floor Cleaner

Pluronic L-60	3%
Sodium metasilicate, cryst.	42%
Sodium sulfate	55%
	100%

38. General Purpose Cleaner

Dodecylbenzene sodium sulfonate (LAS), 40%	15%
Sodium tripolyphosphate	20%
Sodium metasilicate	5%
Sodium sesquicarbonate	25%
Sodium sulfate	35%
	100%

39. Heavy-Duty Alkali Cleaner

Nonionic surfactant	6%
Sodium metasilicate, cryst.	30%
Sodium tripolyphosphate	40%
Sodium bicarbonate	24%
	100%

40. Rug Shampoo (U.S. Patent 3,240,713). Example II—without dye

Maple wood flour	26.0
Water	40.5
Hydrocarbon solvent (naphtha)	20.5
Trichloroethylene	11.2
Coconut diethanolamide (emulsifier)	0.88
Trisodium phosphate (detergent)	0.21
Tetrapotassium pyrophosphate (detergent)	0.38
Hydrogen peroxide (bleach)	0.35

41. Metal Cleaner (U.S. Patent 3,242,093). A preferred formulation comprises a four (4) gram tablet consisting of the following ingredients, said ingredients being specified in parts by weight.

	Parts
Diammonium citrate	2
Sodium bicarbonate	7
Citric acid	7
Surfactant (Pluronic F-68)	1

In the practice of this invention a four (4) gram tablet of the composition listed above is dissolved at room temperature in from about 120 to 240 cm^3 of water and the rusted or corroded object is immersed in the water solution.

42. Hard Surface Cleaners (Brit. Patent 1,045, 691)

Composition	1	2	3	4	5
$Na_3PO_4 \cdot 12H_2O$	20	20	20	20	20
$Na_5P_3O_{10}$	20	20	20	20	20
$Na_3H(CO_3)_2 \cdot 2H_2O$	53.6	52.7	51.7	48.9	52.7
NH_4Cl	4.1	—	—	—	—
$(NH_4)_2SO_4$	—	5.0	—	—	—
$(NH_4)HCO_3$	—	—	6.0	—	—

42. Hard Surface Cleaners (Cont'd)

Composition	1	2	3	4	5
$(NH_4)H_2PO_4$	–	–	–	8.8	–
$(NH_4)_2HPO_4$	–	–	–	–	5.0
Detergent granules	2.3	2.3	2.3	2.3	2.3

VI. CLASSIFICATION AND PROPERTIES OF THE COMPONENTS OF DETERGENT FORMULATIONS

A chemical classification of the materials used in detergent and emulsifier products has been made by McCutcheon (27). Following is a list of these chemical categories:

(1) alkanolamides; (2) alkyl sulfonates; (3) sulfated and sulfonated amines and amides; (4) betaine derivations; (5) diphenyl sulfonate derivatives; (6) ethoxylated alcohols; (7) ethoxylated alkylphenols; (8) ethoxylated amines, amides; (9) ethoxylated fatty acids; (10) ethoxylated fatty esters and oils; (11) fatty esters, other than glycol, glycerol, etc.; (12) fluorocarbons; (13) glycerol esters; (14) glycol esters; (15) imidazoline-type products; (16) isethionates; (17) lanolin-based derivatives; (18) lecithin and lecithin derivatives; (19) lignin and lignin derivatives; (20) monoglycerides and derivatives; (21) phosphate derivatives; (22) polyaminocarboxylic acids and related sequestering agents; (23) protein derivatives; (24) quaternary germicide-type products; (24A) quaternary derivatives (special purpose); (25) salicylanilide, brominated; (26) sarcosine derivatives; (27) silicone and siloxane derivatives; (28) sorbitan derivatives; (29) succinates sulfo-derivatives; (30) alcohol sulfates; (31) ethoxylated alcohol sulfates of ethoxylated alcohols; (32) sulfonates of naphthalene and alkylnaphthalenes; (33) sulfated ethoxylated alkylphenols; (34) sulfated fatty esters; (35) sulfated and sulfonated oils and fatty acids; (36) benzene, toluene, xylene sulfonates; (37) condensed naphthalene sulfonates; (38) dodecyl- and tridecylbenzene sulfonated and free acids; (39) petroleum sulfonates; (40) taurates; (41) tertiary amine oxides; (42) thio- and mercapto derivatives; (43) vinyl and other polymeric resins; (44) α-olefin-sulfonates; (45) miscellaneous; and (46) sucrose esters.

A classification of some of the inorganic builders used in detergent formulations including the phosphates, silicates, carbonates, and other components, such as abrasives, borax, perborate, and the chloroisocyanurates, was also made by McCutcheon (28). Also included in the classification are discussions of physical properties, forms supplied, uses, and suppliers.

The reader is referred to these references as a comprehensive source for general information on the components of detergent formulations. For more detailed information, the technical bulletins of the suppliers should be referred to.

GENERAL REFERENCES

N. K. Adam, The Physics and Chemistry of Surfaces, 3rd ed., Oxford Univ. Press, London, 1941.

A. W. Adamson, Physical Chemistry of Surfaces, Wiley-Interscience, New York, 1960.

A. E. Alexander and P. Johnson, Colloid Science, Vol. I, Oxford Univ. Press (Clarendon), London, 1949.

K. Durham, ed., Surface Activity and Detergency, Macmillan Co., London, 1961.

J. C. Harris, Detergency Evaluation and Testing, Wiley-Interscience, New York, 1954.

J. W. McBain, Colloid Science, D. C. Heath & Co., Boston, 1950.

A. M. Schwartz and J. W. Perry, Surface Active Agents, Wiley-Interscience, New York, 1949.

A. M. Schwartz, J. W. Perry, and J. Berch, Surface Active Agents and Detergents, Vol. 2, Wiley-Interscience, New York, 1958.

TEXTUAL REFERENCES

1. A. M. Schwartz and J. W. Perry, Surface Active Agents, Wiley-Interscience, New York, 1949, pp. 349-350.
2. R. D. Stayner, Mechanism of Cotton Detergency, Oronite Chemicals Co., 1958, p. 12.
3. C. Z. Draves and R. G. Clarkson, Amer. Dyest. Rep., 20, 201-208 (1931).
4. C. Z. Draves, ibid., 28, 421, (1939); AATCC Test Method 17-1952, AATCC Technical Manual, Howes Publ. Co., New York, 1967, p. B-166.
5. H. Seyferth and O. M. Morgan, Amer. Dyest. Rep., 27, 525-532 (1938).
6. L. Shapiro, Ibid., 39, 38-45, 62 (1950).
7. AATCC Test Method 27-1952T, op. cit., p. B-168.
8. J. Ross and G. D. Miles, Oil and Soap, 18, 99-102 (1941).
9. W. D. Harkins and H. F. Jordan, J. Amer. Chem. Soc., 52, 1751-1752 (1930).

10. E. A. Boucher, T. M. Grinchuk, and A. C. Zettlemoyer, J. Colloids and Interface, 23, 600-603 (1967).

11. F. E. Bartell and G. B. Hatch, J. Phys. Chem., 39, 11-23 (1935).

12. J. C. Harris, J. Amer. Oil Chem. Soc., 35, 670 (1958).

13. J. C. Harris, Detergency Evaluation and Testing, Wiley-Interscience, New York, 1954.

14. A. M. Schwartz, J. W. Perry, and J. Berch, Surface Active Agents and Detergents, Vol. 2, Wiley-Interscience, New York, 1958, pp. 489-512.

15. A. M. Schwartz and J. W. Perry, ibid., pp. 349-360.

16. R. E. Wolfrom, J. Amer. Oil Chem. Soc., 35, 652-664 (1958).

17. F. G. Villaume, ibid., 35, 564-566 (1958).

18. H. W. Zussman, ibid., 40, 695-698 (1963).

19. P. S. Stensby, Soap Chem. Spec., 43, Pt. I-V, 41-44 and 98-103, April; 84-86, 90-92, and 130-132, May; 80 and 85-88, July; 94 and 97-102, Aug.; and 96 and 132-138, Sept. (1967).

20. Detergents and Emulsifiers Annual, John W. McCutcheon, Inc., Morristown, N. J., 1967.

21. J. H. Wilson (to Lever Brothers Co.), U.S. Pat. 3,150,098 (1964).

22. J. Reich and H. R. Dallenbach (to Lever Brothers Co.), U.S. Pat. 2,994, 665 (1961).

23. W. M. Bright (to Lever Brothers Co.), U.S. Pat. 3,156,655 (1964).

24. B. H. Gedge, III (to Procter & Gamble Co.), U.S. Pat. 3,356,613 (1967).

25. R. K. Flitcraft and W. B. Satkowski, P. J. Schauer, and R. L. Liss (to Monsanto Chemical Co.), U.S. Pat. 2,925,390 (1960).

26. R. P. Laskey (to Procter & Gamble Co.), U.S. Pat. 3,081,267 (1963).

27. See (20), pp. 10-19.

28. Ibid., pp. 119-127.

Chapter 3

LAUNDRY SOILS

William C. Powe
Whirlpool Research and Engineering Center
Benton Harbor, Michigan

I. INTRODUCTION

The purpose of this chapter is to examine and explain the sources of the major laundry soils, what is known about their chemical composition, and some of their properties that enhance or retard their removal. It is not a historical account of the research done on soils and will not cite many of the people who have contributed to our present knowledge of these soils. You will find more references to these workers in Chaps. 5, 6, 8, and 9.

The human skin is the source of most of the soil found on laundry. These soils are: protein as skin fragments, lipids secreted by the sebaceous glands of the skin and those liberated by breakdown of skin cells; and perspiration residues from the eccrine and apocrine sweat glands of the skin. In addition, particulate soils are picked up from the environment and such things as food residues, and materials from cosmetics and medical preparations.

II. ANATOMY AND FUNCTION OF THE SKIN

The skin constitutes about 10% of the body weight, and in an adult is approximately 1.8 m^2 (1). It has three major functions; by covering the body, it protects the deeper tissues from drying, injury, and invasion of foreign organisms. It is an important factor in heat regulation and it serves as a sensory organ. In addition to its major functions the skin has limited excretory and absorbing powers.

Skin is composed of three layers, the lowest is the subcutaneous or adipose tissue. Above this subcutaneous layer is the dermis, which is mostly connective tissue and serves as a physical support and nutritional supply for the top layer, the epidermis. Small nerves extend into the lower layers of the epidermis, but it lacks a blood supply and nutrients are supplied by diffusion from the dermis (2).

The epidermis varies in thickness from 0.1 to 0.3 mm and has four layers. Skin cells formed in the basal or germinal layer are normal cuboidal epithelial cells. As the cells migrate to the prickel or Malpighian layer, they begin an irregular flattening process. In the third or granular layer the nuclei and cytoplasm of the cells begin to disintegrate and the cytoplasmic proteins become keratin fibers. This process is called keratinization (1-3).

When the cells reach the outer layer, or stratum corneum, they are mostly the tough, horny material keratin, which owes its strength to the disulfide cross-linkages present in its amino acids. The lower part of the horny layer adheres tightly to the rest of the epidermis but the outer layer sloughs off as microscopic scales (3). For the skin cells to migrate from the basal layer and reach the surface and become shedding corneum takes about 27 days in the adult (2).

III. SKIN AS A SOIL

How many cells are lost from the skin surface per unit time is difficult to determine. Mattoni and Sullivan (4) have reported a loss of 3 g of skin a day; Marples (1) lists values of 6 to 14 g a day. In any event, a tremendous number of epithelial cells, which average 25-30 μ in diameter, and fragments of keratinized cells are entrapped in garments in contact with the skin, such as underwear, bed linen, and hosiery. These cells are present in large numbers in wash water and can be identified under the microscope by adding a small amount of picric acid solution to dirty wash water. Skin cells and other protein materials will stain yellow.

One of the problem areas in laundering is the heavily soiled area on the inside of shirt cuffs and collars. These are the main areas where constant abrasion occurs between fabric and skin. The brown stains in these areas result from dirt particles, lipids from the skin surface, and pigments in the skin cells.

When these stains are extracted with solvents to remove the lipids, most of the color remains. If the insides of the collars and cuffs are treated with a protein stain, such as Brilliant Blue, the dye is preferentially adsorbed in the brown areas (5). This probably indicates that most of the color in these stains are pigments in the skin cells. The main skin pigments are carotene, oxyhemoglobin, melanoid, and melanin (3).

Figure 3.1 is a scanning electron micrograph of a cotton-polyester shirt collar showing skin cells wedged between fibers. Figure 3.2 shows a skin cell wrapped around a fiber. The mechanical entrapment of the skin particles and their flexibility, which allows them to conform to the surface of the fiber, are a strong indication of why they are difficult to remove.

FIG. 3.1. Scanning electron micrograph of epithelial cells on the inside of a cotton-polyester shirt collar. Notice that one skin cell is wedged between two fibers (X 2500). Reproduced by courtesy of "Stereoscan," Kent Cambridge Scientific, Inc.

IV. EPIDERMAL APPENDAGES AND THEIR SECRETIONS

The inorganic salts and some of the nitrogenous materials found in laundry soil are primarily from sweat. (The terms perspiration and sweat are used interchangeably in this chapter.) Man produces three kinds of sweat: insensible, eccrine, and apocrine.

A. Insensible Sweat

One of the cooling mechanisms of the body is the release of water from the lungs, and evaporation of water directly through the epidermis. This water loss is called insensible perspiration because it is water vapor and is not visible as sweat on the skin. Kuno (6) states that under resting conditions the insensible water loss from the lungs and skin averages

FIG. 3.2. Scanning electron micrograph of skin cells partially wrapped around fibers on the inside of a cotton-polyester shirt collar (X 5000). Reproduced by courtesy of "Stereoscan," Kent Cambridge Scientific, Inc.

about 23 g/hour for each square meter of body surface. Because this water is not generated by the sweat glands, it does not contribute chloride or other materials to the skin residues.

B. Eccrine Sweat

Eccrine sweat glands are the major thermoregulatory organs of the human body and over 500 cal are lost for each liter of sweat evaporated from the skin surface. Eccrine glands are coiled tubular glands deep in the dermis. The sweat ducts extend upward through the epidermis to the surface of the skin, and are not associated with the hair follicles. These glands are stimulated to secrete sweat by the autonomic nervous system and in response to stimulation of the thermoregulatory center of the hypothalamus in the brain. They may also respond to emotional stimuli (6).

Eccrine glands are distributed all over the body but their concentration is variable. The palms of the hands and the soles of the feet are especially richly supplied with these glands. It has been estimated that the body contains over two million eccrine glands. Their total volume has been calculated to be only about 40 cm^3, yet the maximum amount of sweat secreted may amount to over 10 liters per day (6).

The amount of eccrine sweat produced by an individual, of course, varies tremendously, depending on the temperature and type of activity. A hard-working laborer may secrete 700 ml of sweat per hour, while a person working in a laboratory on a hot summer day might secrete only 130 ml of sweat in the same time (6).

Eccrine sweat is our most dilute secretion, being roughly 99% water and 1% solids. Reports in the literature on the composition of sweat are not in good agreement because of the difficulty in obtaining samples uncontaminated by other materials on the skin. About half of the solids in sweat are inorganic and half are organic (3).

Sodium chloride accounts for almost 50% of the inorganic material in eccrine sweat. Other electrolytes present are calcium, copper, iodine, iron, magnesium, phosphorus, potassium, and sulfur.

Urea is approximately half of the organic material. Small amounts of creatinine, creatine, ammonia, glucose, lactate, and lactic, pyruvic, and uric acids are also present. Arginine and histidine are the major amino acids reported, with smaller amounts of isoleucine, lysine, phenylalanine, threonine, tryptophane, tyrosine, and valine. Small quantities of some of the vitamins are also present (3).

C. Apocrine Sweat

An apocrine sweat gland is a coiled structure that lies deep in the dermis. The duct is narrow and straight and opens into the upper level of the hair follicle near the opening of the sebaceous gland. In contrast to the eccrine glands, the apocrine glands are found only in certain areas of the body. Some of these sites are the anogenital region, the mammary areola, the ear canal, and the glands of Moll in the eyelid. The axillae or armpits are especially abundantly supplied with apocrine glands. Apocrine glands begin to function at puberty and become less active in old age (7).

Apocrine sweating is a process independent of apocrine secretion. A contractile force empties the gland of preformed apocrine sweat. This force is supplied by the myoepithelium, which is the smooth muscle layer that surrounds the apocrine tubules. Stimulation for contraction of the myoepithelium is emotional; apocrine sweat is produced in response to fear, stress, anxiety, and pain, but not to heat (7).

Apocrine sweat is a turbid viscous material in contrast to eccrine sweat, which is a clear watery secretion. The pH of apocrine sweat ranges from 5.0 to 6.5, eccrine sweat varies from 4.0 to 6.0. Apocrine sweat contains protein, carbohydrate, ammonia, and a small amount of ferric iron. Fat cells have been identified within the glands by histological staining (7).

After an apocrine gland has been stimulated to release apocrine sweat, further stimulation of that gland will not cause sweating for 24 to 48 hours. Apparently this amount of time is necessary for the gland to produce more sweat. The amount of apocrine sweat secreted is very small when compared to the output of the eccrine glands. Hurley and Shelley (7) estimate that about 0.001 ml of apocrine sweat results from one stimulation of a gland.

Apocrine sweat is important in soiling of laundry because it is the source of most of the body odor, particularly in the axillae where the glands are most numerous. This typical odor is produced by bacterial activity on the apocrine sweat. Hurley and Shelley (7) demonstrated that apocrine sweat is sterile and odorless as it emerges from the gland and that bacteria are necessary to produce the underarm odor. It is assumed that this odor is the result of bacterial action on short-chain fatty acids. Bacterial growth in eccrine sweat can produce a mild odor but it is not offensive.

D. Sebum

The most important epidermal appendage, as far as laundering is concerned, is the sebaceous gland because it produces the most difficult soil to remove. Morphologically these glands are generally appendages of the outer sheath of the hair follicles and collectively are known as the pilosebaceous apparatus. The sebaceous gland is usually a multilobed gland that opens through its duct into the upper half or third of the hair follicle (3).

As the cells of the sebaceous glands grow and mature, they move toward the center of the gland where the fat-filled cells disintegrate. The fat and the cellular debris are released into the sebaceous duct and flow up the follicular canal and on to the horny layers of the skin. This fat and cellular debris released by the sebaceous glands is called sebum. The force of the continuous release of sebum pushes it to the surface of the skin. It is doubtful if nervous impulses control much of the activity of the sebaceous gland (3).

Sebum melts at about 30°C. This low melting point is due to the fact that sebum is a complex mixture of lipids quite different from body fat, which is primarily mixed triglycerides of palmitic, stearic, and oleic acids. The major lipids present in sebum are paraffinic hydrocarbons, squalene, wax esters, cholesterol esters, cholesterol, triglycerides, diglycerides, monoglycerides, and free fatty acids. Minor constituents are still being isolated and identified (2).

Sebaceous glands are found on all surfaces of the body except the palms of the hands and the soles of the feet. However, sebum can be found on these surfaces because it is spread by the water on the skin surface. The concentration of the glands varies on different parts of the body. The greatest concentration of sebaceous glands is in the scalp and forehead regions. The trunk produces a considerable amount of sebum and the limbs less (1).

This lipid film helps maintain the pliability of the horny layers and delays adsorption of foreign materials through the skin. The short-chain fatty acids present have fungicidal activity and the longer-chain fatty acids are bacteriostatic (3).

Sebum is usually in an emulsified state on the skin surface and can be either an oil-in-water or water-in-oil emulsion, depending on the amount of sweat present at a given time (8). This emulsion formation helps the sebum to flow with the sweat and penetrate the fiber bundles and cover fiber surfaces not in direct contact with the skin.

The analysis and chemistry of sebum are discussed in the next section because the composition of sebum reported by medical workers obtained by extraction of sebum from the skin surface is very similar to the data obtained by extraction of fresh skin fat from clothing.

V. ANALYSIS OF LAUNDRY SOILS FROM THE SKIN SURFACE

A. Introduction

T-shirts are ideal garments to use in collecting skin soils for analysis. Because they are in direct contact with areas of the skin that produce above-average quantities of sebum and sweat, large amounts of typical material can be obtained for study in a short time. Also the size of the garment is convenient for extraction.

Since man is bilaterally symmetrical, a T-shirt can be cut in half after soiling for comparison of the soils present before and after laundering with the reasonable expectation that the amount and composition of the soil present were the same on both halves before laundering. For these reasons much of the data in this chapter have been obtained from soiled T-shirts.

When a soiled garment is extracted with a solvent system of the proper polar-nonpolar balance, an extremely complex mixture of materials is obtained. These include inorganic salts, lime soaps, surfactant residues, lipids from sebum and skin cells, also nitrogenous materials from sweat residues and skin cells. Oxidation products from unsaturated components in the mixture may be present and perhaps residues from cosmetics and ointments.

The information obtainable from these extracts was limited until the development of the various forms of chromatography for separation of closely related materials, and infrared, ultraviolet, and mass spectrography for identification of the chromatographic isolates.

Before the late 1950s, accurate separation and identification of lipid mixtures were very difficult. Saponification was used to determine the total amount of fatty acid and the unsaponifiable fraction. The water-soluble, lime soap, and free fatty acid fractions in clothes soil could be separated from the rest of the fat. Iodine numbers were used to determine unsaturation and acetylation numbers were used to characterize long-chain alcohols. Separation of materials with different carbon chain lengths was made by fractional distillation and low-temperature crystallization. Paper chromatography was about the only chromatographic method in common use.

There is now a much clearer understanding of the composition of clothes soil as a result of the "instrumentation explosion" and the work of many people in developing methodologies for separating and quantitating complex lipid mixtures.

B. Water-Soluble Soils

Bey (9) has published the most extensive data on soils extracted from clothing. He found that for eight people doing light work under normal temperatures, the water-soluble material from unwashed T-shirts extracted with benzene-ethanol 9:1 averaged 1.6% of the total extractable material. The range of this water-soluble soil was 0.5-4.5%. In contrast, the water-soluble portion of soil extracted from the T-shirts of five people doing hard work in hot weather averaged 13.5%, with a range of 5.5-31.7%. These data illustrate the variation in the amounts of sweat residues, which are mostly sodium chloride and urea, that are found on clothing because of the occupation and environment of the wearer.

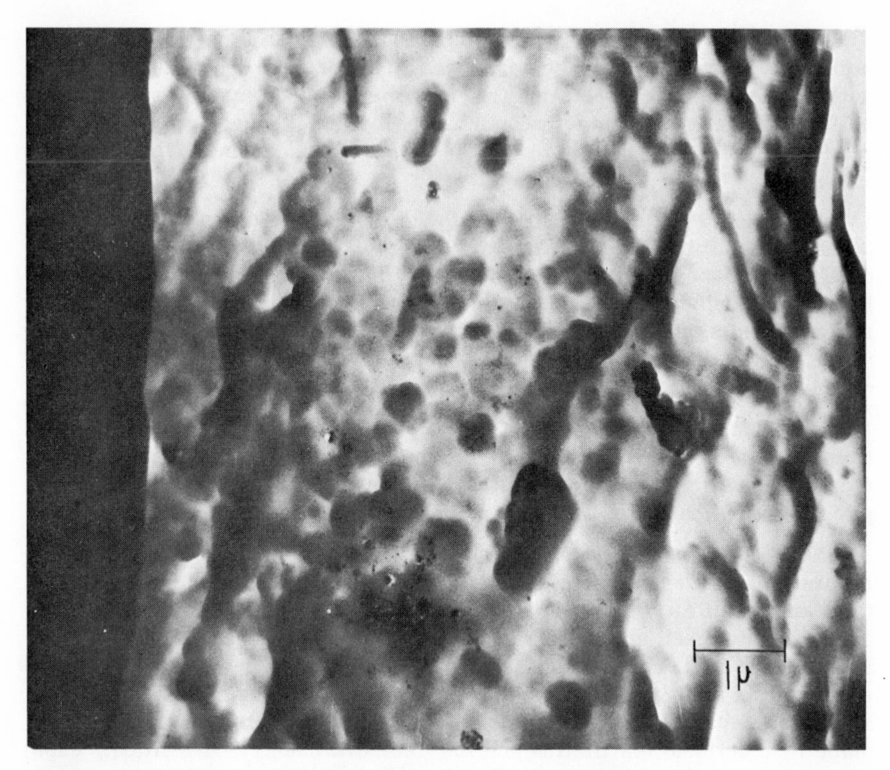

FIG. 3.3. Electron micrograph of a cotton fiber heavily coated with sebum (X 10,000). Reprinted from Ref. 32, p. 881, by courtesy of the Textile Research Journal.

Information on the nitrogenous material derived from protein in the water-soluble portion of clothes soil is very sketchy. Albumin, a water-soluble protein, has been mentioned as being present in the water-soluble fraction. If it is present, it is probably there in small quantities. Tomiyama and Iimori (10) extracted the stained areas of cotton cloth, worn under dress collars, which were mostly skin cells and consequently very rich in proteins and found only 58.8 mg/% of water-extractable nitrogen in the material obtained from 56 g of the stained cloth. See Table 9.21.

C. Sebum

By weight sebum is the major soil present on laundry and its removal is important because unremoved fat acts as a matrix to hold particulate soils already on the fiber and increases deposition of particulates from the wash bath. In addition, the unsaturated compounds present in sebum oxidize to contribute to yellowing of the clothing (11, 12). Because of sebum's importance as a soil, it is essential to understand its composition and chemistry. Figure 3.3 is an electron micrograph of a cotton fiber covered with sebum with particulates embedded in the fat.

In this discussion of clothes soil, sebum will be used as the term for all the lipids extracted from garments even though a small amount of fat has come from skin cells trapped on the clothing or is fat that was liberated during the keratinization process and was not a product of sebaceous glands. In Table 3.1 are data on the types of lipids found in fresh sebum and a range of the values obtained in our laboratory for each lipid class (13). These values fall in the same general range reported by Bey (9).

Sebum is about 70% saponifiable, indicating that 30% is something other than fatty acid. Except for the free fatty acid, the rest of the fatty acids are part of glyceryl, cholesteryl, or wax esters. A very small amount of fatty acid may be present in phospholipids from ruptured cells. Lipids present that are not combined with fatty acid are hydrocarbons, squalene, and free cholesterol (3).

TABLE 3.1

Typical Range in Composition of Fresh Adult Sebum Extracted from T-Shirts[a,b]

Lipid class	Percent
Hydrocarbons	1-3
Squalene	10-12
Wax esters	12-16
Cholesteryl esters	1-3
Cholesterol	1-3
Triglycerides	30-50
Mono-diglycerides	5-10
Free fatty acids	15-30

[a]T-shirts worn 24 hours before extraction with chloroform-methanol 1:1. Average weight of sebum freed from skin fragments, urea, and sodium chloride was 1.3 g for 10 different T-shirts.

[b]From Powe (13).

The free fatty acids in the skin surface lipids come from the triglycerides produced by the sebaceous glands. Sebum in the sebaceous glands does not contain free fatty acids, or di- or monoglycerides. However, hydrolysis of triglyceride starts when sebum is forced into the sebaceous duct and free fatty acid is a major component of the skin lipid film. The skin and/or bacteria produce esterases that break down part of the triglyceride to free fatty acid. Small amounts of di- and monoglycerides are also present since they are the intermediate degradation products in the complete hydrolysis of triglyceride to free fatty acid and glycerol (3, 13, 14).

Free fatty acid reaches a maximum of about 30% in clothes soil derived from skin lipids. I have rarely obtained values higher than this nor have seen much higher values reported by others. Because the free fatty acid is produced by the hydrolysis of triglyceride, a high value for free fatty acid results in a lower value for triglyceride. Nicolaides (15) estimates that the free fatty acid and that combined as triglyceride represent about 70% of the total fatty acids present in sebum.

Table 3.2 presents typical data for the composition and distribution of the fatty acids in fresh sebum according to lipid class. These data are in good agreement with data reported by Bey (9). Small amounts of acids down to C_7 and up to C_{22} or higher are present but are not reported here. Also, small amounts of other branched acids are known to be present.

The fatty acid composition of sebum differs from body fat in that appreciable quantities of odd-numbered carbon chain acids are present and in that there is a considerable amount of branched chain acids. Palmitoleic acid is one of the major acids present in contrast to the small amounts found in human fat. The wax ester fraction contains an especially large amount of palmitoleic acid. Roughly 50% of the fatty acid in sebum is unsaturated.

Besides triglycerides and their hydrolysis products, the only other major source of fatty acid in sebum is the wax ester fraction. The cholesteryl fraction is only 1-3%. A wax ester is a long-chain alcohol esterified with a long-chain fatty acid.

The composition of the wax ester fraction of sebum is very complex. Haahti (17) reported that the major wax esters range from C_{26} to C_{42} with over 50 different peaks identifiable by gas chromatography and each peak representing several different esters with the same carbon number. In a later study Haahti et al. (18) separated the wax ester fraction into four fractions by thin-layer chromatography on silver nitrate-impregnated silica gel plates. These fractions were saturated, mono-unsaturated, diunsaturated, and some unknown polyunsaturated waxes.

TABLE 3.2

Fatty Acid Distribution from Skin Lipids in Fresh Clothes Soil[a,b]

Carbon chain length	Triglyceride, %	Diglyceride, %	Monoglyceride, %	Fatty acid, %	Wax ester, %	Cholesteryl ester, %
12:0	1.0	1.2	0.5	1.0	Trace	Trace
12:1	0.5	0.4	0.3	0.3	Trace	Trace
ISO 13:0	Trace	Trace	—	Trace	—	—
13:0	0.5	0.3	0.4	0.2	0.2	Trace
13:1	0.2	0.3	Trace	Trace	Trace	0.5
14:0	10.5	9.4	10.8	8.3	3.8	3.1
14:1	5.2	3.9	3.6	2.0	9.0	8.2
ISO 15:0	3.0	2.0	2.5	1.8	2.0	1.6
15:0	6.1	5.6	4.3	5.2	4.0	4.6
15:1	2.9	2.0	2.4	2.3	4.9	3.7
16:0	21.4	23.1	22.7	33.0	10.0	14.3
16:1	22.4	22.5	15.4	16.0	39.0	30.0
ISO 17:0	4.0	3.5	2.0	2.5	3.0	2.5
17:0	2.0	2.4	1.3	3.5	5.1	7.1
17:1	2.3	2.2	1.4	2.4	4.6	2.7
18:0	2.2	4.2	4.6	5.0	1.7	2.6
18:1	12.0	14.5	25.4	13.5	8.5	12.7
18:2	3.2	2.0	1.2	1.2	1.5	2.2

[a]Extracted with chloroform-methanol 1:1 from T-shirt worn 1 day.

[b]From Powe (16).

Table 3.3 shows the approximate amount of each alcohol chain length in the wax ester fraction of fresh sebum and the amount of the alcohols that are unsaturated. The wax ester linkage is resistant to hydrolysis and only very small amounts of free long-chain alcohols are found in sebum. If large amounts of fatty alcohols are present on analysis of clothes soil, they must come from some source other than sebum, such as cosmetic creams or medicinal ointments.

TABLE 3.3

Alcohols[a] from Wax Esters in Sebum Extracted from T-Shirts[b]

Chain length	Total, %	Saturated, %	Unsaturated, %
12	0.8	0.5	0.3
13	0.2	0.2	Tr.
14	6.0	5.8	0.2
15	2.5	2.3	0.2
16	8.0	7.4	0.6
17	6.5	5.0	1.5
18	14.0	13.5	0.5
19	5.2	0.7	4.5
20	22.0	8.0	14.0
21	6.7	0.5	6.2
22	10.1	4.1	6.0
23	4.2	1.0	3.2
24	8.7	2.7	6.0
25	1.9	1.2	0.7
26	Trace	Trace	Trace
27	Trace	Trace	Trace

[a]Determined by chromatographing as acetates.

[b]From Powe (16).

It is not certain whether the paraffinic hydrocarbons isolated from sebum are produced by the skin or are there as a contaminate (15, 17). Characterization of this material in our laboratory shows that it is very similar to petrolatum (19). In the analysis of clothes soil any large amount of hydrocarbon is from some source other than sebum.

Squalene is an unsaturated acyclic liquid hydrocarbon containing six isoprene units. It has the empirical formula $C_{30}H_{50}$. Squalene was first isolated from the unsaponifiable fraction of shark-liver oil. Its name comes from Squalidae, the family of small sharks called dogfish. Because squalene has six double bonds, it is very reactive and changes structure on standing in contact with oxygen. More will be said about this property later.

Squalene is formed enzymatically from mevalonic acid and is a precursor of various cyclic triterpenoids, cholesterol, and related steroids. Squalene is produced by the sebaceous glands in man, but the free cholesterol found in sebum is derived mostly from the horny layer of the skin (3).

Sebum adsorption by clothing in an average wearing period is 1-2 percent of the weight of the garment. This is the typical amount of fat extracted from T-shirts after one wearing, which is normally the largest garment in direct contact with the skin for the most hours a day. A medium T-shirt weighs about 110-120 g. Bowers and Chantrey (20) reported that after a 24-hour wearing period, they extracted an average of 1.2 g of sebum from cotton T-shirts and 1.1 g from T-shirts made of 65/35 polyester-cotton. These data agree with results we obtained with cotton T-shirts that averaged 54 mg of sebum adsorption per hour of wearing time (21).

VI. REMOVAL OF SOILS ORIGINATING FROM THE SKIN

A. Removal of Fresh Lipid and Water-Soluble Soils

In Table 3.4 is a comparison of the differential removal of soils originating from the skin. These data were obtained by extracting T-shirts with chloroform-methanol 1:1 to remove all residual soluble soil and then wearing the T-shirt usually for 24 hours. The garments were cut in halves and one half was washed. Then both parts were extracted separately and analyzed to determine the differences in the soil content of the washed and unwashed samples. As would be expected, most of the water-soluble soil was removed and much of the free fatty acid was neutralized to soluble soaps by the alkaline builders in the detergent, but less than 50% of the neutral lipid soil was removed.

TABLE 3.4

Removal of Fresh Soil from a T-Shirt[a,b]

	Before washing, mg	After washing, mg	Percent removal
Water soluble	261	35	87
Neutral lipid	599	279	48
Free fatty acid	100	25	75

[a]Washed and rinsed in a home wash load in 0.15% nonionic detergent at 120°F in soft water.

[b]From Powe (22).

It is impossible to remove all the fat from clothing by laundering in an aqueous system under the best conditions, and under unfavorable conditions a large amount of residual fat can accumulate (23). By unfavorable conditions, I mean excessively hard water, too-little detergent, and water that is not hot enough.

Bey (24) showed that by washing cotton undershirts at 30, 60, and 90°C in a tetrapropylene benzene sulfonate-based detergent in softened water, an average of 2.1, 1.4, and 1.3% residual fat, respectively, was left on the garments.

B. Composition of Residual Soil

In Table 9.23 are some representative values for the amount of fatty soil that accumulated on cotton garments after several years normal use. Total fatty soil ranged from about 2 to 8% and in one extreme case 38% of the fatty soil was lime soaps. The residual free fatty acids in these samples averaged about 3% (23).

Since approximately 50% of the fatty acids in sebum are unsaturated, they tend to oxidize if they remain on fabrics to form yellow and brown polymeric materials (11, 12). These polar materials are eluted from silicic acid and aluminum oxide chromatography columns by acetone and methanol. Data in Table 9.24 reflect this loss of unsaturated acids as the analysis for palmitoleic and oleic acids in the aged soil shows much lower values than the analysis for fresh soil in Table 3.2 in this chapter (23).

Table 3.5 is an analysis of the lipid classes found in aged residual soil from a cotton T-shirt. The main differences in this analysis and the data given in Table 3.2 are that squalene has practically disappeared, the amount of free fatty acid is very low, lime soaps are now present, and over a third of the residual soil is dark-colored oxidation products (25).

Part of the oxidation products come from the unsaturated fatty acids as mentioned above but the oxidation of squalene also contributes much of the material. Oldenroth (26) recognized that the oxidation of squalene contributed to the yellowing of fabric. He found that the peroxide number of squalene increased from 4 to 3000 in 8 days. Wagg et al. (27) also found that squalene contributes to yellowing of fabrics and tried using antioxidants to reduce the oxidation of squalene. However, the antioxidants they used were not very effective.

Table 3.6 reveals how rapidly squalene is oxidized to other products. In 14 days almost 70% of the squalene became a resinous, brown, polar material and none of the original squalene remained (28).

The free fatty acid in sebum, which averages 15 to 30% of the fresh clothes soil, is the only class of lipids that can react with the metal ions in the wash bath. If laundering is done in hard water, it is a matter of

TABLE 3.5

Residual Aged Fatty Soil from a Cotton T-Shirt[a]

Fraction	Percent
Hydrocarbons	3
Squalene	1
Wax and cholesteryl esters	12
Cholesterol	3
Triglycerides	23
Di- and monoglycerides	13
Free fatty acids	2
Lime soaps	7
Oxidation products	36

[a]From Powe (25).

TABLE 3.6

Oxidation of Squalene on Cotton at 25°C in the Dark[a,b]

Eluted from silicic acid with	Percent eluted by number of days stored				
	0	7	14	21	28
Petroleum ether (PE)	100	1	–	–	–
1% Diethyl ether in PE	–	45	–	–	–
25% Diethyl ether in PE	–	18	8	–	–
Diethyl ether	–	22	26	20	12
Acetone	–	14	66	80	88

[a]Cotton swatches soiled to 1% level with purified squalene dissolved in chloroform. Stored in laboratory bench drawer.

[b]From Powe (28).

chance whether the fatty acid will react with calcium or magnesium to form lime soaps or will come in contact with potassium or sodium from the alkaline builders in the detergent to become soluble soaps (13, 24, 29). If lime soaps are formed in any quantity, they help to degrade the appearance of the garment because when wet, lime soaps are gelatinous and act as a matrix to hold particulate soil. When dry, they impart a harsh feel to the fabric.

Since under normal conditions the triglycerides and other esters are not saponifiable during laundering, they are for all practical purposes chemically inert and along with the hydrocarbons are removed by emulsification, solubilization, or the roll-up phenomena, or a combination of these (13, 29, 30).

VII. PARTICULATE SOILS

Particulate soils are mostly windblown fractions of topsoil or incomplete products of combustion from homes, industry, or various forms of transportation. Most soiling occurs by brushing or rubbing against a surface where these particles have settled. Reumuth (31) published a monograph on soils that is richly illustrated with examples of various soils encountered in laundering and hard surface cleaning.

The difficulty in removing particulate soils is based primarily on how small they are, although the type of fiber and the amount of fat present also may have an effect.

A. Siliceous Soils

Figure 3.4 is a comparison of the various types of particles found in topsoil and their size relationship to a cotton fiber. Sand, which ranges from 2000 to 50 μ, can be shaken or brushed from clothing and is certainly removed during laundering. The same can be said for the silt fraction, which includes particles from 50 μ down to about 2 μ. An exception might be particles 10 μ and under which sometimes become entrapped between fibers and may not be removed.

It is the clay fraction which starts at 2 μ and extends down to approximately 0.02 μ that is difficult to remove, and is responsible for most of the redeposited soil on cotton (32, 33). The clay fraction may typically represent 15-20% by weight of a topsoil.

One-micron particles are very small and border on the limits of resolution with an optical microscope. As a comparison, many of the smaller bacteria are 1-2 μ long and some of the larger viruses are about 0.1 μ in diameter. A soil particle that is 0.1 μ and is in direct contact with a fiber is practically impossible to remove by ordinary washing techniques. Other soils that are in this size range are soot and colloidal iron oxide. Figure 3.5 shows an electron micrograph of clay particles in a sample of topsoil. Individual particles and larger clumps containing hundreds of particles are present.

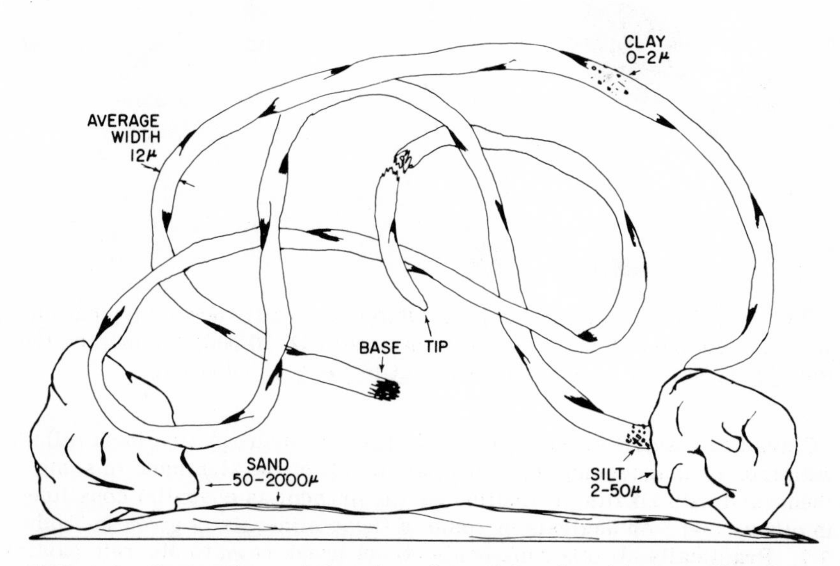

FIG. 3.4. Comparison of the size of particles found in topsoil and a cotton fiber. Clay particles are very difficult to remove because of their small size.

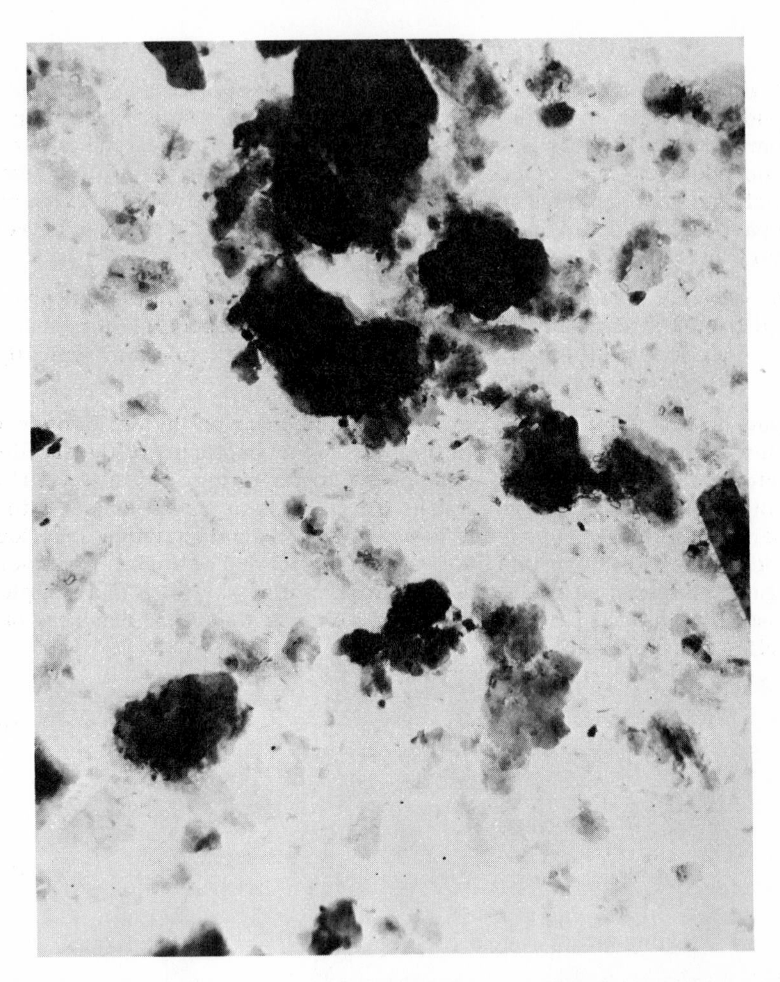

FIG. 3.5. Electron micrograph of Ohio glacial till showing typical clumps and discrete particles of clay found in topsoil (X 10,000). Reprinted from Ref. 32, p. 881, by courtesy of the Textile Research Journal.

Clays are a small group of minerals that are hydrous aluminum silicate with iron or magnesium replacing part or all of the aluminum in some of them, and with alkalis or alkaline earths present as essential constituents in others (34). An analysis of some of these minerals is given in Table 3.7. Practically all other minerals do not break down to discrete particles as small as 1 or 2 μ so they would not readily stick to a fiber surface.

Since the colloidal range extends from ca. 3 to 200 mμ, most clay minerals fall in this range in at least one, if not both dimensions. They have

TABLE 3.7

Typical Analyses of Clay Minerals[a]

	Kaolinite	Montmorillonite	Attapulgite	Illite
SiO_2	45.4	51.2	55.0	51.2
Al_2O_3	38.5	19.8	10.3	25.0
Fe_2O_3	0.8	0.8	3.5	4.5
FeO	—	—	—	1.7
MgO	0.1	3.2	10.5	2.8
CaO	0.1	1.6	—	0.2
K_2O	0.1	0.1	0.5	6.1
Na_2O	0.7	Trace	—	0.2
TiO_2	0.2	—	—	7.5
H_2O	14.2	24.0	19.3	1.0

[a]Reprinted from Ref. 32, p. 883, by courtesy of the Textile Research Journal.

an active surface and are capable of ion exchange, and because they are mostly thin plates, they have a large surface in proportion to their mass. The combination of all these properties make clay minerals ideal particles to become firmly attached to a fiber surface (32). Figure 3.6 is an electron micrograph of kaolinite particles on a cotton fiber. The mechanisms of retention of clay particles on fiber surfaces are discussed in Chap. 6. Figure 6.3 is the structure of a typical clay mineral.

Fortunately most of the clay particles on clothing are there as aggregates and not as individual particles. Evans and Camp (33) have shown by electron micrographs that the larger the size of the aggregates originally on the fiber, the easier the removal. Figure 3.7 shows some clumps of clay from a sample of wash water, also present are some fragments from cotton fibers. Figure 3.8 is a scanning electron micrograph of dirt particles and fat on cotton fibers.

B. Carbonaceous Soils

Although graphite has been used for many years as a model soil, I don't believe it is an important or a common soil except in a few limited cases such as graphite stains in some textile mill operations or stains incurred

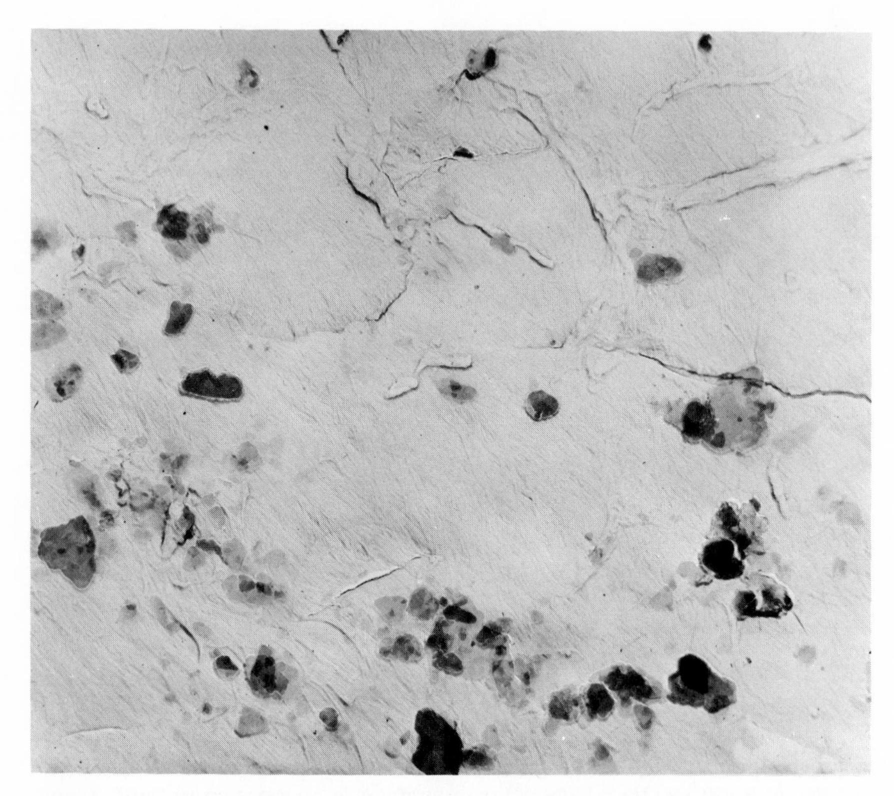

FIG. 3.6. Electron micrograph of kaolinite particles on cotton (X 10,000). Reprinted from Ref. 32, p. 880, by courtesy of the Textile Research Journal.

by accidental contact with a graphite-lubricated surface. Graphite particles are flat and have a crystalline structure with very little organic matter as part of their surface (35). The only contact the average person has with graphite is the "lead" in pencils.

Soot is a form of carbon that is still with us as a soil although it is diminishing as an atmospheric pollutant. There is not much soft coal burned in the United States anymore to heat homes, and the flocs of soot that used to darken the snow several days after it fell are no longer much in evidence. Even in areas of heavy industry, the amount of soot present in soil is probably overestimated.

Laws are being enacted to lower the limits for all types of particulates produced by industry, automobiles, and trucks. There may come a time when the chief sources of carbonaceous soil are tobacco smoke and carbon black particles from abraded rubber. The average automobile tire contains about 4 lb of carbon black (36).

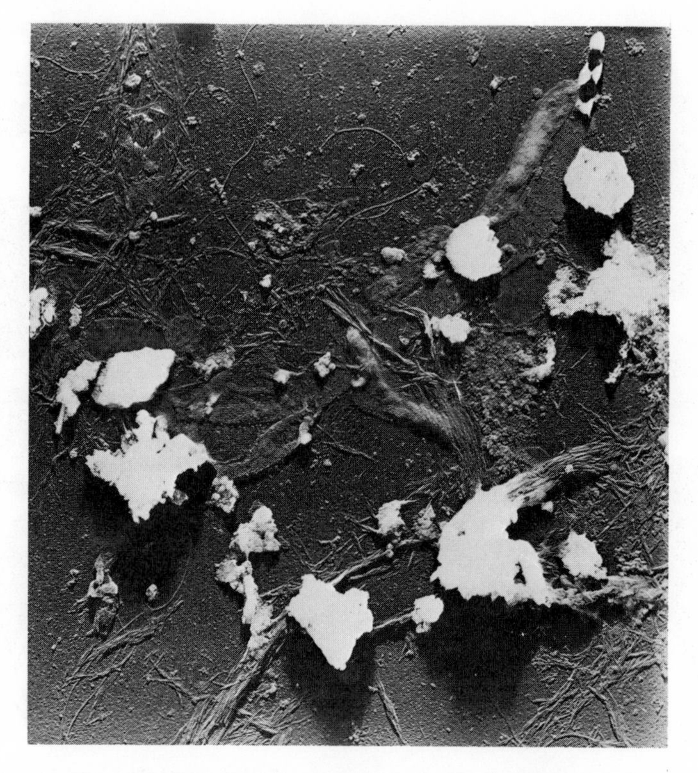

FIG. 3.7. Electon micrograph of clumps of clay particles in the residue of wash water. Cotton fibrils are also present (X 10,000).

Figure 3.9 is an electron micrograph of a soot stain on cotton that was not removed by washing. These spherical particles are joined in chains, which is typical for thermal carbon black. The individual particles range from 40 to 200 mμ but average about 80 mμ. This size range is closely duplicated by Furnace Combustion Black (SRF) which is a semireinforcing carbon black and is 99% carbon. Some industrial carbon blacks have as much as 17% volatile matter (36). Soot probably has a higher organic content than thermal carbon blacks used for industrial purposes.

C. Materials Contributing Color to Soils

I think that the amount of carbon in soils is overestimated because black is equated with free carbon. So much data on detergency have been generated using artificial soils containing carbon black or graphite that, if a natural soil is black, it is assumed to have a large amount of free carbon present.

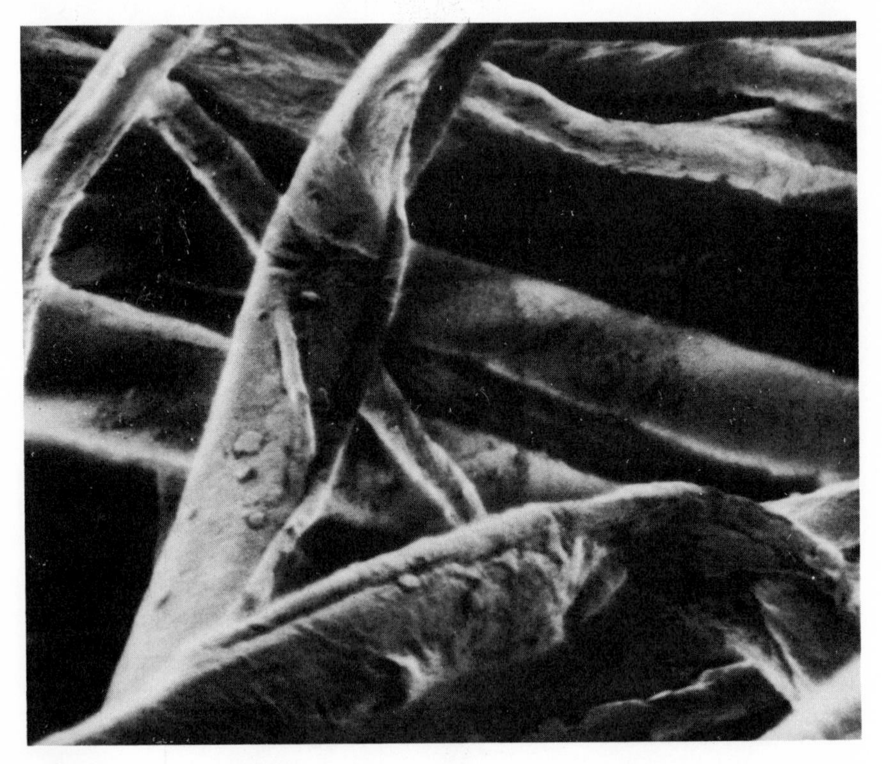

FIG. 3.8. Scanning electron micrograph of dirt particles and fat on cotton fibers (X 2500). Reproduced by courtesy of "Stereoscan," Kent Cambridge Scientific, Inc.

The black color of natural soils is far more likely to be caused by organic material rather than free carbon. Humus, the partially decayed fraction of topsoil, is dark brown or black. Because it is collodial, it will coat mineral particles to give them a black appearance. Only small amounts of organic matter are needed to color clay particles black or dark gray (34).

Inorganic materials can also contribute black coloring to a soil. Our laboratories are in a semirural area, yet the dirt under 100 μ that collects in our air filters is very dark and although it contains clays and other windblown components of topsoil, there is also a substantial amount of black magnetic iron oxide. (See Table 3.10.) Three miles upwind from our air intake is a ferrous foundry.

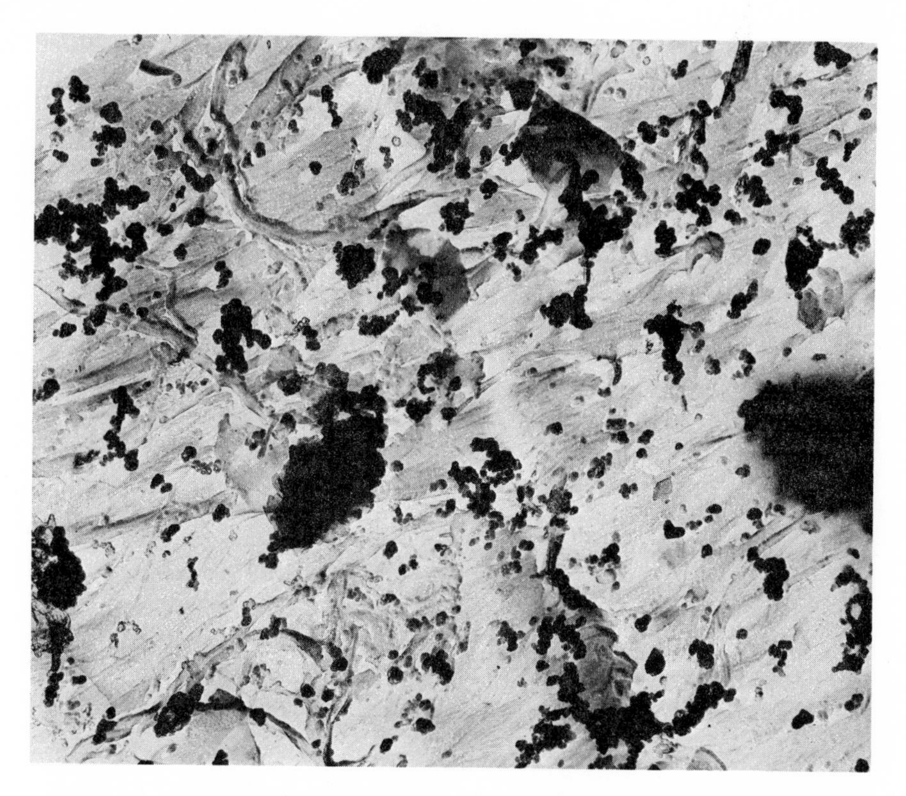

FIG. 3.9. Electron micrograph of soot particles on a cotton fiber (X 10,000). Reprinted from Ref. 32, p. 880, by courtesy of the Textile Research Journal.

VIII. ANALYSIS OF PARTICULATE SOILS

Twenty years ago Sanders and Lambert (37) published an analysis of street dirt from six American cities that showed the composition of the soil to be similar in all the cities. An average of some of their data is given in Table 3.8.

Recognizing that the large amount of carbon in their analyses was mostly of organic origin, they determined how much carbon black would produce the same amount of color. By using 10 mg or less of a large thermal carbon black similar to soot, they tinted 5 g of zinc oxide paste to the same shade of gray produced by adding 1 g of street dirt to 5 g of zinc oxide. From this ratio was developed a "Carbon Black Equivalent" which averaged about 0.6% for the street sweepings.

TABLE 3.8

Average Composition of Street Dirt from Six American Cities[a,b]

	Percent
Water soluble	14.4
Ether soluble	8.3
Total carbon	25.7
Ash	53.5
SiO_2	24.5
R_2O_3 (mostly iron)	10.4
CaO_2	7.0
pH of 10% slurry	7.4
Particle size 4 μ or less	53.0
Carbon Black Equivalent	0.6

[a]Pittsburgh, Detroit, Cleveland, Buffalo, St. Louis, Boston.
[b]From Sanders and Lambert (37).

These data point up the fact that a very small amount of carbon if evenly distributed could produce a large decrease in the reflectance of a white garment, since only 0.2% carbon reduced the reflectance of a white material, zinc oxide, to the color of street dirt. To my knowledge, there is no accurate method to measure small amounts of free carbon in complex mixtures like street dirt.

McCarthy and Moore (38) developed a method for determining free carbon in atmospheric dust using 70% nitric acid to oxidize the organic material and subsequent loss of weight on ignition to determine the elemental carbon. However, the smallest amount of carbon used in a test mixture was 10%. They also cautioned that the method would be inaccurate if a large amount of clay were present because the loss of bound water from the clay structure would give false high values for carbon.

Walter (39) published a determination of elemental carbon in an artificial soil containing the following: 35% kaolin, 25% sodium silicate, 5% palmitic acid, 10% paraffin, 20% sugar, and 5% lampblack. The soil was washed and centrifuged a number of times to remove the organic material and various acid and alkali treatments were used before ignition. The final value for the 5% carbon was 40% high.

In the firing of clays oxidation of the organic material begins at 200-300°C; in the 400-500°C range oxidation of sulfides occurs; from 400-700°C the hydroxyl structural water from the lattice is lost; and at 900°C the clay mineral structure is destroyed (34).

Table 3.9 contains data obtained by thermogravimetric analysis showing the differences of heating soot and graphite in air (40). Notice that soot is completely destroyed before the graphite begins to lose any weight. If the soot were present with clays, both materials would lose their oxidizable material in the same temperature range. The relatively low temperature necessary for destruction of most of the soot is indicative of its high organic content in contrast to the crystalline graphite, which is resistant to oxidation.

Data in Table 3.10 were collected to determine how the reflectance of several soil samples known to be mostly clay compared in reflectance with some natural soils currently being used for detergency studies (41). Also, some bleaching experiments with sodium hypochlorite were run to determine if the more carbonized soils such as soot would remain black and give some indication of the amount of carbon soil present.

TABLE 3.9

Effect of Temperature on the Weight Loss of Soot and Graphite[a,b]

Percent weight loss	Temperature °C	
	Soot	Graphite
10	442	702
20	494	730
40	536[c]	760
60	570	810
80	590	815
100	600	820

[a]Determined by thermogravimetric analysis of a 10 mg sample in air, at a heating rate of 10°/minute.

[b]From Miranda (40).

[c]Ignition of sample.

TABLE 3.10

Some Properties of Dark-Colored Particulate Soils[a]

Soil	% Wt. of garment	Reflectance[b] Rd	Reflectance[b] +b	Appearance after bleaching[c]	% Wt. loss on ignition	% Iron
Particulates from cotton-polyester blouse	1.5	29	3.6	White	30	—
Particulates from cotton T-shirt	2.5	19	2.8	White	33	1.7
Vacuum cleaner dirt, < 100 μ	—	15	6.0	Light brown	42	2.0
Laboratory air filter, < 100 μ	—	19	3.5	Brown	38	8.5
Dark-brown clay, < 100 μ	—	25	5.3	White	12	—
Silica, < 100 μ	—	85	0.7	—	—	—
Silica plus 0.25% soot	—	17	2.5	White	—	—
Dark-brown clay plus 0.25% soot	—	23	5.2	White	—	—
Dark-brown clay plus 1% soot	—	13	3.2	White	—	—

[a]From Powe (41).

[b]Gardner Color Difference Meter.

[c]48 hours in 5% sodium hypochlorite at 25°C.

Early in our studies of clothes soils we found, as had been reported by others, that the removal of large quantities of highly colored partially oxidized fat by solvent extraction did not improve the reflectance of a heavily deposited cotton garment more than a few reflectance units. The only method that was effective for removing the particulate soil for study was to soak the garments in a very hot and very strong soap solution for several days. When built synthetic detergents were used under the same conditions, the results were very poor (42).

By soaking cotton garments in potassium stearate or behenate at or near the boiling point for several days and by adding sodium hydroxide to keep the pH above 11, the reflectance of a T-shirt that was about 65 to 70 could be raised to 85. If this T-shirt was then placed in a fresh alkaline soap solution at a high temperature and sodium perborate was added, more colored particulate soil was removed and the final reflectance of the garment was around 90, which is as high as a new T-shirt. The formation of oxygen bubbles near or under some of the smaller dirt particles probably supplied the additional energy needed to pry them from the fiber surface. This method is very similar to the European method of laundering cottons except more synthetic detergents are now used than soap.

The particulate soil was recovered by acidifying the soap solution and extracting with perchloroethylene to remove the fatty acids. The particulate dirt collected at the solvent-water interface and was recovered for analysis. Before the soap treatment, the reflectance of the blouse in Table 3.10 was 75 and the T-shirt 60. These data show that it is possible to accumulate in several years 1-2% of particulate soil on a cotton garment that cannot be removed easily.

X-ray diffraction of the particulate material from the T-shirt in Table 3.10 showed the presence of alpha quartz and magnesium aluminum silicates (clays) (43). Emission spectrographic analysis revealed the major elements to be Al, Si, Fe, Ti, Mg, and Na, in that order. The values reported for these elements are very similar to the data given in Table 3.11 for the same sample analyzed by flame-emission spectrography and atomic absorption (44).

The original colors of the natural soils as indicated by their reflectances in Table 3.10 were in the same general range as the sample prepared with silica and 0.25% soot. This was almost the identical amount used by Sanders and Lambert when they mixed zinc oxide and carbon black. The addition of 0.25% soot to the dark clay did not appreciably lower the reflectance of the soil but blending 1% soot with the clay produced the darkest soil studied.

The decolorization rate of the samples varied but the dark clay was white in less than 1 hour. One hundred mg of soil of each type was bleached with 40 ml of 5% sodium hypochlorite at room temperature. The clay containing 1% soot was the slowest artificial soil to decolorize. Since all

TABLE 3.11

Major Elements Present in Particulate Soil Removed from a Cotton T-Shirt[a]

Element	Weight percent[b]	Detection method	
		Flame emission	Atomic absorption
Aluminum	4.4	X	
Silicon	3.0		X
Iron	1.3		X
Titanium	1.0	X	
Magnesium	0.6	X	
Sodium	0.5	X	
Calcium	0.4		X
Zinc	0.1		X
Potassium	0.1	X	

[a]From Blickensderfer (44).

[b]Analysis made on ashed sample but percentages calculated on original sample weight.

the samples lost their original black or dark-gray color, no information was obtained by this method that might indicate how much free carbon was present in the particulate soil samples from the garments, or the vacuum cleaner, or filter soils. That soot could be decolorized under these conditions was an unexpected result.

These facts leave one on the horns of a dilemma when trying to determine how much, if any, free carbon is present in a sample of soil containing siliceous materials.

IX. SOILS ON SYNTHETIC FIBERS

Most of the discussion in this chapter has been about the analysis of soils from cotton and not much has been said about the soils on synthetic fibers. However, all the fibers are exposed to the same soils and it is beyond the scope of this chapter to discuss the soiling propensities of various synthetic fibers other than for a few general remarks.

It has been our observation, as well as that of many others, that in model systems clean nylon and polyester surfaces do not have much attraction for hydrophilic siliceous soils. Hydrophobic soils such as carbon black will adhere to clean surfaces of these fibers.

Unfortunately, polyester does have a strong affinity for fatty soils and if enough detergent is not used, will actually scavenge fat from the wash bath. This fatty surface will then increase the adsorption of dirt particles present in the wash water. Bowers and Chantrey (20) have shown that the loss in reflectance caused by deposition of soil from the wash bath is proportional to the amount of fat on the polyester fiber. Bey (24) did not find any selective retention of lipid classes in his study of sebum removal from polyamide, polyester, and polyacrylonitrile.

We found on several occasions that most of the larger particles trapped by fat buildup on nylon and polyester were released by the extraction of the fat. In two instances these particles had the appearance and infrared spectra of talc, $[H_2Mg_3(SiO_3)_4]$, an acid magnesium silicate which is a major component of many types of powders. In other samples the infrared spectra of the particulates were typical of aluminum magnesium silicates or silicon dioxide (45).

REFERENCES

1. M. J. Marples, The Ecology of the Human Skin, Thomas, Springfield, Ill., 1965, Chap. 1.
2. N. Nicolaides and R. E. Kellum, J. Amer. Oil Chem. Soc., 42, 685 (1965).
3. S. Rothman, Physiology and Biochemistry of the Skin, Univ. of Chicago Press, Chicago Ill., 1954.
4. R. H. Mattoni and G. H. Sullivan, MRL TRD 62-68, Wright-Patterson Air Force Base, Ohio, June 1962.
5. W. C. Powe, unpublished work, 1964.
6. Y. Kuno, Human Perspiration, Thomas, Springfield, Ill., 1956.
7. H. J. Hurley and W. B. Shelley, The Human Apocrine Sweat Gland in Health and Disease, Thomas, Springfield, Ill., 1960.
8. M. B. Sulzburger, in The Human Integument (S. Rothman, ed.), Publ. No. 54, Amer. Ass. Advan. Sci., Washington, D.C., 1959, p. 150.
9. K. Bey, Fette, Seifen, Anstrichm., 65, 611 (1963).
10. S. Tomiyama and M. Iimori, J. Amer. Oil Chem. Soc., 42, 449 (1965).
11. O. Oldenroth, Fette, Seifen, Anstrichm., 61, 1220 (1959).
12. E. Walter, ibid., 61, 188 (1959).

13. W. C. Powe, Div. of Colloid and Surface Chem., 148th National Meeting, Amer. Chem. Soc., Chicago, Ill., Sept. 1964.

14. N. Nicolaides, J. Amer. Oil Chem. Soc., 42, 691 (1965).

15. N. Nicolaides, ibid., 42, 708 (1965).

16. W. C. Powe, unpublished work, 1960.

17. E. Haahti, Scand. J. Clin. Lab. Invest., 13, Suppl. 59, 76-84 (1961).

18. E. Haahti, T. Nikkari, and K. Juva, Acta Chem. Scand., 17, 538 (1963).

19. W. C. Powe, unpublished work, 1961.

20. C. A. Bowers and G. Chantrey, Text. Res. J., 39, 1 (1969).

21. W. C. Powe, unpublished work, 1961.

22. W. C. Powe, unpublished work, 1964.

23. W. C. Powe and W. L. Marple, J. Amer. Oil Chem. Soc., 37, 136 (1960).

24. K. Bey, Fette, Seifen, Anstrichm., 66, 579 (1964).

25. W. C. Powe, unpublished work, 1961.

26. O. Oldenroth, Fette, Seifen, Anstrichm., 66, 704 (1964).

27. R. E. Wagg, D. M. Meek, and A. L. D. Willan, Proc. 4th Int. Congr. Surface Activity, Brussels, 1964, Vol. 3, p. 269.

28. W. C. Powe, unpublished work, 1963.

29. B. A. Scott, J. Appl. Chem., 13, 133 (1963).

30. W. C. Powe, J. Amer. Oil Chem. Soc., 40, 290 (1963).

31. H. Reumuth, "Der Schmutz in seiner ganzen Vielfalt," Reiniger-Revue Chem. Reinigung Färberei, Baden-Baden, Heft 5, 6, 7, 8, 9 (1965).

32. W. C. Powe, Text. Res. J., 29, 879 (1959).

33. W. P. Evans and M. Camp, Proc. 4th Int. Congr. Surface Activity, Brussels, 1964, Vol. 3, p. 259.

34. R. E. Grim, Clay Mineralogy, McGraw-Hill, New York, 1953.

35. S. B. Seeley and E. Emendorfer, Encyclopedia of Chem. Technology, Interscience Encyclopedia, Inc., New York 1954, Vol. 3, p. 86.

36. W. R. Smith, ibid., Vol. 3, pp. 34-65.

37. H. L. Sanders and J. M. Lambert, J. Amer. Oil Chem. Soc., 27, 153 (1950).

38. R. McCarthy and C. E. Moore, Anal. Chem., 24, 411 (1952).

39. G. Walter, Wäscherei-Tech. und -Chem., Heft 11, 886 (1958).

40. T. J. Miranda, Whirlpool Corp., unpublished work, 1971.

41. W. C. Powe, unpublished work, 1971.

42. W. C. Powe, unpublished work, 1959.

43. I. E. Levine, California Research Corp., private communication, 1961.

44. P. W. Blickensderfer, Whirlpool Corp., unpublished work, 1970.

45. W. C. Powe, unpublished work, 1959, 1964.

Chapter 4

THEORIES OF PARTICULATE SOIL ADHERENCE AND REMOVAL*

W. G. Cutler
Whirlpool Corporation
Benton Harbor, Michigan

R. C. Davis
Whirlpool Corporation
Benton Harbor, Michigan

H. Lange
Henkel and Cie GmbH
Duesseldorf, West Germany

*(Editor's Note: This chapter is an abridgment of Chapter 4, "Physical Chemistry of Cleaning Action," by H. Lange, in Solvent Properties of Surfactant Solutions, K. Shinoda, ed., Dekker, New York, 1967.)

I. GENERAL REMARKS ON SOILING AND CLEANSING

A. Definitions

Dirt has sometimes been defined as "matter in the wrong place" (1). To define soil in an appropriate manner, certain restrictions must be introduced, at least when speaking about the physical chemistry of cleansing action. One of these restrictions is concerned with the geometric extension of the undesired material. This extension shall be so small in one or two or three dimensions that the material cannot be removed by rough mechanical means, e.g., by lifting with a spatula. Therefore, we are concerned only with small particles or thin coatings. Furthermore, the material characterized as dirt shall adhere sufficiently to the object to be cleaned as not to be removable by such simple mechanical means as beating, brushing, or wiping with a duster. Finally, materials readily soluble in water shall not be regarded as dirt in the sense used here (except when we discuss dry cleaning).

Dirt may be solid or liquid at any given temperature. Furthermore, it may be present in some state intermediate between the solid and the liquid, e.g., plastic or viscoelastic. Solid dirt, or—as it often is named—soil, is present in most cases as discrete particles small in all three dimensions or as fiber fragments. Liquid (oily) dirt is mostly present as a thin, more or less evenly distributed layer on the object to be cleaned, or "substrate." Dirt usually consists of an inhomogeneous mixture of solid and liquid components. The substrate as well as the solid particles is wetted by the liquid component in this case.

B. Adhesion of Soil Particles

As mentioned, the term "dirt" connotes a certain degree of adhering strength at the substrate, and the nature of the adhesive forces will therefore be considered. The strength of adhesion may be expressed quantitatively in a relatively clear way for liquid dirt as the free energy of adhesion at solid-liquid interfaces, i.e., at the interface between a liquid dirt and a substrate. When taking account of the adhesion of solid particles at a substrate the problem is always much more complicated. This is because solid-solid adhesion rarely occurs evenly over an extended region of the solid surface. Solids generally adhere at relatively few points or, more exactly, in very small regions of the zone of adhesion. The size of these regions is determined not only by the surface geometry of the particle and substrate but also by their elastic and plastic properties since the adherents deform each other at the region of contact (2-4). This is shown schematically in Fig. 4.1.

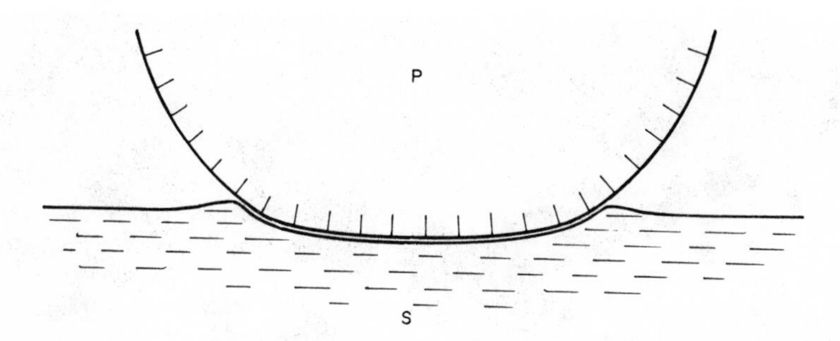

FIG. 4.1. Adhesion of a particle P on a substrate S with deformation by attractive forces.

Generally, the main cause of adhesion is dispersion forces or van der Waals forces. Any additional forces of adhesion due to electrical contact potentials are usually much lower than the van der Waals forces. Electric excess charges that may be generated by the frictional electricity enhance the speed of soiling by catching dust from the air but do not increase the strength of adhesion (2, 4, 5).

More detailed information about the strength of adhesion has been obtained recently (6, 8, 9) by direct measurements using the centrifuge method (6, 7). Forces of adhesion showing broad distribution curves between about 0.01 and 0.5 dyn have been observed for spherical gold particles having diameters of 3-8 μ on substrates of different types. The force is approximately proportional to the diameter of the particle and higher at smooth surfaces of the adherents than at rough ones. Regarding the hardness of the adherents it has been observed that the strength of adhesion is highest if the degree of deformability of the adherents is very different. The strength of adhesion increases with the time of contact. In a humid atmosphere the strength of adhesion is higher than in a dry one. The distribution curves are in accord with theoretical considerations based on the statistics of the surface roughness (2-4).

In certain cases soil particles are not only held by forces of adhesion at the substrate but also may be occluded in microscopic holes or crevices. This kind of case has been discussed sometimes, especially for textile material (10, 11). However, investigations by Götte et al. (12), with the aid of an electron microscope, have shown that no considerable amounts of soil can be occluded in the interior of the fibers. Almost all the soil is deposited at the surface of the fibers. An example is shown in Fig. 4.2, which is a picture taken with the electron microscope of an iron oxide soiling on cotton (12).

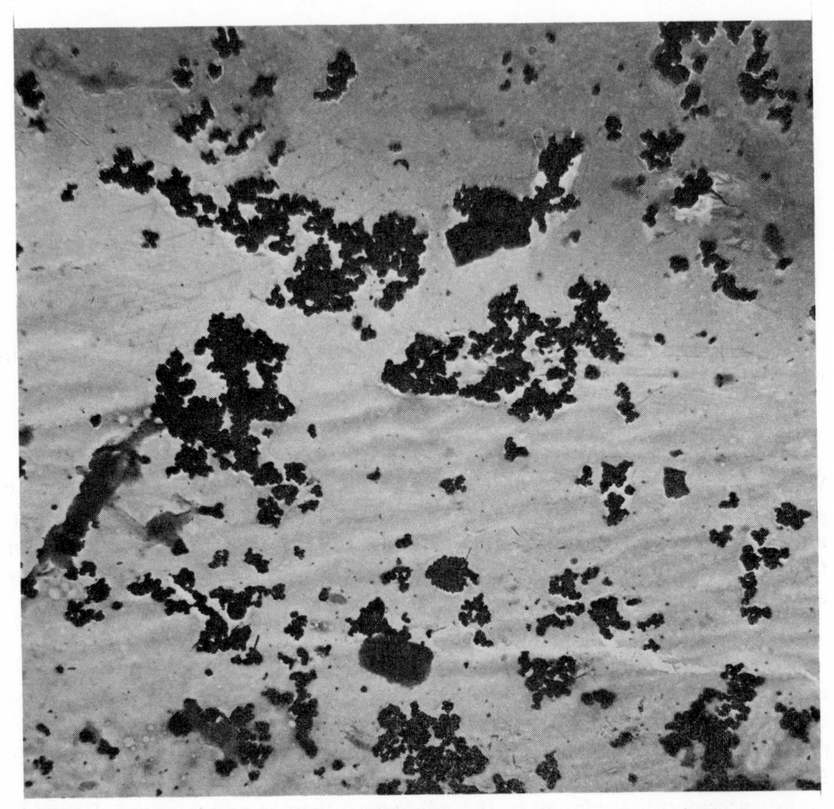

FIG. 4.2. Cotton fiber soiled with iron oxide (magnification 5000 ×).

II. ATTACHMENT OF SOILS TO SUBSTRATES

A. Adsorption of Surfactants as a Condition of Cleansing Action

Surfactants are characterized by their property of being capable of adsorption at interfaces of all types. When the cleansing action of surfactants is considered, their adsorption on the substrate and on the soil is important. The mechanical, electrical, and chemical properties of the substrate-liquid and soil-liquid interfaces are altered by the adsorption. These alterations are the most important condition for the efficacy of surfactants in the cleansing process. The degree of adsorption depends on the type and concentration of the surfactant, on other materials present in solution, on the temperature, and particularly on the type of the adsorbent, i.e., the substrate or soil.

Some work on the adsorption of surfactants from aqueous solutions on soil pigment of different types has been published. Weatherburn et al. (13, 14) have measured the adsorption of soaps and other anionic as well as cationic surfactants on carbon black as a function of the concentration. For the homologous series of the sodium alkyl sulfates it was found that the adsorption isotherms shift to lower concentrations with increasing chain length. With increasing concentration, the adsorption strongly rises up to the critical micelle concentration (cmc). Above this concentration adsorption occurs to a much smaller extent or not at all. The adsorption increases by the addition of sodium sulfate, and decreases by raising the temperature.

Furthermore, the adsorption of surfactants is largely influenced by the type of soil pigment. Its hydrophily and, in certain cases, its ability to exchange ions are especially important in this connection. Moillet (15) has pointed out that an adsorption with reversed orientation, i.e., with the hydrophilic group of the surfactant molecule pointing toward the soil particle surface and with the hydrophobic group oriented toward the aqueous medium, takes place in certain cases. The polar, hydrophilic group then may be bonded either by electrostatic forces or by ion exchange. When investigating the adsorption of ionic surfactants on certain polar particles such as, e.g., alumina and ion-exchange resins, Tamamushi and Tamaki (16) found that a superposition of an adsorption caused by electrostatic attraction or by ion exchange and an "ordinary" adsorption, i.e., an adsorption caused by van der Waals attractive forces, occurs. Adsorption isotherms of sodium dodecyl sulfate on soil pigments of different types (17) are shown in Fig. 4.3.

The adsorption of surfactants on the substrate is generally governed by the same laws as those on soil particles. Extended work has been done by several authors with textile materials as substrate. However, for textiles, it must be realized that adsorption in its proper sense, i.e., accumulation on the substrate-water interface, is largely swamped in most cases by sorption in the interior of the fibers and cannot be experimentally separated from it. Weatherburn et al. (18, 19) have measured the sorption of surfactants by different natural and man-made textile fibers. The lowest sorption of anionic surfactants has been observed on cotton, the highest one on wool.

Sometimes the sorption of surfactants on fibers is mainly caused by certain accompanying materials present in them. Ginn et al. (20) have stated that cotton nearly completely loses its ability of sorbing anionic surfactants if the waxy components have been removed from it. Schwarz et al. (21, 22) have pointed to the role of calcium ions accumulating on the cotton during repeated washes in hard water. Some sorption isotherm of sodium dodecyl sulfate on different fibers (17) are shown in Fig. 4.4.

An extended review of the adsorption of surfactants on soil and substrate has been given by Harris (23). Kölbel and Hörig (24) have found some

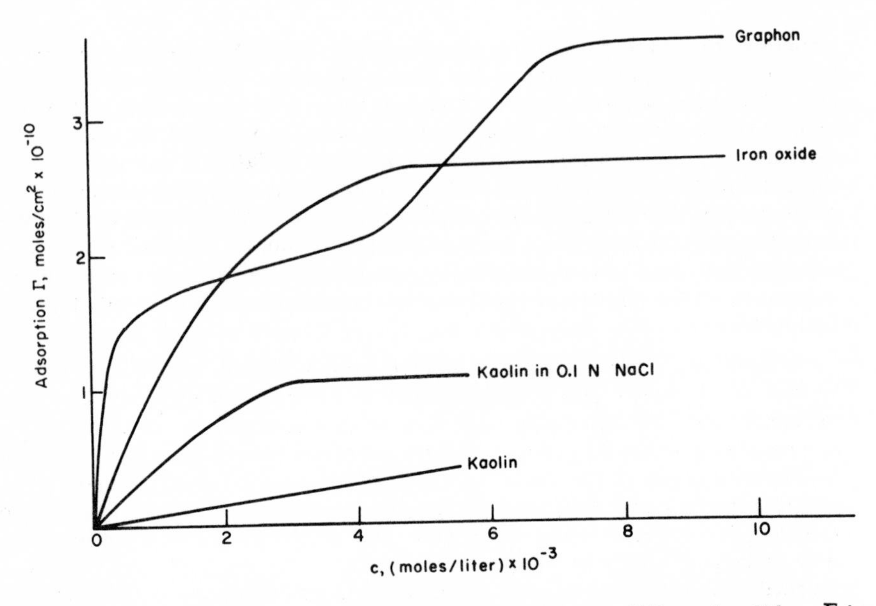

FIG. 4.3. Adsorption of sodium dodecyl sulfate on different solids: Γ is the amount adsorbed per unit surface area; c is the equilibrium concentration.

correlations between the extent of sorption by different types of fibers and their chemical structure.

B. Solid Soil Particles

1. Correlation to Problem of Stability of Hydrophobic Colloids

As explained in Sec. I,B, soil particles generally adhere to the substrate by van der Waals forces. When such a system is immersed in water, the energy of adhesion is altered. A new van der Waals constant A results from the constant A_{12} for the dry system and the constants A_{00}, A_{10}, and A_{20} for the mutual interaction of the water molecules and for the interaction of these with the substrate and the soil particles according to an equation derived by Hamaker (25):

$$A = A_{12} + A_{00} - A_{10} - A_{20} \tag{4.1}$$

Exact numerical values of the single constants are scarcely known. Generally, A will have values between 10^{-13} and 10^{-12} erg. Some values obtained by different methods have been collected in Table 4.1. Additional values are given by Lyklema (25a).

Measurements by the centrifuge method have shown that the force of adhesion of small solid particles at a solid substrate is greatly diminished after immersion in water (33, 34). This is clearly demonstrated by the example shown in Fig. 4.5. The interaction between solid particles and substrate surrounded by an aqueous medium is also greatly influenced by the outer diffuse parts of the electric double layers formed at the particle-liquid and substrate-liquid interfaces. As the sign of a charge of the double layers is the same for particle and substrate (without regard to exceptional cases) the mutual interpenetration of the layers results in a repulsion. This electrical repulsion is superimposed on the van der Waals attraction. The Born repulsion, having a much smaller range than the other types of interaction mentioned, is important in this connection only insofar as it determines the smallest possible distance between particle and substrate.

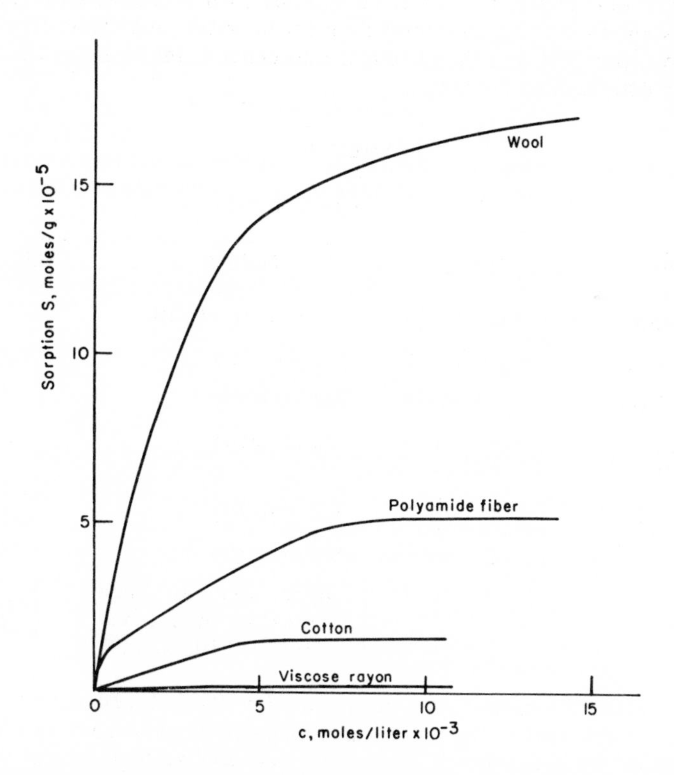

FIG. 4.4. Sorption of sodium dodecyl sulfate on different textile fibers; S is the amount adsorbed per gram of fiber material. (The curve for wool is based on data given by M. Selck, dissertation, Dresden, 1935.)

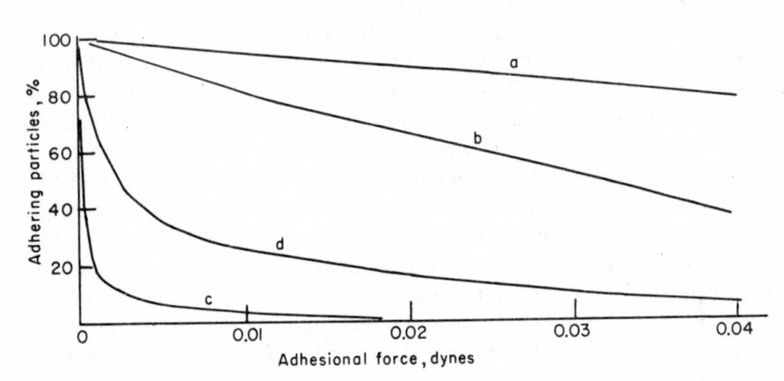

FIG. 4.5. Adhesion of spherical gold particles (3 μ diameter) on a polyester foil (a) in air, 65% relative humidity; (b) in water, solid-solid contact established prior to immersion; (c) in water, adherents first immersed separately and then brought into contact; (d) as in (c) but 12 days after establishing contact.

TABLE 4.1

Values of Some Constants

Substance	ergs A	Method	References
Polystyrene	$\sim 5 \times 10^{-13}$	From optical data	(26)
Paraffin wax	$\sim 1.6 \times 10^{-13}$	Sol stability	(27)
Arachidic acid	$\sim 10^{-14}$	Sol stability	(28)
Silicic acid (amorphous)	$2\text{-}3 \times 10^{-14}$	Sol stability	(29)
Silver iodide	$\sim 5 \times 10^{-13}$	Sol stability	(30)
Selenium	$\sim 2 \times 10^{-13}$	Sol stability	(31)
Platinum	$\sim 3 \times 10^{-12}$	Direct force measurement by the method of crossed filaments	(32)
Gold	$\sim 4 \times 10^{-12}$		

To understand the cooperation of the van der Waals force and the electrical interaction at the cleansing process it is necessary to consider the forces or the corresponding potential energies as functions of the distance between the particle and the surface of the substrate. This problem is very similar to that of the theory of the stability of hydrophobic sols as developed by Derjaguin and Landau (35) as well as by Verwey and Overbeek (36) on the basis of the cooperation of electrical and van der Waals interactions (37-40).

The calculations published by the authors of (35, 36), however, hold only for the interaction between two parallel platelike particles or between two spherical ones. The plate-plate model is important for the theory of the cleansing process only insofar as soil particles adhere to the substrate surface with plane faces as, for example, in the case of crystal faces. This is the case, in particular, for the platelike crystals of clay minerals. On the other hand, for particles more approaching the shape of a sphere or having curved surfaces the model of a sphere and a plate is more reasonable.

As will later be explained in detail, the calculations for the potential energy of the van der Waals attraction V_A, the double-layer repulsion V_R, and the resultant of them $V = V_A + V_R$ as functions of the distance x in the plate-plate or the sphere-plate model lead to curves the general appearance of which is shown in Fig. 4.6. The initial parts of the curves are indefinite for the present. The meaning of the distance δ will be explained later. The curve for V first rises steeply with increasing x, passes a maximum of the height E, and finally falls approaching the abscissa axis. Here, the potential energy at infinite distance is arbitrarily chosen as the zero level of the ordinate axis. As a zero point of the abscissa axis, the distance between particle and substrate when directly adhering one to the other is chosen, i.e., the minimum distance allowed by the Born repulsion. A remote particle has to surpass the energy barrier E when approaching the substrate surface. The height of E largely depends on the potential ψ_0 of the electrical double layer. Starting from this view, Durham (37, 41) has correlated the amount of redeposition of soil to the value of ψ_0.

As will be explained later, the curves can be calculated when some approximations are introduced. However, the starting point of the curves cannot be calculated in this way. Apart from other reasons, such calculations cannot be made because the particle directly adheres at the substrate in the starting state. The process of the penetration of the liquid between particle and substrate requires special considerations.

2. Energetics of Removal of Soil Particles

The case of soil particle having a plane face, e.g., a crystal face, and adhering with a likewise plane substrate surface may be considered first. In the beginning state substrate and particle may be wetted by the surrounding liquid but no liquid may have penetrated between the two in the zone of contact—state I in Fig. 4.7. The complete removal of the particle from the substrate may be divided into two steps. Step 1 comprises the penetration of a thin liquid layer between substrate and particle. The state reached after step 1, i.e., state II in Fig. 4.7, may correspond to a distance δ that must be at least as high as to allow the formation of layers of solvation or adsorption of any substances dissolved in the liquid, especially surfactants and the electrical double layers. On the other hand, δ must be so small that V_A and V_R are still quite high. Apart from that,

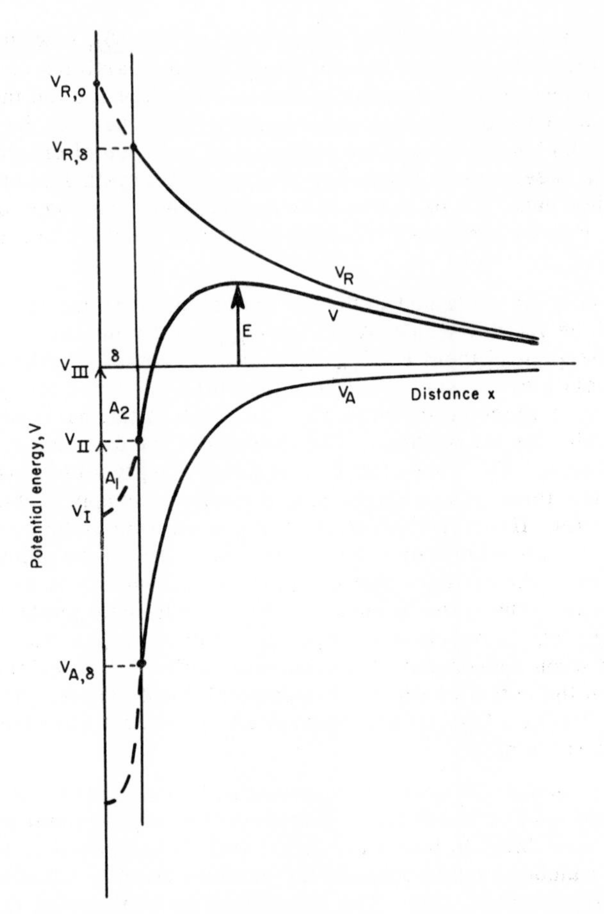

FIG. 4.6. Potential energy $\underline{V}$ of a particle-substrate system vs. distance $\underline{x}$; $\underline{V}_R$ is the electrical double-layer repulsion energy, $\underline{V}_A$ is the van der Waals attraction energy.

the value of δ may be quite arbitrary. Step 2 comprises the final removal of the particle from the substrate, i.e., the separation up to a distance so large that no forces of interaction are effective—state III.

When considering how the total work necessary for the removal of a particle from the substrate is composed of the amounts of work connected with both steps, the scheme of potential energies given in Fig. 4.7 may be used; $\underline{V}_I$, $\underline{V}_{II}$, $\underline{V}_{III}$ are the potential energies in the states I, II, and III, respectively, for 1 cm^2 of initial contact area. The scheme has been drawn for $\underline{V}_{III} > \underline{V}_{II} > \underline{V}_I$. This order is not necessary; the following considerations do not depend on that assumption.

Step 1 may be considered first. It will be divided into two partial steps. First the soil particle adhering on the substrate is removed to the distance δ without liquid penetrating between soil and substrate. For this process an amount of work $\underline{W}_1$ must be done against the van der Waals attraction. Thereafter the liquid is considered to penetrate between substrate and particle. In this process an amount of work equal to the sum $\underline{J}^*$ of the

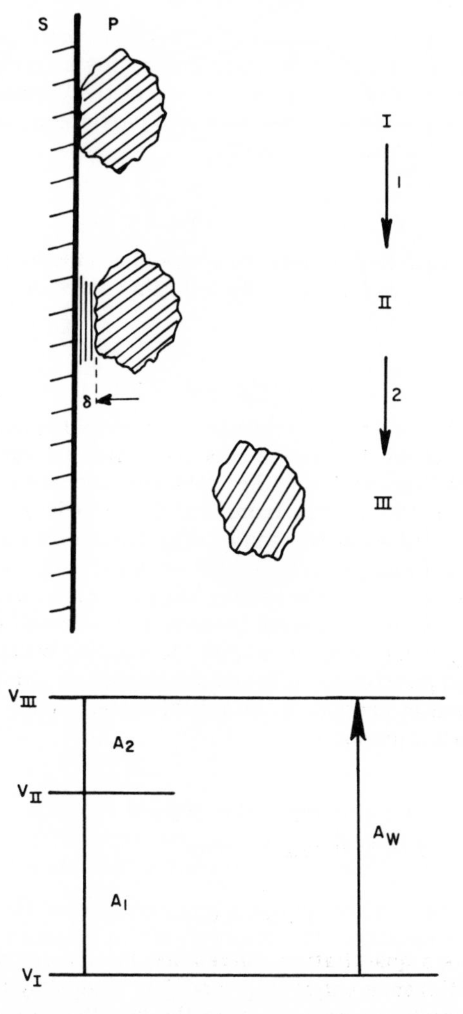

FIG. 4.7. Stepwise removal of a particle $\underline{P}$ from a substrate $\underline{S}$ in a liquid $\underline{L}$. Diagram of the corresponding energy levels.

wetting tensions $\underline{j}_S^*$ and $\underline{j}_P^*$ of the liquid at the substrate and the particle is gained. The asterisks of $\underline{j}_S^*$ and $\underline{j}_P^*$ mean that these terms differ from the corresponding terms in the free liquid on account of the interactions existing between substrate and particle at the small distance δ.

Hence, the work to be done for step 1 is given by

$$\underline{A}_1 = \underline{W}_1 - \underline{J}^* \tag{4.2}$$

The work $\underline{A}_2$ for step 2 is composed of the contributions of the interactions caused by the van der Waals attraction and by the electrical double layers; $\underline{A}_2$ is equal to the total potential energy $\underline{V}_\delta$ at the distance δ if state III where no interaction between substrate and particle is chosen as the reference state for the potential energy:

$$\underline{A}_2 = -\underline{V}_\delta \tag{4.3}$$

The work to be done for the total process of removing an adhering particle $\underline{A}_W$ is equal to the sum $\underline{A}_1 + \underline{A}_2$ and, therefore, according to Eqs. (4.2) and (4.3),

$$\underline{A}_W = \underline{W}_1 - \underline{J}^* - \underline{V}_\delta \tag{4.4}$$

To use Eqs. (4.2) and (4.4) practically, one should know something about $\underline{J}^*$, especially about the difference between it and the sum $\underline{J} = \underline{j}_S + \underline{j}_P$ of the "ordinary" wetting tensions. For this purpose one may imagine step 2 carried out in a different way from that described. First, the thin liquid layer that had penetrated at the end of step 1 is removed while the distance δ between substrate and particle is kept constant. The work $\underline{J}^*$ must be expended for this purpose. Thereafter the particle is brought to a large distance without any liquid flowing between particle and substrate. For this, the work $\underline{W}_2$ must be done against the van der Waals attraction. Finally, the liquid penetrates between substrate and particle and wets the faces that had been in contact in the initial state. Hence, the total work to be done for step 2 amounts to

$$\underline{A}_2 = \underline{J}^* + \underline{W}_2 - \underline{J} \tag{4.5}$$

From Eqs. (4.3) and (4.5) it follows that

$$\underline{J} - \underline{J}^* = \underline{W}_2 + \underline{V}_\delta \tag{4.6}$$

Equation (4.6) is a quantitative expression for the consideration given above that the difference between $\underline{J}^*$ and $\underline{J}$ is caused by the interaction at the distance δ, which can be seen from the fact that $\underline{W}_2$ is equal to the difference of the potential energies for the distance δ and for a very large distance in the unwetted state and $-\underline{V}_\delta$ for the wetted state.

Now $\underline{A}_W$ can be expressed without making use of the term $\underline{J}^*$. From Eqs. (4.4) and (4.6) it follows that

$$\underline{A}_W = \underline{W}_1 + \underline{W}_2 - \underline{J}$$

or, in a simplified form,

$$\underline{A}_W = \underline{W} - \underline{J} \tag{4.7}$$

where $\underline{W} = \underline{W}_1 + \underline{W}_2$. The correlations given by Eqs. (4.2)-(4.7) can be clearly demonstrated by the diagram shown in Fig. 4.8.

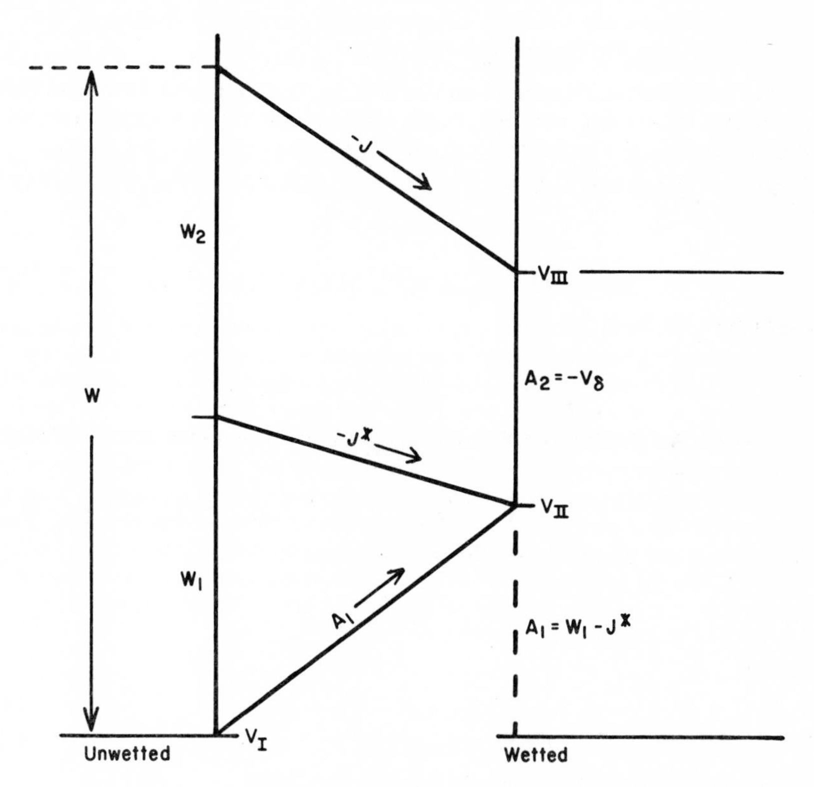

FIG. 4.8. Diagram illustrating Eqs. (4.2)-(4.7).

Now it may be considered in what manner the magnitudes contained in Eqs. (4.2), (4.3), and (4.7) are influenced by any substances dissolved in the water. $\underline{W}_1$ and $\underline{W}$ depend only on the material of substrate and soil. Hence, they may be disregarded in this connection. $\underline{J}^*$, $\underline{V}_\delta$, and $\underline{J}$ remain to be considered.

Some information about J is obtained by expressing the wetting tensions j_S and j_P, the sum of which is J, by the corresponding surface tensions γ_S and γ_P of substrate and particle as well as by the interfacial tensions between them and the liquid:

$$j_S = \gamma_S - \gamma_{SL} \quad \text{and} \quad j_P = \gamma_P - \gamma_{PL} \tag{4.8}$$

When there is a film adsorbed from the solution at the interfaces, γ_{SL} and γ_{PL} are smaller than the interfacial tensions γ^0_{SL} and γ^0_{PL} in the absence of the adsorbable solute by the spreading pressures π_{SL} and π_{PL}, respectively, of the films. Hence

$$\gamma_{SL} = \gamma^0_{SL} - \pi_{SL} \quad \text{and} \quad \gamma_{PL} = \gamma^0_{PL} - \pi_{PL} \tag{4.9}$$

If the adsorbed substance is an electrolyte and therefore forms an electrical double layer, π_{SL} and π_{PL} can be divided each into a nonelectrical or thermodynamical component π'_{SL} or π'_{PL}, respectively, and a component $\Delta\pi_{SL}$ or $\Delta\pi_{PL}$, respectively, caused by the energy of the double layer—cf., e.g., (42, 43):

$$\pi_{SL} = \pi'_{SL} + \Delta\pi_{SL} \quad \text{and} \quad \pi_{PL} = \pi'_{PL} + \Delta\pi_{PL} \tag{4.10}$$

From Eqs. (4.8)-(4.10)

$$j_S = \gamma_S - \gamma^0_{SL} + \pi'_{SL} + \Delta\pi'_{SL}$$

as well as the analogous equation for j_P results. So, one finally obtains

$$J = \gamma_S + \gamma_P - (\gamma^0_{SL} + \gamma^0_{SP}) + \pi'_{SL} + \pi'_{PL} + \Delta\pi_{SL} + \Delta\pi_{PL} \tag{4.11}$$

Hence, one can divide J into three components:

$$J = J^0 + J' + \Delta J \tag{4.12}$$

Herein

$$J^0 = \gamma_S + \gamma_P - (\gamma^0_{SL} + \gamma^0_{PL}) \tag{4.13}$$

is the contribution of the pure solvent (water) and

$$J' = \pi'_{SL} + \pi'_{PL}) \tag{4.14}$$

is the nonelectrical contribution and

$$\Delta J = \Delta\pi_{SL} + \Delta\pi_{PL} \tag{4.15}$$

is the electrical contribution of the adsorbed film. In these relations the contribution of any substances dissolved in water (surfactants and other compounds) to J is expressed in a manner to be explained in detail later on.

From Eq. (4.7) it follows that an increase of J causes a diminution of the total work A_W necessary for the removal of an adhering particle. This means with regard to Eq. (4.12) that a surfactant contributes to the diminution of A_W via both the nonelectrical part and the electrical part ΔJ. As may be seen from the foregoing explanations, the total potential energy V at the distance x is composed of the terms V_A and V_R for the van der Waals and the double-layer interactions, respectively

$$V = V_A + V_R \tag{4.16}$$

The amount of V_R is given by the difference of the free energies F_x and F_∞ of the double layers on both sides for the distances x and ∞:

$$V_R = 2(F_x - F_\infty) \tag{4.17}$$

Here, for the sake of simplicity, it has been assumed that F_x and F_∞ are equal at both sides, i.e., at the substrate and the particle. This, presupposes that the interfacial potential is equal at both sides. When substrate and particle are covered by a nearly saturated monolayer of a surfactant, this condition will be fulfilled to a good approximation. Moreover, in this case Eq. (4.15) is simplified to

$$\Delta J = 2\ \Delta\pi \tag{4.18}$$

The electrical component $\Delta\pi$ of the interfacial tension lowering is equal to the decrease of the free energy of the system caused by the formation of the free double layer, i.e., the double layer not interacting with another one. Hence.

$$\Delta\pi = -F_\infty \tag{4.19}$$

From Eqs. (4.17)-(4.19)

$$V_R = 2(F_x + \Delta\pi) = 2F_x + \Delta J \tag{4.20}$$

follows for the distance δ, correspondingly,

$$V_{R,\delta} = 2(F_\delta + \Delta\pi) = 2F_\delta + \Delta J \tag{4.21}$$

is obtained.

These correlations are demonstrated by Fig. 4.9. The free energy of the double layer F_x as a function of the distance x is plotted there. The curve shown has been calculated by using the theory of the electrical double layer (36) for a special example, viz., for the potential $z = 4$ (the

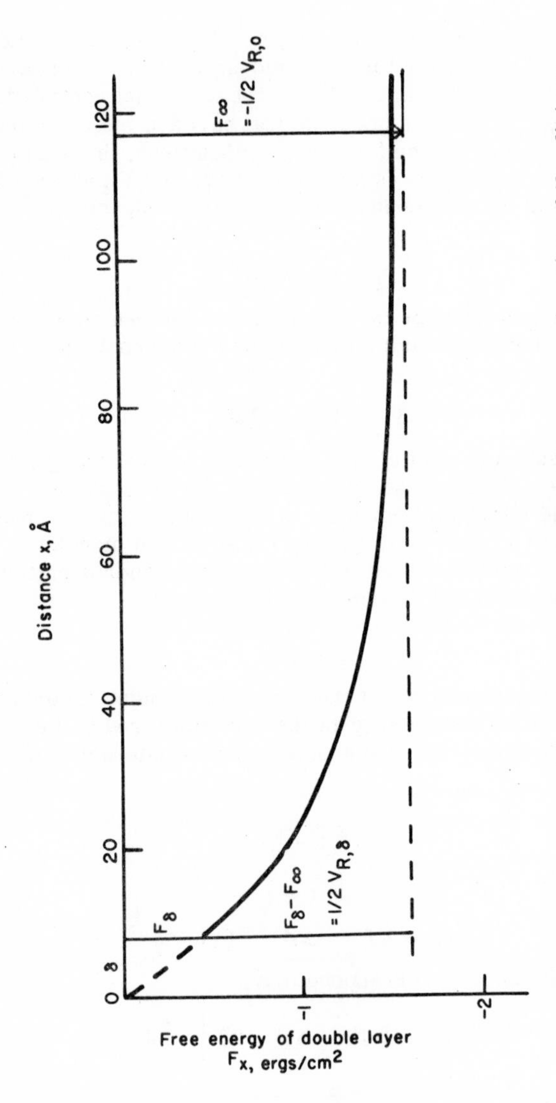

FIG. 4.9. Free energy F_x of the electrical double layer as a function of the distance x. Calculated curve; see text.

meaning of z as a measure of the interfacial potential is explained in the following section), and a concentration of 10^{-2} moles/liter of a 1-1 valent electrolyte.

F_x must be zero for $x = 0$. This simply follows from the fact that no electrolyte solution can be present between two adhering plates, and therefore no double layer can be formed. With increasing x the curve falls down and asymptotically approaches F_∞. As the double layer is spontaneously formed, F_∞ must be negative. The fact that the curve only gradually descends to F_∞ is caused by the interaction of both double layers. It means, at the same time, that some electrical repulsion occurs at finite distances.

$V_{R,\delta}$ is always smaller than ΔJ by the amount of $2F_\delta$, as can be seen directly from Fig. 4.9. As $V_{R,\delta}$ is the electrical part of V_∞, this means with regard to Eqs. (4.3) and (4.7) that A_2 is diminished by the electrical double layer, but that the diminution of A_2 is less by $2F_\delta$ than that of A_W. The nonelectrical part J' does not influence A_2.

As A_1 is the difference between A_W and A_2, it follows from the foregoing that the electrical double layer contributes to the diminution of A_1 by the amount of $2F_\delta$. The nonelectrical part J' is completely effective in diminishing A_1. The same result is obtained by substituting J^* in Eq. (4.2) by using Eq. (4.6) and setting $V_\delta = V_{A,\delta} + V_{R,\delta}$.

To sum up, any substances dissolved in water, especially surfactants, can influence the work A_W necessary for the complete removal of adhering particles as well as the partial amounts A_1 and A_2 in the manner listed in Table 4.2.

It may be quite possible that there are systems in which $J^* > W_1$ for a small region of δ. Then A_1 is negative according to Eq. (4.2). In this case the first penetration of a liquid layer will occur spontaneously.

When the adhering particles have curved surfaces, the situation is different in many respects. The model of a spherical particle at a flat

TABLE 4.2
Plate-Plate Model

Type of influence	A_1	A_2	A_W
Nonelectrical part J' of the spreading pressure	Decreased	Uninfluenced	Decreased
Electrical double layer influencing ΔJ as well as $V_{R,\delta}$	Decreased	Decreased	Decreased

substrate surface may be considered as the simplest case. However, the following considerations are also valid for any other curved surface. From the standpoint of pure geometry there is no contact area at all between a body having a curved surface and a second body having a flat surface. Nevertheless, in reality there are always contact areas of small, but finite dimensions between them because an elastic or plastic deformation of the adhering bodies at the contact region occurs on account of their mutual attraction (2, 3). The case may be considered in which the contact area generated by the deformation with the radius r_1 is much smaller than the region with the radius r_2 where a perceptible interaction between substrate and particle still occurs; see Fig. 4.10. There, d is half the

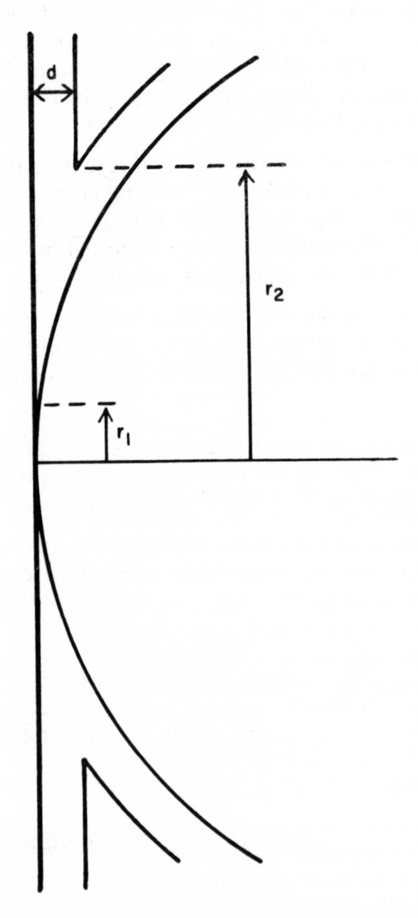

FIG. 4.10. Contact between sphere and plate with zone of interaction.

distance of the latter region, i.e., as it were, the "reach" of the interaction From $r_1 << r_2$, and therefore even more from the fact that $r_{12} << r_2^2$, it follows that the process of the first penetration of liquid into the very small zone of proper contact, i.e., the area with the radius r_1, is not important energetically. Therefore, it is not necessary in this model to consider this process as a separate step when calculating the total work for the removal of the particle. Instead, the removal of the particle may be considered as a single process not to be separated into any steps occurring one after the other. It begins at a distance given by the Born repulsion. This distance may be chosen as the zero point of the distance scale. The work A_W is the sum of the components of the van der Waals attraction and the double-layer repulsion only:

$$A_W = -(V_{R,0} + V_{A,0}) = -V_0 \tag{4.22}$$

The difference between the plate-sphere model and that of two parallel plates is given mainly by the fact that for the former the interaction occurs in a region in which the distance continuously enlarges in the outward direction. It is to be expected, therefore, that geometrically less simple systems in which the distance also continuously grows in the surroundings of a very small area of contact exhibit a behavior very similar to that of the plate-sphere system.

The actual geometry of the adhesion of real particles and substrates is much more complicated, as already mentioned in Sec. I,B. However, zones of direct contact will always alternate with regions where the distance continuously enlarges. Depending on the fraction of zones of direct contact, the behavior of the systems is better described by the model of parallel plates or by the plate-sphere model.

In principle, the equilibrium between particles adhering at the substrate and removed particles should be determined by the magnitude of A_W. The kinetics of the removal of the particles from the substrate as well as of the inverse process, i.e., the deposition or redeposition of the particles, is additionally controlled by the height of the activation barrier E. The kinetic retardation of the deposition is quite analogous to the stability of hydrophobic colloids caused by a similar energy barrier set up by the electrical double layer (35, 36).

3. Calculation of Curves of Potential Energy as a Function of Distance between Particle and Substrate

To obtain some information on the magnitude of the terms A_1, A_2, A_W, and E, it is useful to consider the potential energy V as a function of the distance x between substrate and particle on the basis of the theory of the electrical double layer (35, 36). For a system consisting of a substrate having a plane surface and a particle with a plane face parallel to the substrate surface at a distance $\underline{x}$ and immersed in an aqueous electrolyte solution, the model of two parallel plates can be used. The calculation (36, 38, 39) leads to an equation of the general form

$$\underline{V}_{\underline{R}} = \frac{\kappa}{\nu^2} f(\kappa \underline{d})_{\underline{z}} \tag{4.23}$$

provided that ions of one valency ν only are present. κ is the term known from the Debye-Hückel theory

$$\kappa = \left(\frac{8 \pi \eta}{\epsilon \underline{k} \underline{T}} \nu \underline{e} \right)^{\frac{1}{2}} \tag{4.24}$$

and $\underline{d}$ is equal to $\underline{x}/2$. $\underline{z}$ is defined by

$$\underline{z} = \frac{\nu \underline{e} \psi 0}{\underline{k} \underline{T}} \tag{4.25}$$

where k is Boltzmann's constant, $\underline{T}$ is the absolute temperature, $\underline{n}$ is the concentration of electrolyte expressed by the number of ions of one sign per cubic centimeter, ϵ is the dielectric constant, and $\underline{e}$ is the charge per electron. The value of pure water will be inserted for ϵ in the following calculations. Indeed, ϵ is influenced in the double layer by dielectric saturation on account of the high field strength occurring there. However, this effect does not much influence the measurable properties of the double layer, as has been shown by Grahame (44). ψ_0 is the potential drop in the diffuse part of the double layer and is given the short name "interfacial potential" in the following discussion; $\underline{z}$ is a convenient measure of ψ_0. For $\nu = 1$ and 25°C, $\psi_0 = 25.6\underline{z}$ mV.

Numerical values of the function $f(\kappa d)_{\underline{z}}$ in Eq. (4.23) can be taken from a table given by Verwey and Overbeek (36) for 25°C and different values of $\underline{z}$. Some simplifying assumptions are involved in the calculation. The Gouy-Chapman model (45, 46) of the electrical double layer is regarded as valid. That implies the simplifying assumption of a charge homogeneously distributed over the surface. However, neglecting the "discreteness of charge effect" is of little importance for the structure of the diffuse part of the double layer (47, 48).

For the calculation of the van der Waals attraction energy an approximate formula valid for $\underline{d} \ll \underline{D}$ (where $\underline{D}$ is the thickness of the plate) is sufficient because the interaction is negligible at distances comparable with the dimensions of the particles or substrates. In this case the potential energy of attraction is given by

$$\underline{V}_{\underline{A}} = - \underline{A}/12\pi \underline{x}^2 \tag{4.26}$$

The total work per square centimeter needed for bringing two parallel plates from infinite distance to the distance $\underline{x}$ is given by the sum $\underline{V} = \underline{V}_{\underline{R}} + \underline{V}_{\underline{A}}$. Shown in Fig. 4.11 are calculated curves for $\underline{V}_{\underline{R}}$, and $\underline{V}_{\underline{A}}$, and $\underline{V}$

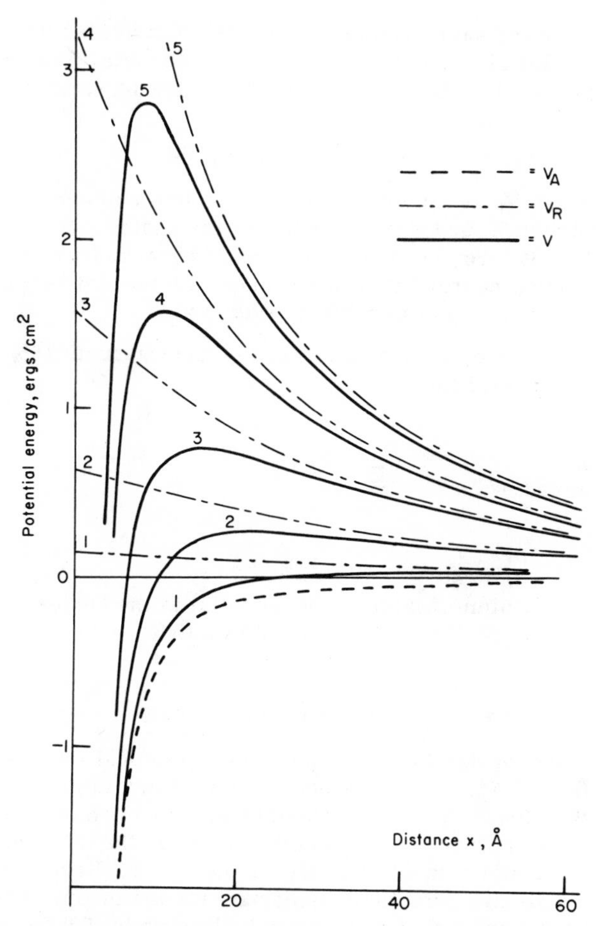

FIG. 4.11. Potential energies $\underline{V}_A$, $\underline{V}_R$, and $\underline{V}$ of two parallel plates for values 1 - 5 of $\underline{z}$ as indicated by the figures.

as functions of the distance $\underline{x}$ for several values of $\underline{z}$. A value of $\underline{A} = 2 \times 10^{-13}$ erg and a concentration of 10^{-2} mole/liter of a 1-1 electrolyte has been assumed there. The curves for $\underline{V}$ have the same general shape as the schematic curve shown in Fig. 4.6. The height of the maximum $\underline{E}$ strongly rises with increasing potential $\underline{z}$.

Before discussing the applications of the $\underline{V}$, $\underline{x}$ curves to the theory of the cleansing action, the sphere-plate model may be considered. Derjaguin (49) has derived some equations enabling us to calculate the interaction energy between two bodies with curved surfaces if this energy is known as a function of the distance for two parallel plates of the same material and immersed in the same medium. This method can also be used to pass over from the electrical interaction energy in the plate-plate model to that in the sphere-plate model. For this the following equation holds (38):

$$V_R = 2\pi a \int_{H_0}^{\infty} V_H \, dH \tag{4.27}$$

In this equation, V_H is the interaction energy in the distance H, H_0 is the shortest distance, and a is the radius of the sphere. By comparison with the analogous formula for two spheres of equal radius (36) it follows that the interaction energy between a sphere with the radius a and a plate is equal to that of two spheres with the radius $2a$.

V_H can be expressed, according to Verwey and Overbeek (36), by an equation of the general form

$$\frac{\nu^2}{\kappa} V_H = f(u, z), \tag{4.28}$$

Here u is defined by

$$u = \nu e \psi_d / kT \tag{4.29}$$

where ψ_d is the minimum value of the potential in the center between the plates. From Eqs. (4.27) and (4.28) it follows that

$$\frac{\nu^2}{2a} V_R = \pi \int_x^{\infty} f(u, z) \, d(\kappa H) \tag{4.30}$$

Numerical values of this function obtained by graphical integration for different values of κx, and z have been collected in a table by Verwey and Overbeek (36). Hence, V_R can be obtained as a function of x for different values of z or ψ_0, respectively, a and n. However, the table contains only data for potentials equal to or higher than $z = 2$. Nevertheless, smaller potentials are sometimes important for the theory of the cleansing action. In these cases an equation given by Derjaguin (50) for two equal spheres can be used; after fitting to the sphere-plate model, it reads

$$V_R = \epsilon a \psi_0^2 \ln(1 + e^{-\kappa x}) \tag{4.31}$$

The van der Waals attraction energy can be evaluated according to Hamaker (25) by

$$V_A = - Aa/6x \quad \text{for } x \ll a \tag{4.32}$$

For the total potential energy $V = V_R + V_A$ of the sphere-plate system as a function of the shortest distance x one obtains curves very similar to those shown in Fig. 4.11 for the plate-plate model (38, 40).

4. Influence of Surfactants

Now we shall consider in what manner surfactants influence the quantities discussed in the two preceding sections. Recently it has been shown directly by experiment that the force of adhesion of small solid particles at a solid substrate in an aqueous medium is greatly diminished by the presence of surfactants (33, 34). An example, Fig. 4.12, is the influence of sodium dodecyl sulfate and of dodecyl octaglycol ether on the force of adhesion of carbon-coated spherical gold particles on a foil of polyester in an aqueous medium.

Surfactants are more or less adsorbed from aqueous solutions by most solid surfaces. The adsorbed film has a spreading pressure π_{SL} or π_{PL}, respectively, which contributes to the amounts of work A_1 and A_W for the plate-plate model. The spreading pressure can be calculated if the adsorption isotherm of the surfactant at the solid under consideration is known. For ionic surfactants the adsorption isotherm must be known at constant total electrolyte concentration. Then the equation

$$\pi = RT \int_0^a \frac{\Gamma}{a} \, da \qquad (4.33)$$

is applicable (51). In this equation R is the gas constant, Γ is the amount adsorbed per unit area, and a is the activity. For very low concentrations c the activity a can be replaced by c.

As an example, it may be mentioned that for sodium dodecyl sulfate at a concentration of 8×10^{-3} mole/liter on graphitized carbon (Graphon) a

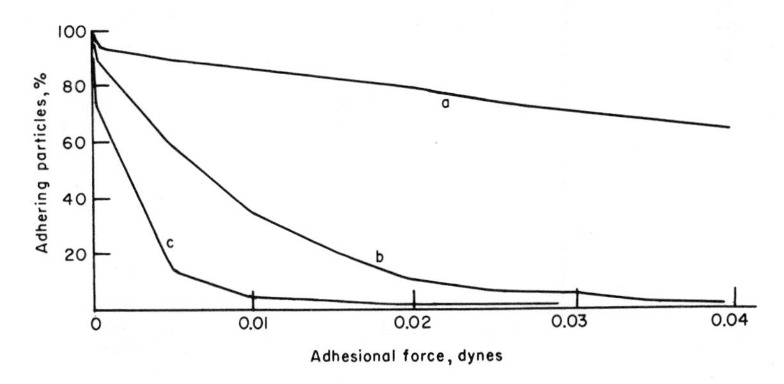

FIG. 4.12. Adhesion of spherical gold particles (3μ diameter) on a foil of regenerated cellulose (a) in water; (b) in 0.5 g/liter octaoxyethylene dodecanol; (c) in 4 g/liter sodium dodecyl sulfate. Solid-solid contact was established prior to immersion.

value of π = 29 dyn/cm has been found (17) as calculated from the adsorption isotherm by Eq. (4.33). From that one can conclude that the contributions to $\underline{J}$ given by the spreading pressure of adsorbed films of surfactants can be quite high and may play an important role in the removal of solid soil particles.

The variation of the interfacial potential ψ_0 produced by the adsorption of ionic surfactants on the particles and the substrate greatly influences the potential curves and consequently the amounts of $\underline{A}_1$, $\underline{A}_2$, and $\underline{A}_W$. As a reasonable approximation, ψ_0 can be regarded as being equal to the electrokinetic or ζ potential because, as already mentioned, ψ_0 means the potential drop in the diffuse part of the double layer.

The influence of an anionic, a nonionic, and a cationic surfactant on the ζ potential of dispersed carbon black is demonstrated in Fig. 4.13 (52). Admittedly, not the ζ potential proper but the directly measurable electrophoretic mobility of the particles is plotted there. Of course, the

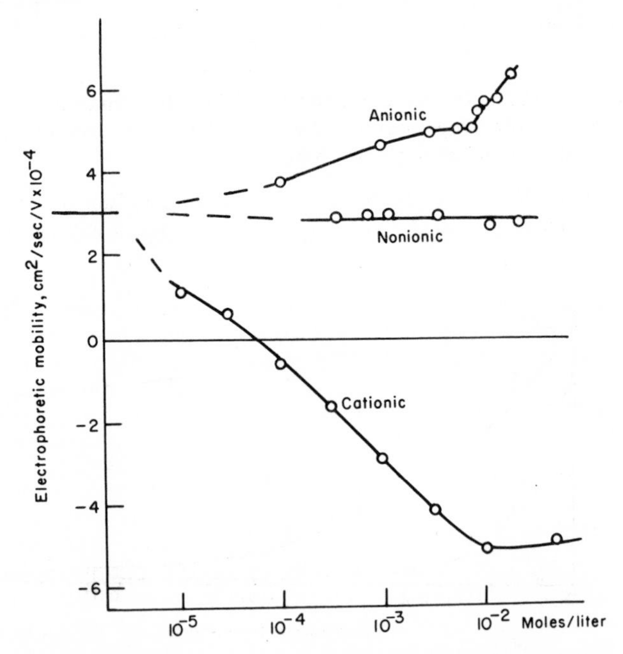

FIG. 4.13. Electrophoretic mobility of carbon-black particles in aqueous solutions of sodium dodecyl sulfate (anionic), polyoxyethylene dodecanol (nonionic), and dodecyl pyridinium chloride (cationic) as a function of surfactant concentration.

ζ potential cannot be quantitatively calculated from u for particles of a geometrically irregular shape but, even so, u can be regarded as a qualitative measure of ζ. Even in pure water, the carbon-black particles migrate to the anode; i.e., their charge, and hence their interfacial potential, is negative. The same is true for most other solid particles in pure water. By addition of an anionic surfactant the negative potential is increased. This is due to the adsorption of the negative long-chain ions. The increase is exceptionally steep immediately after the critical micelle concentration is surmounted. Long ago, Urbain and Jensen (53) observed an increase of the negative charge of solid particles dispersed in water by the addition of sodium or potassium salts of fatty acids, i.e., ordinary soaps. On the addition of nonionic surfactants the potential does not vary sensibly. This is to be expected because of the absence of ionized groups in the surfactant molecule. Cationic surfactants, on the other hand, decrease the negative potential of the particles strongly at very low concentrations as a result of the adsorption of the positively charged long-chain ions and cause an inversion of the sign of the potential at higher concentrations. Quite similar curves have been obtained with particles different from carbon black.

The ζ potential of fibrous materials is best determined by electroosmotic measurements. Long ago, Karrer and Schubert (54) stated that the most common textile fibers also take up a negative interfacial potential in pure water. According to more recent investigations by Stackelberg et al. (55) and Benzel (56), the ζ potential of textile fiber is influenced by anionic, nonionic, and cationic surfactants in the same general manner as that of carbon black and other particles.

Cationic surfactants cannot generally be used for cleansing purposes, at least as far as textile materials are concerned, because of the diminution or canceling of the negative charge of particles substrate. In certain regions of concentration they may cause, according to Götte (57), an "inversion" of the cleansing action, i.e., a cleansing action less than that of pure water.

Regarding the influence of anionic surfactants on the potential curves it must be recognized that these compounds are electrolytes and, therefore, contribute to the total ionic strength of the solution. The influence of the alteration of the V, x curve caused by an anionic surfactant on the energy quantities responsible for the removal and redeposition may be considered first. Some reviews of this topic can be found in a number of publications (40, 58, 59). It may be assumed, for example, that by the addition of an anionic surfactant at a constant ionic strength of 10^{-2} mole/liter (univalent ions only) the interfacial potential is increased from $z = 2$ to $z = 4$, i.e., from about 50 to 100 mV. These figures are of a reasonable order of magnitude. The distance may be arbitrarily assumed to be 6 Å. From

Fig. 4.14 the values of $\underline{V}_\delta$ and $\underline{E}$ can be taken for both potentials in the plate-plate model. One can immediately see that $\underline{E}$ strongly increases with rising potential. Hence, the redeposition of soil particles is the more inhibited the higher the potential $\underline{z}$.

In the manner of presenting the V, x curves as chosen in Fig. 4.14 the zero points of $\underline{V}$, i.e., the potential energies at infinite distance between particle and substrate, are arbitrarily fixed at the same level for both curves. This method of presentation is reasonable when the energy during the mutual approaching of particle and substrate is considered—as it has been here. However, it must be borne in mind that one really chooses reference levels for both curves having different absolute values

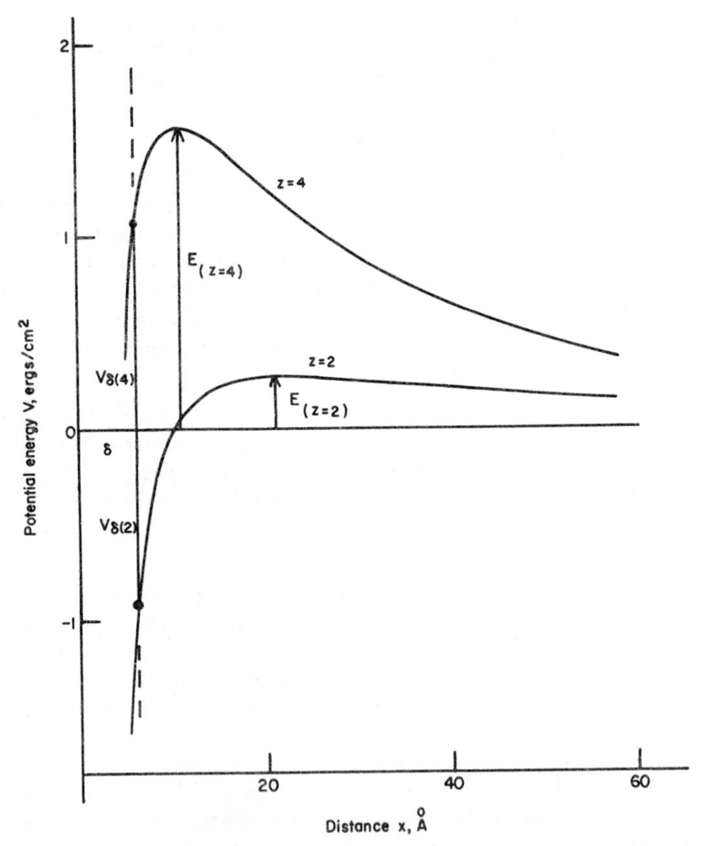

FIG. 4.14. V, x curves of the plate-plate model for z = 2 and z = 4, ionic strength 10^{-2} mole/liter. For the van der Waals constant a value of 2×10^{-13} erg has been assumed.

of the potential energy. This is because different values of the free energy $\underline{F}_\infty$ of the double layer are involved—cf. Fig. 4.9. The same is true for the curves shown in Fig. 4.11.

Consequently, this choice of standard levels of $\underline{V}$ is unreasonable and misleading when the removal of an adhering particle rather than its deposition is considered. When a particle adheres at a substrate, no electrical double layer at all can exist between the two. Hence, the potential energy in this state is independent of the potential and ionic strength.

To present both curves in Fig. 4.14 in such a manner that they are correlated to the same potential energy for the state of adhesion, the curves must be shifted so that the difference of the values of $\underline{V}_{R,0}$ is canceled. In this way, the diagram shown in Fig. 4.15 is obtained from Fig. 4.14. Indeed, the level of each of the curves in Fig. 4.15 corresponds to the potential energy at infinite distance, but the curves are mutually shifted along the ordinate axis by the difference $\Delta \underline{V}_{R,0}$ of the levels for $\underline{z} = 2$ and $\underline{z} = 4$. Either level corresponds to the value $\underline{V}_{III}$ in Fig. 4.8 and is denoted as $\underline{V}_{III(2)}$ and $\underline{V}_{III(4)}$, respectively.

The absolute height of the potential energies for the state of adhesion, as corresponding to the level $\underline{V}_1$ in Fig. 4.8, cannot be determined even in this way, but at least it is equal for either value of $\underline{z}$. In Fig. 4.15 a point for $\underline{V}_1$ is marked on the ordinate axis at arbitrary height.

The points of intersection of the curves with the vertical drawn at a distance δ correspond to the levels $\underline{V}_{II}$ in Fig. 4.8; they are denoted as $\underline{V}_{II(2)}$ and $\underline{V}_{II(4)}$, respectively, in Fig. 4.15. The height of these points above $\underline{V}_I$ is equal to the work $\underline{A}_I$ required for the first step of the removal of a particle, whereas the height above or below $\underline{V}_{III(2)}$ or $\underline{V}_{III(4)}$, respectively, is equal to the positive or negative value of $\underline{V}_\delta = -\underline{A}_2$.

By inspection of Fig. 4.15 one can see that $\underline{A}_I$ is lower for $\underline{z} = 4$ than for $\underline{z} = 2$. This result is not influenced by ignorance of the absolute value of $\underline{V}_I$, because it is based on the difference of the values $\underline{V}_{II(2)}$ and $\underline{V}_{II(4)}$ only. Furthermore, the diagram shows that $\underline{V}_\delta$ changes from a negative to a positive value and, therefore, $\underline{A}_2$ inversely changes from a positive to a negative value when the potential is increased from $\underline{z} = 2$ to $\underline{z} = 4$. Finally, the height of the maximum above $\underline{V}_{II}$, and even more above $\underline{V}_I$, is lower for $\underline{z} = 4$ than for $\underline{z} = 2$. Taking all these influences together, the removal of a soil particle is thermodynamically and kinetically promoted by increasing the potential.

The influence of the electrolyte concentration at constant potential is shown in Fig. 4.16. By increasing the concentration, $\underline{V}_{III}$ is sensibly depressed. Hence, the total work $\underline{A}_W$ for the removal of a particle is correspondingly diminished. Furthermore, $\underline{V}_{II}$ is depressed and $\underline{V}_\delta$ decreases. Consequently, both partial amounts of work $\underline{A}_1$ and $\underline{A}_2$ are

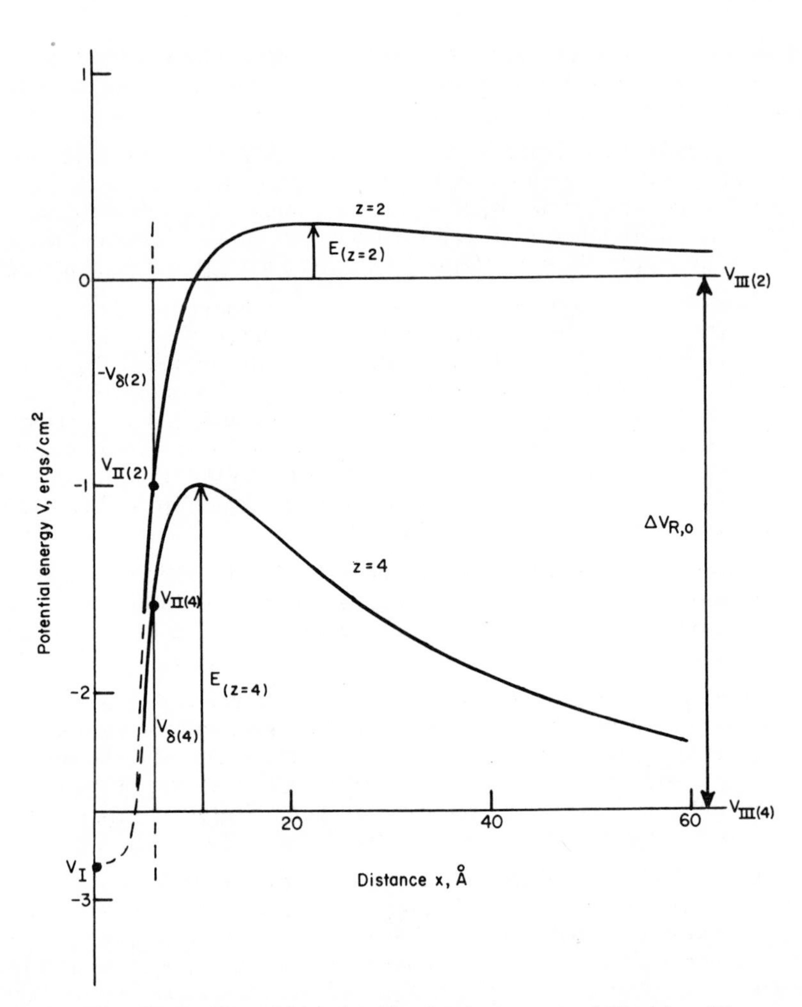

FIG. 4.15. As in Fig. 4.14 but with abscissa axes shifted by $\Delta \underline{V}_{\underline{R},0} = \underline{V}_{\underline{R},0(\underline{z}=4)} - \underline{V}_{\underline{R},0(\underline{z}=2)}$. Numbers on the ordinate axis refer to the level of the upper abscissa axis, i.e., for $\underline{z} = 2$.

diminished. $\underline{E}$ increases, causing a stronger retardation of redeposition. All these influences should favor the cleansing action. It is to be kept in mind, however, that the potential $\underline{z}$ is sometimes diminished in a solution of a given anionic surfactant by the addition of an ordinary salt.

The choice of the value $\delta = 6$ Å is somewhat arbitrary. It is easily shown, however, that the conclusions obtained are not much influenced by an alteration of δ.

The effect of the potential $\underline{z}$ in the sphere-plate model is shown in Fig. 4.17. Here, the abscissa axis is also shifted downward by $\Delta \underline{V}_{\underline{R},0} = \Delta \underline{V}_{\underline{R},0(\underline{z}=4)} - \Delta \underline{V}_{R,0(\underline{z}=2)}$ when going from $\underline{z}$ = 2 to $\underline{z}$ = 4. Indeed, in the sphere plate model the potential energy is influenced even in the state of adhering by the electrical double layer, as may be seen from the zone with the radius $\underline{r}_2$ in Fig. 4.10. Consequently, both curves no longer have a common reference level with regard to the absolute value of the potential energy, but instead have a common level with regard to the potential energy of the particle adhering to thé substrate and therefore being influenced by the double layer.

The influence of the electrolyte concentration $\underline{c}$ in the sphere-plate model is shown in Fig. 4.18. In this model $\underline{V}_{\underline{R},0}$ is independent of $\underline{c}$ for given $\underline{z}$. Hence, there is no reason for shifting the potential-energy level in this case. From the fact that $\underline{V}_{\underline{R},0}$ does not depend on $\underline{c}$, in combination with

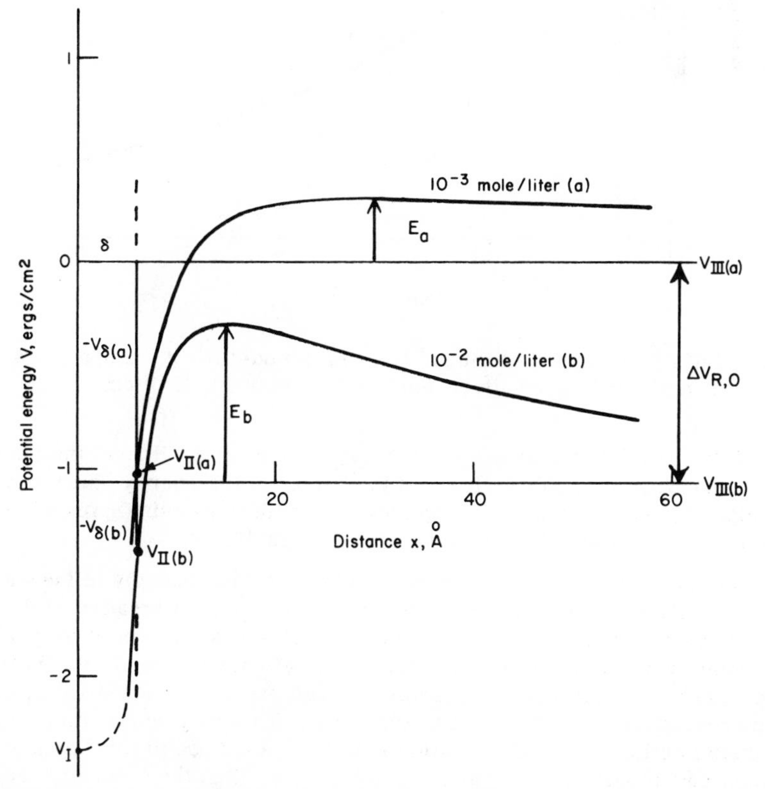

FIG. 4.16. $\underline{V}$, $\underline{x}$ curves of the plate-plate model for 10^{-3} mole/liter (index $\underline{a}$) and 10^{-2} mole/liter (index $\underline{b}$); $\underline{z}$ = 3.

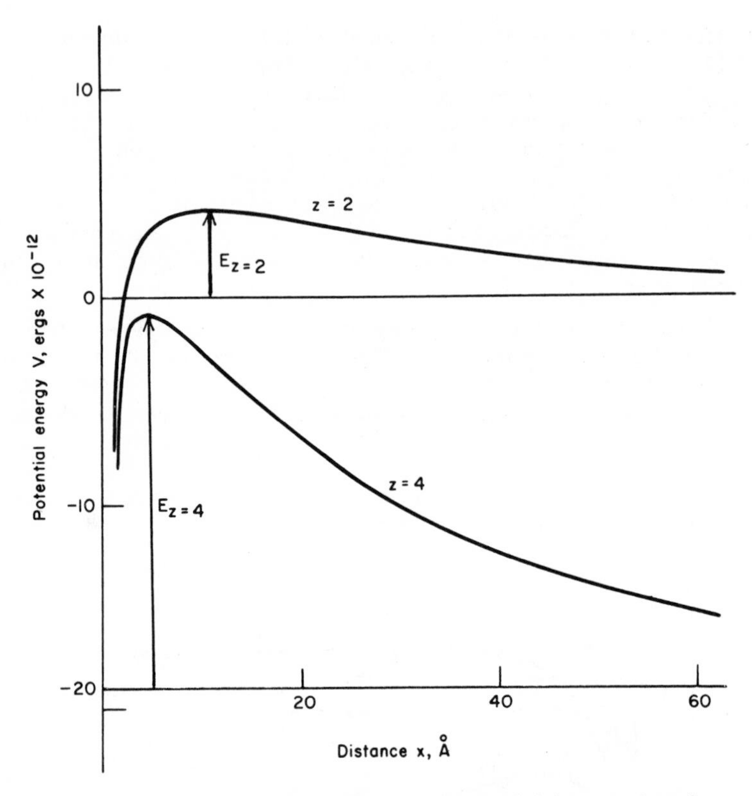

FIG. 4.17. $\underline{V}$, $\underline{x}$ curves of the sphere-plate model for $\underline{z} = 2$ and $\underline{z} = 4$; ionic strength 10^{-2} mole/liter. Sphere diameter 1×10^{-5} cm.

Eq. (4.22), it follows that $\underline{A}_W$ is also independent of $\underline{c}$. Hence, the electrolyte concentration $\underline{c}$ only influences the height of the energy barrier $\underline{E}$ in the plate-sphere model. The lowering of $\underline{E}$ with increasing $\underline{c}$ promotes both the removal and the redeposition of the particles.

It must however be borne in mind in this case also that the influence of the electrolyte concentration $\underline{c}$ is not restricted to an alteration of the $\underline{V}$, $\underline{x}$ curve at constant potential $\underline{z}$ but that $\underline{z}$ itself is influenced by $\underline{c}$; $\underline{z}$ decreases with rising $\underline{c}$ for a given interfacial charge density σ. On the other hand, the adsorption of anionic surfactants, and hence σ and $\underline{z}$, can be increased by the addition of electrolytes. This will be favorable for the cleansing action. The last-mentioned influence will be predominant if without added salt the adsorption is much lower than the saturation adsorption. This is the case for a surfactant with a short hydrophobic chain or at low concentration. On the contrary, when the surfactant has a long hydrophobic chain or is used at high concentration, saturation adsorption occurs even without added salt. Then, addition of salt can only decrease z and therefore lower the cleansing efficiency.

These predictions about the influence of salts on the cleansing action of anionic surfactants have been verified by washing tests with artificially soiled cotton (60). The runs shown in Fig. 4.19 and 4.20 serve as examples. In Fig. 4.19 the degree of whiteness obtained with sodium dodecyl sulfate at four different concentrations has been plotted as a function of the concentration of added sodium chloride. On the left side of the diagram the values obtained with the pure aqueous solutions of the surfactant without sodium chloride have been marked. The washing efficiency increases with rising concentration of the surfactant. The influence of added sodium chloride is very different depending on the concentration of the surfactant. At the lowest concentration, viz., 1×10^{-3} mole/liter, the washing efficiency first rises steeply and finally, after passing through a maximum, falls. A similar, although less pronounced, initial rise is shown by the curve for the next two higher concentrations. At the highest concentration, viz., 1×10^{-2} mole/liter, the maximum washing effect is reached without added salt. The decreasing parts of all the four curves run into a joint curve at about 1 N NaCl. The concentration obviously is sufficient to afford a saturated adsorption layer of the surfactant on the fibers and the soil

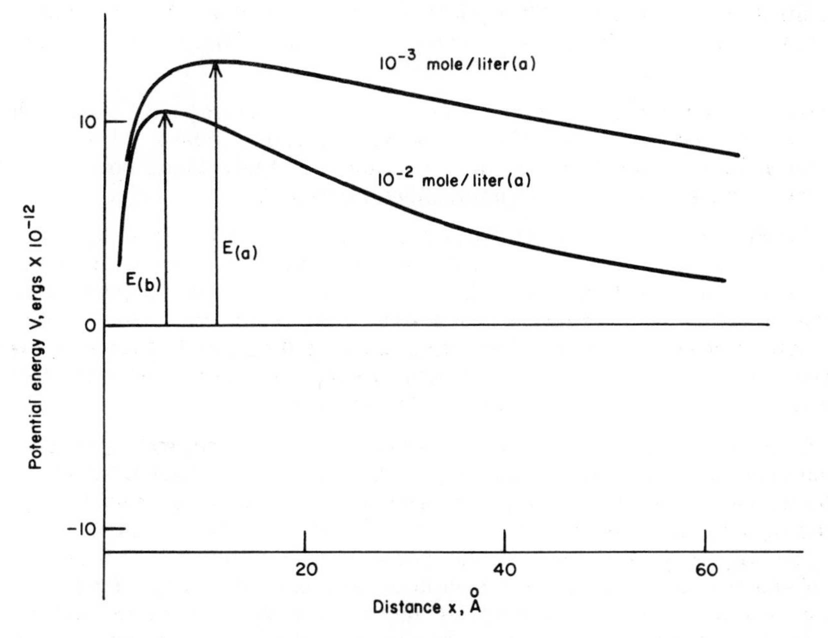

FIG. 4.18. V, x curves of the sphere-plate model for 10^{-3} mole/liter (index a) and 10^{-2} mole/liter (index b); z = 3.

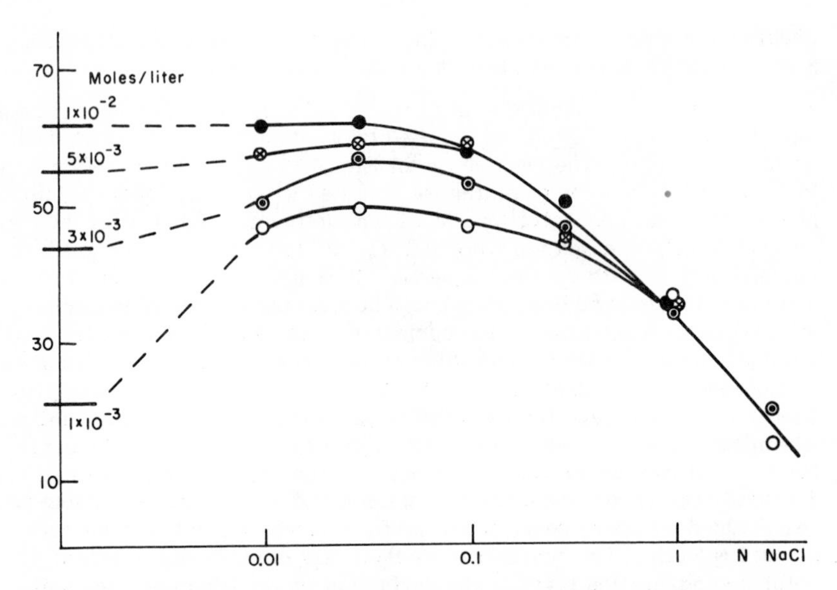

FIG. 4.19. Washing effect of sodium dodecyl sulfate at different concentration as a function of the concentration of added sodium chloride. Values obtained without sodium chloride are marked on the ordinate axis.

particles and, therefore, a constant interfacial charge density even at the lower surfactant concentrations. Hence, a further increase of the salt concentration can only result in a decrease of the interfacial potential and, consequently, a lowering of the washing efficiency.

Shown in Fig. 4.20 are analogous results for a series of washing tests in which sodium alkyl sulfates with different chain lengths were used at a constant concentration of 3×10^{-3} mole/liter. The general appearance of the curves is very similar to that shown in Fig. 4.19. Increasing chain length at constant surfactant concentration has the same influence as rising concentration at constant chain length. Hence, the theoretical prediction mentioned above is also verified in this case.

Up to now, only univalent ions have been taken into account. In fact, however, under practical conditions polyvalent ions also are present. Nevertheless, the influence of the valency of the ions present in the cleansing liquor will not be considered here since the present treatment is especially concerned with the effect of surfactants, and at best these substances contain polyvalent ions in exceptional cases only. Furthermore, for practical applications of surfactants in hard water, polymer phosphates that sequester the polyvalent cations are usually added. The presence of polyvalent anions is of small importance when anionic surfactants are used.

Besides the electric charge of the adsorbed surfactants, their hydration is important. This is particularly true for nonionic surfactants because the electrical mechanism of the cleansing effect does not apply here. Adsorbed surfactant molecules in most cases are oriented in such a manner that the hydrophilic groups are directed toward the aqueous phase. In this way, both the soil particles and the substrate are surrounded by a hydration shell. This can inhibit the redeposition of soil particles because it prevents a too close approach and therefore makes inoperative the van der Waals attraction that is especially strong at shortest distances.

The idea that hydration layers prevent the mutual approach of surfaces coated by them has been verified by experiments on the stabilizing action of surfactants against the coagulation of hydrophobic sols by electrolytes (61). Surfactants can inhibit the coagulation of a sol even if the electrolyte concentration is so high that the electrical interaction between the particles is nearly completely suppressed. Under this condition the stabilizing action can only be caused by the formation of a hydration layer around the surface of the particles. With a sol of paraffin wax the coagulation is most clearly observed by an increase of the turbidity up to a value T_c indicative of the sol in the coagulated state. The stabilizing action expresses itself in the turbidity being lower than T_c by an amount T in spite of the high electrolyte concentration. In Fig. 4.21 is shown ΔT for

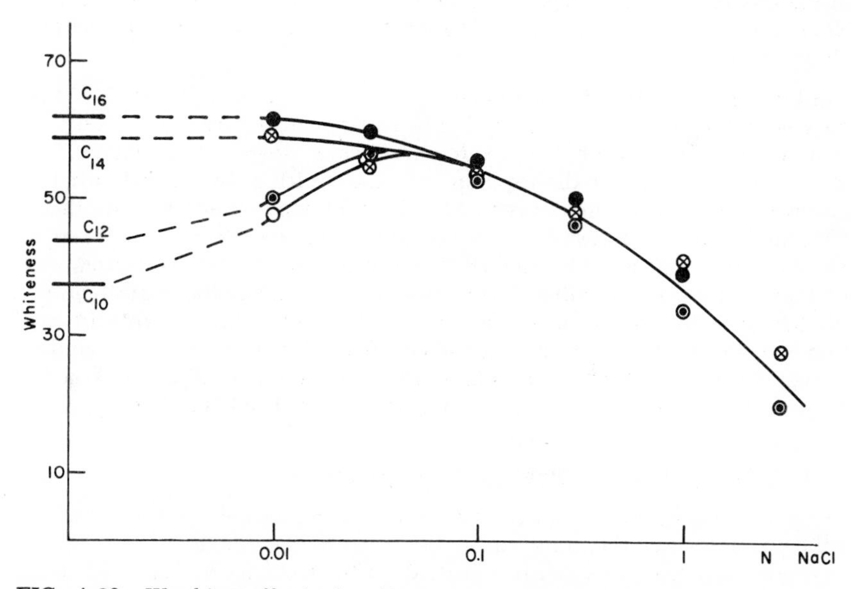

FIG. 4.20. Washing effect of sodium alkyl sulfates with different chain length as a function of the concentration of added sodium chloride. Surfactant concentration 3 x 10^{-3} mole/liter.

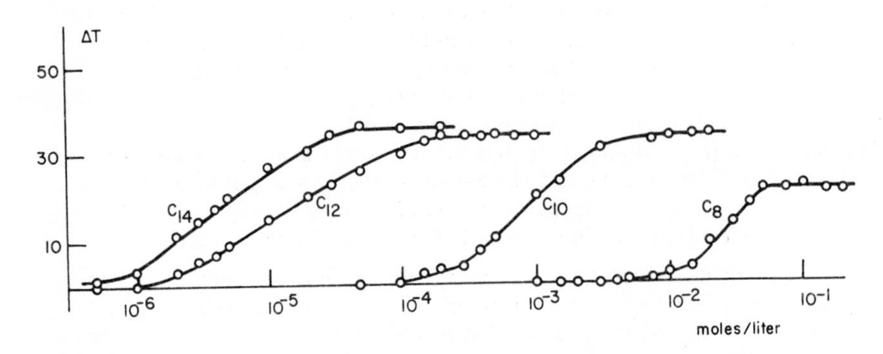

FIG. 4.21. Diminution $\Delta \underline{T}$ of the turbidity of a paraffin sol caused by the addition of sodium alkyl sulfates in the presence of sodium chloride; sum of concentrations 0.5 mole/liter.

sodium alkyl sulfates of different hydrophobic chain length as a function of the concentration. The curves are shifted to lower concentrations with increasing chain length. In Fig. 4.22 analogous results are given for nonionic surfactants, viz., dodecyl polyglycol ethers with different numbers of the ethylene oxide groups, i.e., with variable hydrophilic chain length. Here, the curves are also shifted to lower concentrations with increasing chain length.

In the experiments made with dodecyl polyglycol ethers the shift of the curves may be caused simply by the hydration being enhanced by the increased number of ethylene oxide units, which results in an intensified stabilizing action. For the experiments made with sodium alkyl sulfates, however, the shift of the curves has to be explained otherwise. As the hydrophilic group is always the same, and as the ionic strength is kept constant, the shift must be caused by the variation of the adsorption isotherm of the surfactant at paraffin wax. The higher the chain length, the lower is the concentration for the beginning of an appreciable adsorption and, therefore, of an interfacial concentration of the hydrated sulfate groups sufficiently high for a remarkable stabilization. Analogous results have been obtained for aqueous dispersions of carbon black (17).

5. Influence of Particle Size. Hydrodynamics

Common experience with all types of cleansing processes shows that solid particles are the more easily removed the larger they are. For example, soil particles smaller than about 0.1 μ cannot be removed from textile material at all. This is very likely not caused by the influence of the particle size on the $\underline{V}$, $\underline{x}$ curves.

Direct measurements of the adhesional force performed by the ultracentrifuge method (8, 9) have shown that this force is approximately proportional to the sphere radius. This proportionality also results from Eq. (4.30) for $\underline{V}_R$ and from Eq. (4.32) for $\underline{V}_A$ in the sphere-plate model. In the plate-plate model, $\underline{V}_R$ and $\underline{V}_A$ are proportional to the area of contact. Hence, the same proportionalities are also valid for the quantities formed therefrom by summation or subtraction, as for example, $\underline{V}$, $\underline{V}_\delta$, $\underline{E}$.

Nevertheless, the mechanical force exerted upon the particles in the cleansing process may be much more important in connection with the influence of the particle size. It is well known that, in most cases, a soil consisting of solid particles only cannot be completely removed even by a surfactant solution without mechanical action. This is easily explained from two points of view: first, regarding the necessity of the penetration of a liquid layer (cf. Sec. II,B,2) and second, regarding the activation barrier to be surmounted. The greater the size of a particle, the higher is the probability that the particle is directly hit during the mechanical treatment of a substrate to be cleaned, for example, in a washing machine. Very likely, the force exerted hydrodynamically on the particle is more important. This, however, requires a high gradient of the streaming velocity of the liquid near the substrate surface because there is a very thin liquid layer adhering to the surface and having a zero velocity relative to the substrate surface. Such a high velocity gradient near the surface cannot exist, for hydrodynamic reasons, in a stationary state, but can exist only when abrupt local changes of the streaming velocity occur. The latter is supposed to be the case in a washing machine. In a streaming gradient,

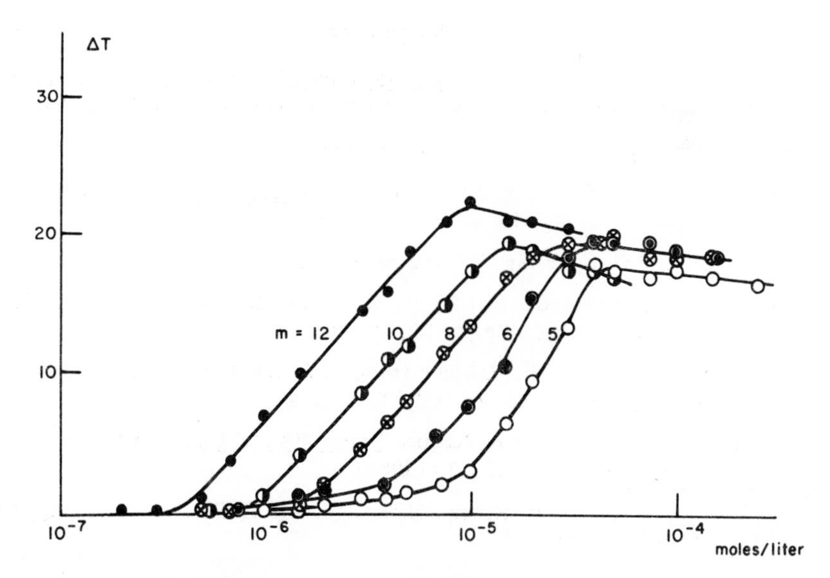

FIG. 4.22. Diminution $\Delta\underline{T}$ of the turbidity of a paraffin sol caused by the addition of polyoxyethylene dodecanols with $\underline{m}$ ethylene oxide units in the presence of 0.5 $\underline{N}$ sodium chloride.

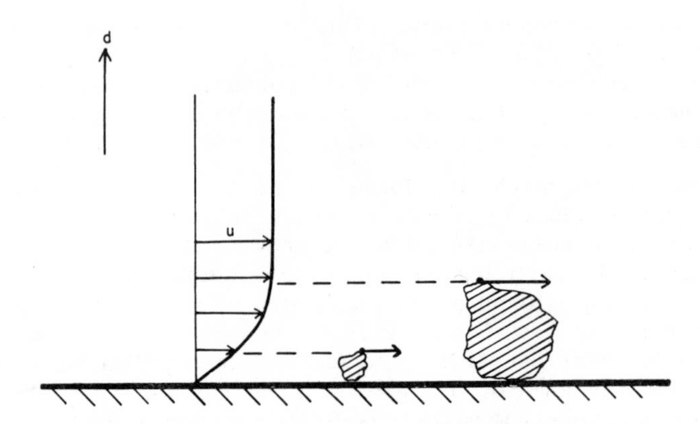

FIG. 4.23. Small and large soil particles adhering at a substrate in a streaming gradient: u is the streaming velocity; d is the distance from the substrate surface.

as shown schematically in Fig. 4.23, the larger particle is attacked by a stronger hydrodynamic force than the smaller one, not only because of the larger area of attack but also because the outer part of it extends to zones of higher streaming velocity.

REFERENCES

1. Quoted, e.g., in Solubilization and Related Phenomena (M. E. L. McBain and E. Hutchinson, eds.), Academic Press, New York, 1955, p. 215.

2. G. Sperling, dissertation, Karlsruhe, 1964.

3. H. Krupp and G. Sperling, Proc. 4th Int. Congr. Surface Activity, Brussels, Vol. II, 447 (1964).

4. H. Krupp and G. Sperling, Z. Angew. Phys., 19, 259 (1965).

5. H. Krupp, G. Sandstede, and K.-H. Schramm, Dechema Monogr., 38, 115 (1960); Chem.-Ing.-Tech., 32, 99 (1960).

6. G. Böhme, K. Krupp, H. Rabenhorst, and G. Sandstede, Trans. Inst. Chem. Eng. (London), 40, 252 (1962).

7. G. Böhme, Z. Egey, and H. Krupp, Proc. 4th Int. Congr. Surface Activity, Brussels, Vol. II, 419 (1964).

8. G. Böhme, W. Kling, H. Krupp, H. Lange, and G. Sandstede, Z. Angew. Phys., 16, 486 (1964).

9. G. Böhme, W. Kling, H. Krupp, H. Lange, G. Sandstede, and G. Walter, Proc. 4th Int. Congr. Surface Activity, Brussels, Vol. II, 429 (1964).

10. H. Reumuth, Wäschereitech.-Chem., 2 (11), 41 (1949).

11. J. Compton and W. J. Hart, Ind. Eng., Chem., 43, 1564 (1951).

12. E. Götte, W. Kling, and H. Mahl, Melliand Textilber., 35, 1252 (1954).

13. A. S. Weatherburn, G. R. F. Rose, and C. H. Bayley, Can. J. Res., 27F, 179 (1949).

14. G. R. F. Rose, A. S. Weatherburn, and C. H. Bayley, Text. Res. J., 21, 427 (1951).

15. J. L. Moillet, J. Oil Colour. Chem. Ass., 38, 463 (1955).

16. B. Tamamushi and K. Tamaki, Proc. 2nd Int. Congr. Surface Activity, London, 1957, 3, 449; Trans. Faraday Soc., 55, 1007, 1013 (1959).

17. H. Lange, unpublished work, 1964.

18. A. S. Weatherburn, G. R. F. Rose, and C. H. Bayley, Can. J. Res., 28F, 51 (1950).

19. A. S . Weatherburn, Text. J. Aust., 28, 826, 888 (1953).

20. M. E. Ginn, F. B. Kinney, and J. C. Harris, J. Amer. Oil Chem. Soc., 38, 138 (1961).

21. W. J. Schwarz, A. R. Martin and R. C. Davis, Text. Res. J., 32, 1 (1962).

22. W. J. Schwarz, A. R. Martin, B. J. Rutkowski, and R. C. Davis, Vortr. Originalfassung Int. Kongr. Grenzflaechenaktive Stoffe, 3, Cologne, 1960, 4, 37; Fette, Seifen, Anstrichm., 64, 57 (1962).

23. J. C. Harris, Soap Chem. Spec. (Nov.-Dec. 1958); (Jan.-Feb. 1959).

24. H. Kölbel and K. Hörig, Angew. Chem., 71, 691 (1959),

25. H. C. Hamaker, Physica, 4, 1058 (1937).

25a. J. Lyklema, Advan. Colloid Interface Sci., 2, 84 (1968).

26. J. H. Schenkel and J. A. Kitchener, Trans. Faraday Soc., 56, 161 (1960).

27. S. N. Srivastava and D. A. Haydon, ibid., 60, 971 (1964).

28. R. H. Ottewill and D. J. Wilkins, ibid., 58, 608 (1962).

29. A. Watillon and Ph. Gerard, Proc. 4th Int. Congr. Surface Activity, Brussels, 1964, Preprints, 4, 28.

30. H. Reerink and J. Th. G. Overbeek, Discuss. Faraday Soc., 18, 74 (1954).

31. A. Watillon and A. M. Joseph-Petit, Vortr. Original fassung Int. Kongr. Grenzflaechenaktive Stoffe, 3, Cologne, 1960, 1, 145.

32. B. V. Derjaguin, T. N. Voropayeva, B. N. Kabanov, and A. S. Tityevskaya, J. Colloid Sc., 19, 113 (1964).

33. G. Böhme, W. Kling, H. Krupp, H. Lange, G. Sandstede, and G. Walter, Z. Angew. Phys., 19, 265 (1965).

34. W. Kling, H. Lange, G. Böhme, H. Krupp, G. Sandstede, and G. Walter, Proc. 4th Int. Congr. Surface Activity, Brussels, 1964, Preprints, 3, 17.

35. B. V. Derjaguin and L. D. Landau, Acta Physicohim. URSS, 14, 633 (1941); see also B. V. Derjaguin, Trans. Faraday Soc., 36, 730 (1940).

36. E. J. W. Verwey and J. Th. G. Overbeek, Theory of the Stability of Lyophobic Colloids, Elsevier, Amsterdam, 1948.

37. K. Durham, J. Appl. Chem. (London), 6, 153 (1956).

38. H. Lange, Kolloid-Z., 154, 103 (1957).

39. H. Lange, ibid., 156, 108 (1958).

40. H. Lange and W. J. Schwarz, Amer. Dyest. Rep., 50, 25 (1961).

41. K. Durham, in Surface Activity and Detergency (K. Durham, ed.), Macmillan, New York, 1961, p. 119.

42. J. T. Davies, Proc. Roy. Soc. (London), A208, 224 (1951).

43. J. T. Davies and E. K. Rideal, Interfacial Phenomena, Academic Press, New York, 1961, p. 231.

44. D. C. Grahame, Z. Elektrochem., 59, 773 (1955).

45. M. Gouy, C. R., 149, 654 (1909).

46. D. L. Chapman, Phil. Mag., 25, 475 (1913).

47. D. C. Grahame, Z. Elektrochem., 62, 264 (1958).

48. H. Brodowski and H. Strehlow, ibid., 63, 262 (1959).

49. B. V. Derjaguin, Kolloid-Z., 69, 155 (1934).

50. B. V. Derjaguin, Trans. Faraday Soc., 36, 203 (1940).

51. Y. Fu, R. S. Hansen, and F. E. Bartell, J. Phys. Colloid Chem., 53, 1141 (1949).

52. W. Kling and H. Lange, Kolloid-Z., 127, 19 (1952).

53. W. M. Urbain and L. B. Jensen, J. Phys. Chem., 40, 821 (1936).

54. P. Karrer and P. Schubert, Helv. Chim. Acta, 11, 221 (1928).

55. M. v. Stackelberg, W. Kling, W. Benzel, and F. Wilke, Kolloid-Z., 135, 67 (1954).

56. W. Benzel, dissertation, Bonn, 1953.

57. E. Götte, Kolloid-Z., 64, 331 (1933).

58. J. C. Harris, Text. Res. J., 28, 912 (1958).

59. W. Kling, Waschereitech.-Chem., 12, 542 (1959); Vortr. Originalfassung Int. Kongr. Grenzflaechenaktive Stoffe, 3, Cologne, 1960, 4, 143; in Tenside, Textilfsmittel, Waschrohstoffe (K. Lindner, ed.), Wissenshaftliche Verlagsgesellschaft m.b.H., Stuttgart, 1964, p. 1225.

60. H. Lange, Fette, Seifen Anstrichm., 65, 231 (1963).

61. H. Lange, Kolloid-Z., 169, 124 (1960); J. Phys. Chem., 64, 538 (1960).

Chapter 5

REMOVAL OF ORGANIC SOIL FROM FIBROUS SUBSTRATES

Hans Schott
School of Pharmacy,
Temple University
Philadelphia Pa.

I. WETTING OUT OF FABRICS

The first step in detergency is the penetration of the wash liquor into the fabric, replacing the fiber-air interface with the fiber-aqueous solution interface. This takes place in two successive stages: first wetting of the surface, and then of the bulk of the fabric.

A. Surface Wetting

The physicochemical basis for wetting of a solid surface by a liquid is as follows: A drop of liquid W placed on a fiber surface F in air A will spread out until it makes a contact angle θ, defined by the plane of the solid surface and the tangent to the liquid-air interface (γ_{WA}); θ is measured in the liquid. According to Fig. 5.1, the equation which describes the equilibrium situation, neglecting gravity effects, is

$$\gamma_{FA} = \gamma_{FW} + \gamma_{WA} \cos\theta \qquad 0° \leq \theta < 180° \tag{5.1}$$

where γ represents surface and interfacial tensions in dynes per centimeter which, for liquids, are numerically equal to surface free energies in ergs per square centimeter. The subscripts designate the phases, but it must be remembered that in practice, the gas phase A contains vapor of the liquid W. Therefore, the interface FA is not the clean fiber surface but is covered by a layer of adsorbed vapor, and

$$\gamma_{FA} = \gamma_{F} - \pi \tag{5.2}$$

where γ_F is the free energy of the clean fiber surface and π the surface or spreading pressure of the film of condensed vapor. Likewise, γ_{WA} refers to the interface between a liquid and the gas phase saturated with vapor of that liquid. Both γ_{WA} and θ can readily be measured directly, but not γ_{FA} or γ_{FW}. A cosine can vary from +1 ($\theta = 0°$) through 0 ($\theta = 90°$) to -1 ($\theta = 180°$). Since $\cos\theta = (\gamma_{FA} - \gamma_{FW})/\gamma_{WA}$, when $(\gamma_{FA} - \gamma_{FW}) \geq \gamma_{WA}$, $\theta = 0°$ and the drop spreads out completely. For $\gamma_{WA} > (\gamma_{FA} - \gamma_{FW}) > -\gamma_{WA}$, the drop makes a finite contact angle. If $(\gamma_{FA} - \gamma_{FW}) \leq -$

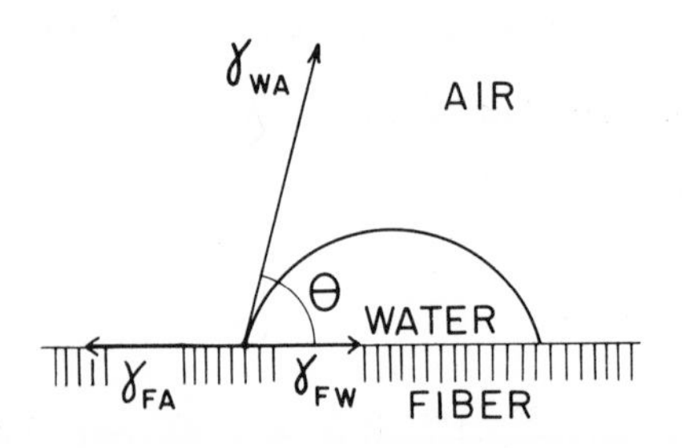

FIG. 5.1 Neumann's triangle of forces at the solid (F)-liquid (W)-air (A) interface.

γ_{WA}, θ would be 180° and the drop should be completely spherical. In practice, a contact angle greater than 90° indicates poor wetting of the solid by the liquid.

Two parameters are useful in wetting relationships, the adhesion tension and the work of adhesion (1, 2). Adhesion tension A of the liquid W for the solid surface F is the difference between the solid-air and solid-liquid tensions:

$$A_{FW} = \gamma_{FA} - \gamma_{FW} = \gamma_{WA} \cos \theta \tag{5.3}$$

The energy required to separate 1 cm^2 of the solid surface from the liquid is measured by the reversible work of adhesion, given by the Dupré equation

$$W_{FW} = \gamma_{FA} + \gamma_{WA} - \gamma_{FW} \tag{5.4}$$

The lower the solid-liquid interfacial tension, the stronger the adhesion. Combining Eqs. (5.1) and (5.4) results in Young's equation

$$W_{FW} = \gamma_{WA} (1 + \cos \theta) \tag{5.5}$$

For complete wetting, $\theta = 0°$, and $W_{FW} = 2\gamma_{WA}$ equals the work of cohesion of the liquid, i.e., the work required to separate a liquid column of 1 cm^2 cross section into two. Despite the fact that $\cos \theta$ cannot exceed unity, it is possible for W_{FW}, the work of adhesion of liquid to solid, to exceed $2\gamma_{WA}$, the work of cohesion of liquid. This corresponds to the nonequilibrium situation where $(\gamma_{FA} - \gamma_{FW}) > \gamma_{WA}$. If the work of adhesion of the liquid to the solid is less than the work of cohesion of the liquid, there is a finite contact angle. This angle is larger, the smaller the adhesion of the liquid to the solid relative to the cohesion of the liquid. Contact angles greater than 90° indicate that $W_{FW} < \gamma_{WA}$. A situation never encountered in practice is for θ to reach 180°, making W_{FW} zero. This would mean no adhesion whatsoever between liquid and solid. On smooth surfaces, contact angles rarely exceed 120 .

From Eqs. (5.3) and (5.4),

$$A_{FW} = W_{FW} - \gamma_{WA} \tag{5.6}$$

i.e., the adhesion tension is the difference between the energy of adhesion of the liquid to the solid surface and the surface tension of the liquid. For details of contact angle measurements see (1-4).

B. Contact Angle Hysteresis, and the Effect of Roughness and Porosity

Among the complications encountered when applying these wetting relationships to practical situations, particularly to fabrics, are hysteresis in contact angle measurements, and roughness and porosity of the surface on which a liquid spreads.

Frequently, a liquid makes a greater contact angle when advancing over a dry surface than when receding from a previously wetted surface. The difference between the advancing and the receding contact angle, called hysteresis, has been reduced or eliminated in some instances by working with very clean solid and liquid surfaces (5). Obviously, a surface contaminant soluble in or displaced by the spreading liquid would produce such hysteresis. A liquid which makes a finite contact angle with a solid surface is less attracted to the dry surface than to itself. Therefore, absorption or penetration of some of the liquid into the solid will enhance the attraction of the solid surface for the liquid, lower the receding contact angle according to Eq. (5.5), and bring about the hysteresis.

Another possible cause of the contact angle hysteresis is a slow reorientation of the surface molecules taking place when in contact with the liquid. Thus, a surface which has been in contact with water may have more hydrophilic groups turned outward, resulting in a smaller contact angle when the water recedes from it, than for the same surface in contact with air or oil when water first advanced over it.

Even in cases where there is no contact angle hysteresis on a smooth and plane surface, the introduction of surface roughness can result in pronounced hysteresis. If there are asperities or grooves on a surface which make an angle of inclination ϕ with respect to the horizontal plane of the surface, the liquid front will not advance perpendicularly to the grooves until the apparent contact angle is at least as large as $\theta + \phi$, θ being the contact angle for the smooth surface. Likewise, the liquid front cannot recede unless the apparent contact angle falls to $\theta - \phi$: the hysteresis is 2ϕ (6, 7). The steeper the grooves or asperities, the greater the contact angle hysteresis (7, 8).

Surface roughness affects the contact angle apart from hysteresis. If the roughness factor $\underline{r}$ is the ratio between the actual surface area and the apparent or geometric area of the smooth surface (9), Eq. (5.1) becomes

$$\underline{r}(\gamma_{\underline{FA}} - \gamma_{\underline{FW}}) = \gamma_{\underline{WA}} \cos \theta_{app} \tag{5.7}$$

or

$$\underline{r} \cos \theta = \cos \theta_{app} \tag{5.8}$$

where θ_{app} is the apparent contact angle the liquid makes with the rough surface. Since $r \geq 1$, the sign of $\cos \theta_{app}$ is not changed by the roughness but its absolute value is increased. Hence, roughness increases the amount by which θ differs from 90°. If $\theta < 90°$, $\theta_{app} < \theta$; θ_{app} may be zero while θ is still finite. The work of adhesion becomes

$$W'_{FW} = \gamma_{WA} (1 + r \cos \theta) \tag{5.9}$$

The energy of adhesion of a liquid to a rough surface will be greater than to a smooth surface of the same material if $\theta < 90°$, and smaller if $\theta >$ 90°

The next step is to consider a liquid spreading on a fabric, which is a porous surface, when the pores remain filled with air. The apparent or geometric solid-liquid interfacial area for the liquid at rest on the fabric surface consists of an area f_{FW} of true solid-liquid interface plus an area f_{FW} of liquid-air interface per square centimeter of apparent solid-liquid interface (10). Therefore, neglecting hydrostatic pressure,

$$\cos \theta_{app} = f_{FW} \cos \theta - f_{WA} \tag{5.10}$$

$$W''_{FW} = \gamma_{WA} (1 + f_{FW} \cos \theta - f_{WA}) \tag{5.11}$$

If the pores become completely filled with liquid, f_{WA} vanishes and Eq. (5.8) results, becuase $f_{FW} = r$. When a liquid has a finite true contact angle θ on a smooth surface, the introduction of pores, i.e., of a liquid-air interface, will always result in an apparent contact angle which is greater than θ and hence in reduced work of adhesion. This is in contrast with liquid spreading on a rough but nonporous surface, where $\theta_{app} > \theta$ only when $\theta > 90°$.

Cassie and Baxter (10) demonstrated that fabrics can be represented by a fine grid of parallel fibers, and have shown how to calculate the apparent contact angle, given the true contact angle of the liquid on a smooth surface of the same material and the geometry of the fiber assembly in terms of $(R + d) / R$. Here R is the fiber radius and $2 (R + d)$ the distance between the centers of two neighboring fibers as shown in Fig. 5.2. Table 5.1 lists the values calculated for θ_{app} using different values of θ and of the openness of the weave, taken from Fig. 3 of Cassie and Baxter's excellent paper (10). Values of θ_{app} measured for water on paraffin-coated gratings of parallel metal wires were in excellent agreement with calculated values (10). A large apparent contact angle due to a porous surface with a geometrically effective structure, rather than a waterproofing agent, is chiefly responsible for the nonwetting of ducks' feathers by water (11)

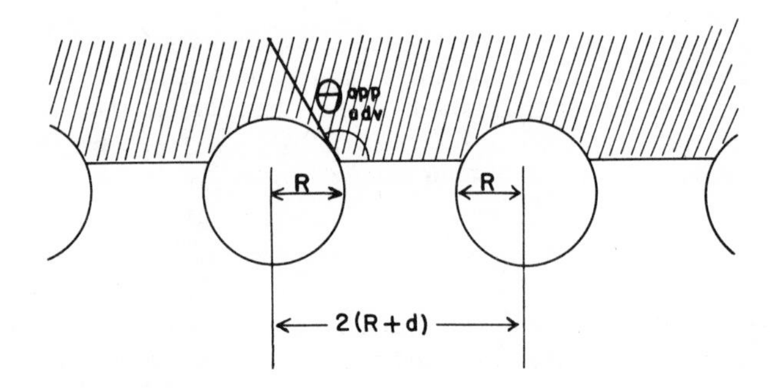

FIG. 5.2. Liquid (shaded area) under zero hydrostatic pressure resting on a grid of parallel cylinders, which represents a fabric surface (10).

TABLE 5.1

Relation between True Advancing Contact Angle θ on a Smooth Surface and Apparent Advancing Contact Angle θ_{app} on Fabric, for Different Values of θ and of Tightness of Weave[a]

	θ_{app} for	
θ	$(R + d) / R = 2.0$	$(R + d) / R = 4.0$
0°	56°	98°
30°	66°	107°
60°	92°	120°
90°	121°	140°
120°	148°	159°

[a]From Cassie and Baxter (10).

C. Penetration of Liquids into Fabrics

In order to study the kinetics of penetration of a liquid into a fabric, the fabric can be treated as a bundle of fibers representing an assembly of parallel capillaries. This is not an oversimplification because liquids readily penetrate the wide openings between the strands of yarn. The much slower process of penetration of the interior of the yarns is largely governed by capillary forces owing to the tight packing of the fibers: the yarns act as wicks.

The rate of advancement of the meniscus of liquid in a capillary (length L per time t) in the absence of inertial forces and of gravitational effects, i.e., in a horizontal capillary, is given (12, 13) by

$$\frac{dL}{dt} = \frac{R_e \gamma_{WA} \cos \theta_{app}}{4L\eta} \tag{5.12}$$

where η is the viscosity of the liquid and R_e the effective capillary radius of the interfiber spaces. Integrating and introducing the roughness factor r,

$$L^2 = \frac{R_e t \gamma_{WA} r \cos \theta}{2\eta} \tag{5.13}$$

Accordingly, Hollies *et al.* (13) found that plots of L^2 versus t for a given fabric are linear; the slope k represents the rate constant for water transport. Effective capillary radii are generally in the 5-20 = μ range; for highly twisted yarns of fibers which swell in water, they may be smaller. According to Eqs. (5.12) and (5.13), an effective wetting agent should reduce θ_{app} to zero while lowering γ_{WA} as little as possible since a high surface tension actually hastens the capillary penetration by the liquid if $\theta_{app} < 90°$. According to Eq. (5.1), γ_{FW} should be strongly reduced by the wetting agent to have a large value for $\gamma_{WA} \cos \theta$. The wetting agent should be adsorbed strongly at the fiber-water interface but weakly at the water-air interface.

Fabric and yarn construction and fiber morphology affect the rate of wetting or wicking far more than does the chemical nature of the fiber material. Synthetic fibers are smooth; their low roughness factors result in apparent contact angles only slightly below the contact angles for perfectly smooth surfaces. Wool fibers have scaly and rough surfaces, but any increase in the rate of water transport because of reduced apparent contact angles is far outweighed by the adverse effect of the random fiber arrangement in the yarn. Owing to crimp, fuzziness, and a wide range of fiber diameters, wool yarns have only short and discontinuous capillary spaces between the fibers. Synthetic fibers are much better aligned, have uniform diameters, and no fuzz. This results in yarns with long, continuous interfiber capillaries and good wicking and wetting (13, 14). Felting of an acrylic fabric, which increased the randomness of the fibers, resulted in proportionally lower wicking rates for yarn and fabric (14). Table 5.2 illustrates the difference in apparent contact angles between fiber and

TABLE 5.2

Apparent Contact Angles of Water on Fibers and Yarns, and Rates of Water Transport in Yarns and Fabrics[a]

Material	Form	Advancing angle	Receding angle	Water transport rate constant *k*[b]
Polyethylene	Fiber	86°	49°	
	Yarn	25°	16°	
Polyhexamethylene adipamide	Fiber	83°	60°	
	Yarn	55°	33°	
	Yarn	50°		0.15
	Fabric			0.16
	Yarn	29°		0.23
	Fabric			0.18
Polyethylene terephthalate	Fiber	79°	74°	
	Yarn	49°	20°	
Cotton	Yarn	33°		0.21
	Fabric			0.20
	Yarn	18°		0.45
	Fabric			0.16
Wool	Fiber	85°	34°	
	Yarn	108°	21°	0.001
	Fabric			0.003

[a] From Hollies *et al.* (13, 15).

[b] $$k = \frac{R_e \gamma_{WA} \cos \theta^{adv}_{app}}{2\eta}, \ \mathrm{cm^2/sec}.$$

yarn. The rather good quantitative agreement for the rates of transport of water in a given yarn and in the corresponding fabric supports the capillary model for wetting out of fabrics (13, 15).

A few of the straight lines in the square of distance versus time plots according to Eq. (5.12) intercept the time axis rather than going through the origin, particularly those lines obtained with tightly woven fabrics. This induction period results from the slower penetration of water through the surface of these fabrics relative to the motion of water in their interior.

Comparison of the magnitude of these intercepts with the rate constants for water transport indicates that surface wetting and water travel in the fabric interior are controlled by the same fabric properties (15).

The wetting of cotton fabric, held in a horizontal position, by sodium dodecylbenzene sulfonate solutions was followed by means of a set of electrodes in contact with the fabric, which measured the rate with which these conductive solutions spread across the dry fabric (16). Wetting times were found to increase with increasing surfactant concentration or decreasing surface tension until the critical micelle concentration was reached. This indicates that the process was one of capillary wetting, for which high rates require as high a γ_{WA} value as possible consistent with a low contact angle.

The effect of hydrostatic pressure on the rate of wetting out of yarns and fabrics has been studied theoretically and experimentally because it is important in designing water-repellent fabrics (17, 18).

When solutions of surfactants are the penetrating liquids, dynamic interfacial tensions often play a greater role than equilibrium values. Surfactants lower the free energy of solid-water and water-air interfaces by being positively adsorbed at these interfaces, forming oriented monolayers. The surfactant concentrations at the interfaces are greater than in the bulk solution. During the wetting of a fabric, new solid-liquid and liquid-air interfaces are continuously being created as the solution displaces air from the fabric (19, 20). Because these newly formed interfaces have initially the same surfactant concentration as the bulk solution, their interfacial tensions are close to those against water. Surfactant molecules diffuse from the bulk solution to these fresh interfaces until they attain the equilibrium excess concentration, thereby lowering the interfacial tensions until the equilibrium or static values are reached. The intermediate values of the two interfacial tensions, called dynamic tensions, are higher, nonequilibrium values which are a function of the age of the interfaces. The sinking times of cotton disks in wetting-agent solutions were found to increase with the dynamic variability of the surface tension. Solutions of those wetting agents which underwent the greatest increase in surface tension during expansion of a solid-liquid interface were the slowest to wet out the fabric (20). The time lag in the adsorption of surfactant at interfaces affects capillary wetting in two ways: it increases the disparity between advancing and receding contact angle, and delays attainment of the equilibrium $\gamma_{WA} \cos \theta$ value. The increase in contact angle hysteresis generally hinders wetting (2).

As mentioned above, a requisite for an efficient wetting agent is that it lower γ_{FW} considerably and γ_{WA} much less. A second requisite is that it produce low dynamic fiber-water interfacial tensions by rapidly diffusing to the interface and readily forming an oriented adsorbed interfacial film. Even though extensive surfactant adsorption by the fiber generally produces low contact angles, the resultant depletion of surfactant from the penetrating liquid may outweigh this advantage and result in slower wetting (19, 21). As an illustration, nonionic surfactants are adsorbed more extensively by cellulose and diffuse more slowly in water than anionic surfactants. During wetting of cotton fabric by surfactant solutions, this results in pronounced depletion of nonionic surfactants from the advancing liquid front, and makes them less suitable as wetting agents than anionic surfactants. Diffusion of surfactant from the bulk solution to the new interfaces formed by immersing cotton skeins in solutions of two nonionic surfactants was shown to determine the rate of wetting (sinking times) (21). In the case of three anionic surfactants, adsorption and diffusion had a negligible effect on wetting rates. For their solutions, the rate of wetting of cotton skeins was shown to depend on the contact angles; the sinking time was an exponential function of cos θ (21).

D. Effect of the Chemical Nature of Surfaces on Their Wettability

While geometrical factors (surface roughness and porosity) have a profound effect on contact angles and wettability, the chemical nature and physicochemical characteristics of the solid surface are also important.

The work of Zisman and co-workers, summarized in (3), has uncovered a linear relationship between the cosine of the contact angles formed on a smooth surface by different liquids and the surface tension of these liquids:

$$\cos\theta = \underline{a} - \underline{b}\,\gamma_{\underline{WA}} \tag{5.14}$$

This linear relationship was found to hold for several organic polymers and low molecular weight compounds despite the wide variety of spreading liquids used. These solids, like most organic substances, have surface free energies below 100 $ergs/cm^2$ and are therefore said to have low-energy surfaces. The data of Fig. 5.3 refer to "equilibrium" contact angles (3); by using thoroughly purified liquids and very clean and glossy-smooth surfaces, and by letting the liquid drops advance or recede so slowly as to approach equilibrium conditions, advancing and receding contact angles were identical.

The surface tension at which the straight lines of Fig. 5.3 reach unit value for cos θ is called critical surface tension for spreading ($\gamma_{\underline{c}}$). Liquids of surface tension equal to and smaller than the $\gamma_{\underline{c}}$ of a solid will spread out indefinitely on its surface, i.e., have contact angle zero, while liquids of surface tension greater than $\gamma_{\underline{c}}$ will make a finite contact angle. As a cosine cannot exceed unity, the linear wetting curves of Fig. 5.3 become horizontal once they reach $\gamma_{\underline{c}}$. Because of this boundary condition, Eq. (5.14) becomes

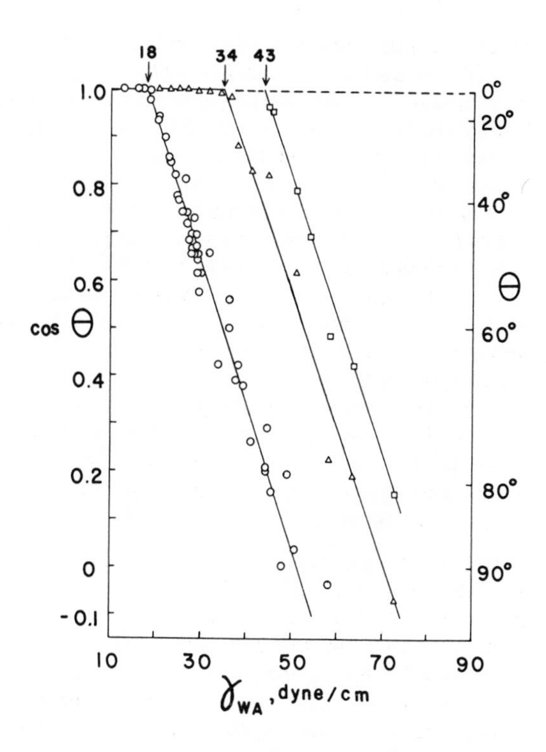

FIG. 5.3. Zisman plots (cos θ versus surface tension) at 20° for three low-energy surfaces: polytetrafluoroethylene (circles), polyethylene (triangles), and polyethylene terephthalate (squares).

$$\cos \theta = 1 + \underline{b} (\gamma_{\underline{c}} - \gamma_{\underline{WA}}) \tag{5.15}$$

The critical surface tension is a characteristic property of the solid. The most inert and least polar solids such as polytetrafluoroethylene have the lowest γ_c values, and very few liquids spread freely on their surfaces. The critical surface tension has been identified with the surface tension of the solid; $\gamma_c = \gamma_F$ (22-24). More likely, γ_F is somewhat greater than γ_c (25).

Critical surface tensions have been correlated with the following bulk properties of the solids: For organic polymers, γ_c increased with increasing cohesive energy density (25) and second-order transition temperature (27). For group VI.A elements, γ_c was found to be proportional to the atomic radius (28). The topic of surface energy and wetting of solids is one of the active frontiers in physical chemistry.

Contact angles of water and of benzene as well as critical surface tensions of several polymers are listed in Table 5.3. The contact angle of water on regenerated cellulose (viscose, cellophane) has been shown to decrease from 32° to zero within 10-15 min contact (29). Evidently, some water of the drops was absorbed by and penetrated into film and fiber, so that the residual drops eventually spread on water imbibed by the solid. Owing to the presence of waxes, raw cotton is quite hydrophobic. Extraction with ethanol, and with ethanol followed by ether, reduced the advancing contact angle of water from 105 to 65° and to 10-30°, respectively (29). Repeated laundering and bleaching of fabrics removes not only these waxes but the entire primary wall of the cotton fiber, i.e., the thin outermost layer which contains most of the noncellulosic constituents, so that the cotton surface becomes hydrophilic. However, calcium salts of the fatty acids from perspiration and of sulfonated surfactants are precipitated onto the fabric by the action of water hardness during washing. These "lime soaps" are quite hydrophobic [the contact angle of water on calcium palmitate is 90° but decreases with increasing pH (33)] and tend to impart this hydrophobicity to the fabric. Fabric softeners have the same effect.

In addition to chemical composition, orientation and crystallinity of the solid can greatly affect its wettability. The contact angles of water on secondary cellulose acetate films which had been cast from solution onto glass or metal surfaces were up to 10° lower on the "air side" than on the side which, during film formation, was in contact with the high-energy substrates. For cellulose nitrate films, the contact angles of water were 7-13° higher on the "air side" than on the "glass side" (34). Apparently, the polar groups in the polymer surface were oriented outward at the "glass side" and inward at the "air side." The critical surface tension of a butadiene-acrylonitrile copolymer was 30 dyn/cm at the "glass side" versus 37 at the "air side." For chloroprene and chlorosulfonated polyethylene, the orientation induced by the glass surface lowered γ_c by only 2 dyn/cm (35). The importance of density and crystallinity of the surface layer is demonstrated by the fact that aggregates of single crystals of linear polyethylene, cold-pressed into a film, had γ_c = 46 dyn/cm. Molding these aggregates above the melting point, which lowers density and crystallinity, reduce γ_c to the value listed in Table 5.3 for low-density polyethylene (36). Heterogeneous nucleation and crystallization of polymer melts in contact with high-energy inorganic surfaces resulted in high surface densities for these polymers and in wettability similar to single crystals. Polymer melts solidified in contact with low-energy surfaces such as polytetrafluoroethylene had surface layers of wettability and density comparable to the amorphous solid (37).

Measurements with aqueous solutions of simple organic liquids gave essentially the same critical surface tension values for polymers as with pure liquids (3, 32); they have the advantage in experimental work of not swelling or dissolving many of the polymer substrates. Not all cos θ

TABLE 5.3

Contact Angles of Water and Benzene on Smooth Surfaces of Fiber-Forming Polymers at 20° (or 25°), and Critical Surface Tensions

Polymer	θ, degrees, for		γ_c, dyn/cm
	Water[a]	Benzene[b]	
Polytetrafluoroethylene	108, 124 (22)[d]	46	18, 16 (22)[d]
Polypropylene	90 (22)	—	29 (22), 29 (27)
Polyethylene	94, 91 (22), 92 (30)	< 5	31, 32 (22), 31.5 (31) 31 (27), 31.5 (32)
Polystyrene	91, 90 (22)	0	33, 30 (22), 36 (27)
Polyvinyl chloride	87, 80 (22), 88 (30)	0	39, 39 (22), 39 (27)
Polyvinylidene chloride	80	0	40, 40 (27)
Polyethylene terephthalate	81	0	43, 42.5 (32)
Polyhexamethylene adipamide	70, 70 (30)	0	46
Polyacrylonitrile	48 (30)	0	44 (27)
Cellulose (regenerated)	0-32[e], 38 (30)	0	45 (23), 44 (29)

[a] γ_{WA} = 72.8 dyn/cm.

[b] γ_{OA} = 28.9 dyn/cm.

[d] Data of Zisman et al. (3) except where other references are given.

[e] See text.

versus $\gamma_{\underline{WA}}$ curves of such solutions were linear, but they were smooth and without discontinuities in slope (3, 38). In the case of aqueous surfactant solutions, the wetting curves generally consisted of two portions. The breaks occurred at surfactant concentrations somewhat below the critical micelle concentrations (cmc) and at surface tensions somewhat above those observed at the cmc, regardless of the nature of the solid surface (3, 38). At concentrations above the cmc, the wetting curves of anionic, cationic, and nonionic surfactants were linear and converged to the same $\gamma_{\underline{c}}$ values as were obtained with pure liquids (3).

These observations led Zisman to conclude that the wetting of low-energy surfaces by aqueous surfactant solutions is not caused primarily by adsorption of the surfactant molecules at the solid-water interface with their hydrocarbon tails attached to the solid surface and the polar moiety oriented outward toward the water, thereby converting the solid surface to one of high energy, as is held by some (5). His preferred wetting mechanism is a lowering of the surface tension of water, caused by the adsorption of a film of surfactant molecules at the water-air interface oriented with the hydrocarbon tails toward the air and the polar groups inside the water. This reduces the free energy of the water surface by covering it with a film of hydrocarbon chains (3). Accordingly, the important factor in the wetting of low-energy surfaces by aqueous surfactant solutions is that the surfactant must lower the surface tension of water below the critical surface tension of the solid.

The difference between the two wetting mechanisms may seem academic because most surfactants, owing to their amphiphilic nature, are preferentially adsorbed at and lower the free energy of both solid-liquid and liquid-air interfaces. Moreover, both wetting mechanisms result in the same final orientation of the surfactant molecules in the adsorbed monolayer at the solid-water interface, viz., the hydrocarbon tail attached to the low-energy solid and the polar head in the water. However, the distinction may be important to a surface chemist selecting or synthesizing a wetting agent.

Zisman's mechanism for wetting by surfactant solutions depends on lowering $\gamma_{\underline{WA}}$, whereas in their studies on penetration of water into fabrics, Hollies et al. (13) and Durham and Camp (16) found the rate to be proportional to $\gamma_{\underline{WA}}$. These results are not incompatible, because the two studies refer to completely different situations. The former deals with equilibrium spreading on smooth surfaces, the latter involves kinetic considerations applied to capillary structures. Fabrics have higher critical surface tensions for wetting than do smooth surfaces of the same material (see Sec. II, B, 1, Table 5.5). As long as the surfactant lowers the surface tension of water below $\gamma_{\underline{c}}$ of the fabric, comparatively high $\gamma_{\underline{WA}}$ values promote high wetting-out rates.

Studies by Zisman (3) and Wolfram (22) showed that the work of adhesion of different liquids on a given polymer substrate varies within a much narrower range than their surface tensions. For instance, the work of adhesion of 57 pure liquids on polytetrafluoroethylene varied between 35 and 54 ergs/cm^2, with a mean of 46 and an average deviation 11% of the mean. The surface tensions ranged from 18 to 73 dyn/cm, with a mean of 32 and an average deviation 27% of the mean (3). The work of adhesion of the solutions of two anionic and two cationic surfactants at their cmc on seven polymer surfaces varied between 36 and 60 ergs/cm^2, with a mean of 50 and an average deviation 11% of the mean (38). According to Eqs. (5.5) and (5.15), the work of adhesion is a quadratic function of the surface tension of the wetting liquid, i.e., it cannot be independent of the surface tension.

The penetration of capillary spaces in a solid by a liquid, such as the wetting out of fabric discussed in the previous section, can conveniently be considered in terms of the adhesion tension (also called wetting tension or spreading pressure) $\gamma_{WA} \cos \theta$ or $W_{FW} - \gamma_{WA}$ (1). For aqueous solutions of alcohols spreading on all polymers investigated, and for surfactant solutions below the cmc spreading on polyolefins and polytetrafluoroethylene, Wolfram reported that the adhesion tension increased linearly with decreasing surface tension (38):

$$\gamma_{WA} \cos \theta = e - c\, \gamma_{WA} \tag{5.16}$$

Since c was found experimentally to be equal to 1, e = W_{FW}, according to Eq. (5.5). The surfactant was adsorbed in equal amounts at the solid-liquid and liquid-air interfaces. When the solid surface came into contact with the liquid surface, it did not affect the orientation and packing density of the monolayer of surfactant adsorbed at the latter (38). The value of e or W_{FW} has been identified with the energy of attraction between liquid and solid at the interface due to London dispersion forces (39). If the solid polymer or the liquid contains polar groups, intermolecular attractions other than dispersion forces, e.g., dipole-dipole interaction or hydrogen bonding, come into play at the solid-liquid interface, and c is no longer unity. In that case, the monolayer of surfactant molecules adsorbed at the solid-liquid interface has orientation and packing density different from those of the monolayer adsorbed at the liquid-air interface (38). For solutions of different surfactants, Eq. (5.16) was found to apply to concentrations below the cmc and Eq. (5.14) or (5.15) to concentrations above the cmc; therefore, there is no conflict between the former and the latter two equations.

E. Conclusions

Is the wetting out of fabrics a serious problem in detergency? Generally it is not because polymers commonly used for textile fibers have critical surface tensions for spreading of 40 dyn/cm or higher (see Table 5.3) and many surfactants lower the surface tension of water to about 30 dyn/cm at their cmc. Furthermore, fabrics have higher γ_c values than smooth films of the same material. The difference was found to be about 20 dyn/cm for fluorocarbon and silicone finishes: The critical surface tension for spreading of cotton fabric treated with two fluorocarbon finishes was 24-25 dyn/cm compared to 5 dyn/cm for glass slides coated with these finishes (40). These soil-repellent finishes probably render the fabric difficult to wet by detergent solutions.

II. REMOVAL OF ORGANIC SOIL

Soil in domestic laundry can be divided into two main categories, organic or greasy soil and particulate, inorganic soil.

A. Nature of Organic Soil

Among the main water-insoluble constituents of organic soil are sebum from perspiration, scales from the skin which are largely proteins (41, 42), foodstuffs, oils from hair preparations and other cosmetic lotions, blood, and mineral and lubricating oils, as well as other occupational dirt.

Human sebum is a complex mixture of widely varying composition. The main components of aged sebum, and their representative levels, are free fatty acids (25%), triglycerides (32%), mono- and diglycerides (12%), waxes (18%), hydrocarbons, chiefly squalene (9%), and cholesterol (4%, about one-half as esters), plus smaller amounts of other sterols and phospholipids. The fatty acids range from n-C_7 (heptanoic) to n-C_{22} (behenic) and include branched and diunsaturated species. Palmitic and oleic are the most prevalent acids (41, 43-45). Extracts from garments and bed sheets were found to contain somewhat more hydrocarbons (46) or unsaponifiable matter (46a).

Freshly secreted sebum contains more triglycerides and less free fatty acids, but enzymes from bacteria on the skin hydrolyze some of the former to produce more of the latter (45). Among the chemical changes that sebum undergoes on the fabric are oxidation and polymerization, possibly catalyzed by traces of iron. This results in stronger odor, deeper color, and greater difficulty of removal. As an example, the central, most heavily soiled portion of a pillowcase contained 7.3% fatty matter. One-half of this fabric was stored in the dark in the absence of fresh air, where it lost only 1.4 reflectance units in a month and another 3.9 units in a year. The other half, kept in the open, lost 8.5 units in a month and an additional 12.6 units in a year (41). Sebum can accumulate in garments despite frequent laundering; solvent-extractable materials amounting to

13% have been found in cotton undershirts and pillowcases (41). The accumulated organic soil consisted largely of triglycerides, lime soaps, and unsaponifiable material; it contained very little free fatty acids. The combined fatty acids were much lower in oleic acid than sebum, probably because of oxidation (47).

Sebum is a soft, yellow to brown solid. A differential thermal analysis curve of fatty material extracted with a 9:1 benzene-ethanol mixture from underwear is shown in Fig. 5.4. The sample had been chilled to 0°C and was then slowly heated. The endothermic melting range extends to 48°, but 90% of the material is molten at 37°. At the usual machine washing temperature of 50 ± 5°, sebum is a rather mobile liquid.

Several recently marketed commercial detergents are recommended for washing the entire domestic laundry load in cold water. As can be seen from Fig. 5.4, sebum is largely solid at 15 or 20°. Solid or semi-solid organic soil is removed from fabrics in a completely different manner from molten organic soil, subsequently referred to as oily soil. The two situations must be discussed separately.

B. Mechanisms of Removal of Molten Organic Soil

1. The Rolling-Up Mechanism

The pioneering work of N.K. Adam (48) has shown that the removal of oily soil from fiber surfaces consists of its displacement by the wash water through a mechanism of preferential wetting. Kling (49) termed it "Umnetzung" or exchange wetting.

Formally, there is an analogy between the displacement of air by water, dealt with in the preceding section, and the displacement of oil by water. When a fiber $\underline{F}$ partly covered with a film of liquid oil $\underline{O}$ is immersed in a very dilute aqueous surfactant solution $\underline{W}$ which does not completely displace the oil from the fiber surface, a state of equilibrium is reached in which both oil and water remain in contact with the fiber and meet along the three-phase boundary line marked $\underline{P}$ in the cross-sectional diagram of Fig. 5.5. The contact angles $\theta_{\underline{W}}$ measured in water and $\theta_{\underline{O}}$ measured in oil, which are supplementary, are finite in this case, which corresponds to the situation of Fig. 5.6(b). Resolving the interfacial tensions parallel to the solid surface, and remembering that the resultant force at $\underline{P}$ is zero because of the condition of equilibrium, gives

$$\gamma_{\underline{FW}} = \gamma_{\underline{FO}} + \gamma_{\underline{WO}} \cos \theta_{\underline{O}} \tag{5.17}$$

or

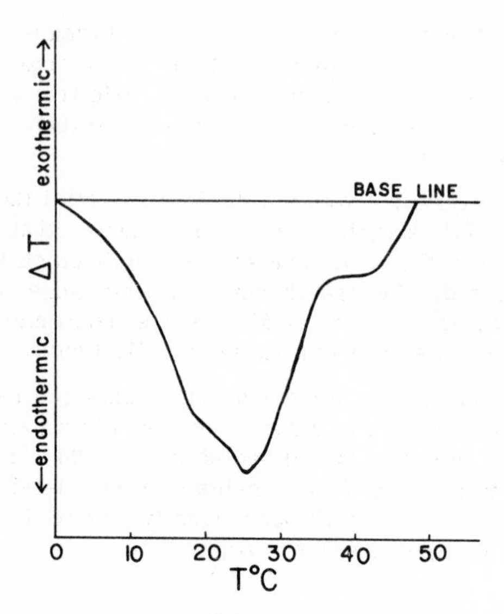

FIG. 5.4. Differential thermal analysis curve of sebum.

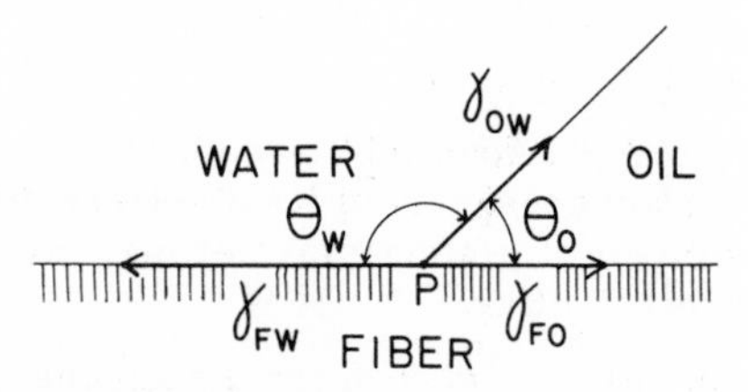

FIG. 5.5. Force diagram at the fiber-water-oil boundary, representing the equilibrium state for finite contact angles.

$$\gamma_{\underline{FO}} = \gamma_{\underline{FW}} + \gamma_{\underline{WO}} \cos \theta_{\underline{W}} \tag{5.18}$$

analogous to Eq. (5.1). Equations (5.17) and (5.18) are equivalent because $\cos \theta_{\underline{O}} = \cos(180° - \theta_{\underline{W}}) = -\cos \theta_{\underline{W}}$. The foregoing assumes that water and oil are mutually insoluble. Otherwise, the equations refer to water saturated with oil and to oil saturated with water.

When a fiber soiled with oil is immersed in pure water, usually very little happens; $\theta_{\underline{O}}$ remains close to zero and $\theta_{\underline{W}}$ is nearly 180° [Fig. 5.6 (a)]. Addition of a very small amount of surfactant to the water lowers $\theta_{\underline{W}}$ and increases $\cos \theta_{\underline{W}}$. Most surfactants used for washing have little or no solubility in the oil phase and will therefore not affect $\gamma_{\underline{FO}}$. Since they are positively adsorbed at the fiber-water and/or water-oil interfaces

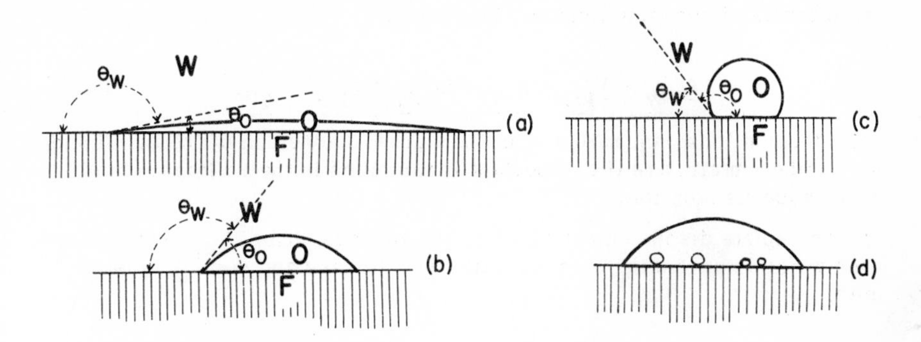

FIG. 5.6. Rolling up of oil on fiber immersed in water: (a) through (c) represent successive stages, corresponding either to increased time of contact or, if equilibrium was attained, to increasing surfactant concentration.

in the form of a monomolecular film, they will lower $\gamma_{\underline{FW}}$ and/or $\gamma_{\underline{WO}}$, so that, according to Eq. (5.18), cos $\theta_{\underline{W}}$ must increase. This situation corresponds to Fig. 5.6(b). The film of surfactant molecules being adsorbed at the fiber surface displaces the oil because it adheres more strongly to that surface (2, 50). Increasing amounts of surfactant progressively lower $\theta_{\underline{W}}$ until it is nearly zero, at which point the oil has been almost completely displaced from the fiber surface and rolled up into droplets [Fig. 5.6(c)]. Even when the surfactant is not positively adsorbed at the fiber-water interface and does not lower $\gamma_{\underline{FW}}$, as is the case with anionic surfactants in distilled water on pure cellulose (51), it can promote rolling up of the oily soil on the fiber, evidently by lowering $\gamma_{\underline{WO}}$. The rolled-up oil droplets are readily sloughed off the fabric and suspended in the aqueous phase by the agitation of the washing machine (48).

The work of adhesion of water and of oil for the fiber is given by the Dupré equations (5.4) and (5.19) referred to the system fiber-liquid-air.

$$\underline{W}_{\underline{FO}} = \gamma_{\underline{FA}} + \gamma_{\underline{OA}} - \gamma_{\underline{FO}} \tag{5.19}$$

The difference between the two works of adhesion is obtained by subtracting Eq. (5.19) from Eq. (5.4). Substituting into Eq. (5.18) and rearranging results in

$$(\underline{W}_{\underline{FW}} - \gamma_{\underline{WA}}) - (\underline{W}_{\underline{FO}} - \gamma_{\underline{OA}}) = \gamma_{\underline{FO}} - \gamma_{\underline{FW}} = \gamma_{\underline{WO}} \cos \theta_{\underline{W}} \tag{5.20}$$

or, in terms of adhesion tension [Eq. (5.6)]:

$$\underline{A}_{\underline{FW}} - \underline{A}_{\underline{FO}} = \gamma_{\underline{WO}} \cos \theta_{\underline{W}} = \gamma_{\underline{FO}} - \gamma_{\underline{FW}} \quad (5.21)$$

The adhesion tensions, rather than the works of adhesion of the two liquids on the fiber in air, are the important parameters in the rolling up of oil by the aqueous solution.

For complete displacement of oil by the aqueous phase, $\theta_{\underline{W}}$ must be zero when water is advancing over the oily fiber surface and cos $\theta_{\underline{W}}$ must be unity:

$$\underline{A}_{\underline{FW}} - \underline{A}_{\underline{FO}} \geq \gamma_{\underline{WO}} \quad (5.22)$$

The oil will be completely displaced from the fiber by an aqueous surfactant solution if the adhesion tension of the aqueous solution to the fiber exceeds the adhesion tension of the oil to the fiber by an amount at least equal to the interfacial tension between the aqueous solution and the oil (1, 48, 50). The difference in adhesion tension can be determined directly rather than being calculated from contact angles, by measuring the force necessary to pull a vertical plate through a water-oil interface (52) or by measuring the angle of inclination at which an oil droplet resting on a horizontal plate immersed in aqueous surfactant solutions beings to move when the plate is being tilted (53).

Equation (5.21) refers to the equilibrium situation at the boundary line between water, oil, and fiber. For the nonequilibrium case where the oil is being displaced by the aqueous surfactant solution, the resultant interfacial tension $\underline{R}$ rolling up the oil into a droplet is given (49) by

$$\underline{R} = \underline{A}_{\underline{FW}} - \underline{A}_{\underline{FO}} - \gamma_{\underline{WO}} \cos \theta_{\underline{W}} \quad (5.23)$$

which reduces to Eq. (5.21) at equilibrium ($\underline{R} = 0$). Kling and Koppe (54) calculated the "work of detergency" $\underline{W}_{\underline{D}}$, that is, the work per unit area required for complete displacement of oil by water. When $\theta_{\underline{W}}$ is just zero,

$$-\underline{W}_{\underline{D}} \cong \underline{A}_{\underline{FW}} - \underline{A}_{\underline{FO}} - \gamma_{\underline{WO}} = \gamma_{\underline{FO}} - \gamma_{\underline{FW}} - \gamma_{\underline{WO}} \quad (5.24)$$

In many practical situations, $\theta_{\underline{W}}$ is not reduced to zero: the rolling-up mechanism stops short of complete displacement of the oil. It is then possible to calculate the "residual work of detergency" $\underline{W}_{\underline{D},\underline{R}}$ (55), i.e., the theoretical work required to reduce $\theta_{\underline{W}}$ from the equilibrium value to zero and thereby to complete the displacement or rolling up of the oil. Since the area of fiber-oil contact decreases as the oil is being rolled up,

$W_{D,R}$ is a function of the volume V of the oil droplet (55):

$$W_{D,R} = \pi(3V/\pi)^{2/3}\left[(4)^{1/3} - (2 + 3x - x^3)^{1/3}\right]\gamma_{WO} \quad (5.25)$$

where $x = \cos\theta_W$ at equilibrium. As required, $W_{D,R}$ becomes zero for $\theta_W = 0°$ (complete displacement of the oil).

A numerical example is given in Fig. 5.7, namely, the displacement of mineral oil from glass and from poly ϵ-caprolactam (6-nylon) film by aqueous solutions of sodium dodecyl sulfate (55). For simplicity, $W_{D,R}$ was calculated for $V = 1/3\sqrt{\pi}$, making the first term in Eq. (5.25) unity. As can be seen from the values of the residual work of detergency, the oil was much more easily displaced from the high-energy glass surface

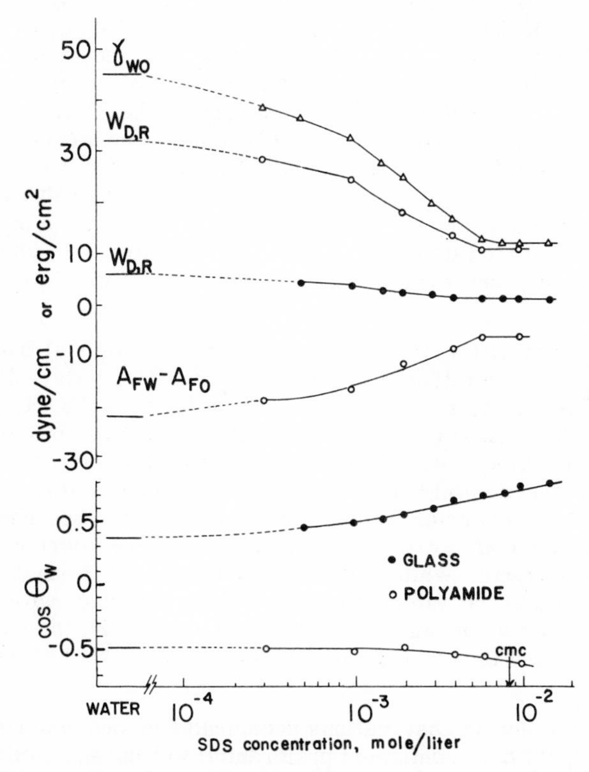

FIG. 5.7. Interfacial tension γ_{WO}, difference in adhesion tensions $A_{FW} - A_{FO}$, and residual work of detergency $W_{D,R}$ in the system aqueous sodium dodecyl sulfate (SDS)-mineral oil on glass and on 6-nylon, as a function of surfactant concentration (55).

than from the low-energy polymer. The residual work of detergency for both substrates, and $\gamma_{\underline{WO}}$, decreased with increasing surfactant concentration, while $\underline{A}_{FW} - \underline{A}_{FO}$ increased. The three parameters leveled off at about 6×10^{-3} mole/liter, which is slightly below the cmc of sodium dodecyl sulfate in the absence of oil (8.1×10^{-3} mole/liter or 2.32 g/liter). Solubilization of oil lowers cmc values.

The high residual work of detergency for mineral oil on 6-nylon indicates that sodium dodecyl sulfate is a poor detergent active, a fact well known to formulators. Table 5.4 combines two sets of data comparing the displacement of mineral oil from three fibers and one film by solutions of two surfactants. The data of Kling and Lange (55) are not directly comparable to those of Stewart and Whewell (30) because the former employed 6-nylon and the latter 66-nylon, and because the two mineral oils used had different interfacial tensions against water. Lissapol N, a surfactant of Imperial Chemical Industries Ltd., is octylphenol$(EO)_{9 \text{ to } 10}$ and has a cmc of 0.1 to 0.2 g/liter. At or just below its cmc, the residual work of detergency of the nonionic surfactant for mineral oil on nylon was almost zero; i.e., the surfactant solution displaced the oil almost completely from the nylon surface. Lissapol N removed mineral oil from polyethylene more effectively than sodium dodecyl sulfate removed it from nylon, despite the greater affinity of mineral oil for polyethylene than for nylon, indicated by the greater $\theta_{\underline{W}}$ value of the oil on polyethylene than on nylon immersed in water. Table 5.4 illustrates the usefulness of the residual work of detergency as a parameter for rating the effectiveness of detergents.

The effectiveness with which a surfactant removes oily soil from fabrics is governed by its adsorption at the fiber-water and/or water-oil interface, raising $\underline{A}_{FW}$ and/or lowering γ_{WO}. Adsorption at the water-air interface, which lowers γ_{WA} and promotes foaming, does not contribute to the detersive action (50), contrary to what most housewives believe. Foaming is in fact harmful because it reduces the mechanical work to which the fabric is subjected in the washing machine and because it lowers the amount of surfactant available for adsorption at the interfaces between water and oil or fabric. While it would be unusual for a surfactant to be strongly adsorbed at the water-oil or fiber-water interface without being positively adsorbed at the water-air interface as well (2), the important factor in detergency is surfactant adsorption at one or both of the first two interfaces.

Extensive microscopic observations documented by photomicrographs (48, 50, 56, 57, 58) have confirmed preferential wetting and rolling up as the major mechanism for the removal of oily soil from hydrophilic and hydrophobic fibers alike by anionic, cationic, and nonionic surfactants. Adam (48) remarked on the large size of the oil globules being rolled up and shaken off the fabric. The oil globules shown in the published photomicrographs vary between 20 and 120 μ in diameter. They are one to

TABLE 5.4

Displacement of Mineral Oil from Polymer Surfaces by Surfactant Solutions at Room Temperature[a]

	Polyethylene		66-Nylon[b]		Viscose		6-Nylon[c]		
	Water	0.1 g/liter L[d]	Water	0.1 g/liter L	Water	0.1 g/liter L	Water	0.58 g/liter SDS[e]	1.73 g/liter SDS
γ_{WO}, dyn/cm	58	9	58	9	58	9	45	24	12
θ_W[f]	128°	112°	86°	33°	36°	< 30°	60.5	60°	54°
$A_{FW} - A_{FO}$, erg/cm^2	-36	-3.3	4	7.5	47	>7.7	-22	-12	-7
$W_{D,R}$, erg/cm^2	—	5.4	—	0.07	—	< 0.05	32	17.5	10

[a]From data of Stewart and Whewell (30) and Kling and Lange (55).

[b]Polyhexamethylene adipamide.

[c]Poly ε-caprolactam.

[d]L is Lissapol N.

[e]SDS is sodium dodecyl sulfate.

[f]Measured in the aqueous phase, which is the advancing position.

two orders of magnitude larger than the droplets in common industrial emulsions. Possibly, the mechanical agitation in the washing machine breaks up such globules into smaller ones. The surfactant, being adsorbed at the water-oil interface, helps to disperse the oil and stabilizes its emulsion in water (50). However, emulsification plays at best a secondary role in the removal of oily soil. Once removed from the fiber surface, the tendency of the oil toward redeposition is much less than its tendency to remain there in the first place, owing to adsorption of surfactant at the fiber-water and/or water-oil interface (40).

An interesting phenomenon observed in party rolled-up oily soil is the appearance of small aqueous droplets inside the half-formed oil drops, which are located directly at the fiber surface as shown schematically in Fig. 5.6(d) (59, 60). The water droplets are formed by osmotic passage of water from the surrounding bath to traces of salt located at or near the fiber surface which is covered by oil. Such salt originated from perspiration, suint, or from previous washes. Evidence for osmosis is the fact that when a fiber with an oil globule adhering to it is placed in concentrated salt solutions, the aqueous droplets inside the oil phase collapse, but when the concentrated salt bath is replaced by water, they re-form (58-60).

The work of Berch et al. (40) illustrates the importance of the chemical nature of the fiber on soilability and washability. Fiber surfaces are commonly classified as varying from low-energy, nonpolar, or hydrophobic to relatively high-energy, polar, or hydrophilic; although compared to many inorganic solids, all organic polymers have low surface free energies. The extremes are represented by polytetrafluoroethylene and cellulose. This attribute of low or high surface free energy refers to the fiber-air interface.

In air, the tendency toward soiling is greatest and the removal of soil most difficult for fibers of the highest surface free energy. Viscose and cotton fibers covered with oily soil are more difficult to wipe clean than polyester fibers, because when the oil wetted the fibers, it reduced the high surface free energy of the cellulose by a larger amount than the lower surface free energy of the polyester fibers. Treatment with a fluorochemical finish lowers the surface free energy of fabrics so much that they become "soil-repellent," because even oily soils of relatively low polarity will not spread and penetrate them. As shown in Table 5.5, treatment with a fluorocarbon finish reduced the critical surface tension for wetting of a cotton fabric from greater than 72 dyn/cm to 24 dyn/cm; this fabric will repel liquids having surface tensions as low as 25 dyn/cm.

Under water, the situation is reversed. The tendency toward soiling by redeposition and redistribution of oily soil in the washing machine (displacement of detergent solution from the fabric by oil) is greatest and the removal of oily soil is most difficult for the most hydrophobic fibers.

TABLE 5.5

Relation between Soil Removal and Critical Interfacial Tensions of Cotton Fabrics Treated with Different Finishes and Contact Angles on the Corresponding Polymer Films[a]

Polymer[b]	γ_c[c], dyn/cm	$\gamma_{c,W \to O}$[d], dyn/cm	θ_O[e], degrees	Soil removal[f], %
None (untreated cellulose)	> 72	< 2.8	117	96
Acrylic	> 72	< 9.3 > 4.3	87	85
Fluorocarbon	< 25 > 24	< 15 > 9.3	66	60
Silicone	< 45 > 38	> 50	32	41

[a]From Berch et al. (40).

[b]As fabric finish or in film form.

[c]Critical surface tension for wetting of finished fabrics in air.

[d]Critical interfacial tension for displacement of water from finished fabrics by organic liquids.

[e]Receding contact angle of mineral oil on polymer films cast from finishing agents, immersed in solution of built anionic detergent in water of 100 ppm hardness.

[f]Removal of mineral oil from finished fabrics by built anionic detergent in water of 100 ppm hardness.

These are the fibers which possess the lowest surface free energy (in air) and therefore the highest interfacial free energy in water, as shown in Table 5.5. The critical surface tension for displacement of water by organic liquids from finished cotton fabrics, which is a measure of the free energy of the fabric-water interface, increased from untreated cotton fabric to one treated with a silicone finish. The contact angles of drops of mineral oil on films prepared from the finishes and immersed in a detergent solution, measured in the oil phase, decreased in the same order. Smaller θ_O values indicate increasing difficulty in rolling up the mineral oil. This was confirmed by the actual data on oil removal from the finished fabrics, which was highest for untreated cotton and lowest for the silicone-treated fabric (last column of Table 5.5).

Comparison of Tables 5.3 and 5.5 shows that fabrics have higher critical surface tensions in air than smooth films of the same surface material. This is a consequence of the capillary structure of the yarn, which tends to promote wetting that would not occur on a plane surface.

A particularly valuable feature of the work of Berch and associates (40) is that, while the chemical nature of the fabric surface was changed through the use of different finishes, yarn structure and fabric construction remained the same. As discussed in Secs. I, B and C, geometrical factors are major determinants of wetting behavior, and extensive corrections would be required for direct comparison of different fabrics. Oily soil was transferred spontaneously from cellophane to polyester film when the two were immersed together in water or in surfactant solutions (60a).

The reason why the fibers of highest surface free energy in air have the lowest interfacial free energy in water and in detergent solutions is connected with the properties of water. Owing to the strong attraction between its molecules, water has a very high internal pressure, nearly five times that of benzene. This produces a strong unbalance of attractive forces at the water-air interface and consequently a strong inward pull on the water molecules in the surface layer. To increase the surface of water against air by bringing water molecules from the bulk (where the attractive forces are balanced out) to the surface requires considerable energy. Therefore, the surface tension of water is high, 2.5 times greater than that of benzene.

The more polar and the more closely spaced the functional groups of the polymer at the fiber-water interface, the more strongly will they attract the water molecules at that interface and the more nearly will they counterbalance the inward pull exerted by the bulk water on the layer of water molecules at the fiber surface. Likewise, interaction with water lowers the free energy of the fiber surface compared to what it was in air. Because this mutual attraction increases as the polarity of the fibers and their surface free energy in air increase, the fiber-water interfacial free energy will be lowest for the most polar fibers. These considerations are true as long as the surface free energy in air of the fibers remains lower than that of water, a condition fulfilled by all common organic fiber-forming polymers.

This behavior can also be interpreted in terms of Eq. (5.1), according to which the fiber-water interfacial free energy, represented by the interfacial tension γ_{FW}, is equal to the difference between the surface free energy of the fiber, represented approximately by γ_{FA}, and $\gamma_{WA} \cos \theta$. As discussed in Sec. I,D, γ_{FA} in turn is approximated by the critical surface tension for wetting, γ_c. Both γ_c and $\cos \theta$ increase with increasing polarity of the fiber (see Table 5.3). Since cos θ increases over a wider relative range and since it is multiplied by the surface tension of water, which is high, $\gamma_{\underline{FW}}$ becomes smaller as fiber polarity and $\gamma_{\underline{FA}}$ increase. To sum up, polarity, free energy (in air), and hydrophilicity of the fiber surface go hand in hand; they are inversely related to the fiber-water interfacial free energy.

The effect of the polarity of the fiber and of the oily soil on soil removal is illustrated by the work of Stewart and Whewell (30) mentioned previously. They measured contact angles of mineral oil and of olive oil on different fibers immersed in solutions of the nonionic surfactant Lissapol N. The aqueous phase was in the advancing position, as it is in detergency. The work of adhesion of the oil to fibers immersed in aqueous surfactant solutions was calculated by Young's equation applied to Fig. 5.5,

$$\underline{W}_{\underline{FO}/\underline{W}} = \gamma_{\underline{WO}} (1 + \cos \theta_{\underline{O}}) \qquad (5.26)$$

and the difference in adhesion tensions $\underline{A}_{\underline{FW}} - \underline{A}_{\underline{FO}}$ was calculated as $\gamma_{\underline{FO}} - \gamma_{\underline{FW}}$ [Eq. (5.21)]. These two energies are plotted on the ordinates of Fig. 5.8 and 5.9. Detergency results obtained with skeins impregnated with the two oils and washed in 0.1 or 0.5 g/liter Lissapol N at 22°, expressed as percent residual oil, are plotted on the abscissas. The detergency data were corrected for the relative surface areas of the fibers.

According to the critical surface tensions or the contact angles of water listed in Table 5.3, the order of increasing surface hydrophilicity or polarity of the fibers shown in Figs. 5.8 and 5.9 is the same as in Table 5.6. This was also the order of increasing removal of mineral and olive oils by the nonionic surfactant solutions, except for the inversion of the polyamide and acrylic fibers. As indicated in Table 5.6, this inversion was caused by the presence of the surfactant. In agreement with the results of Berch et al. (40) and with the data on residual work of detergency of Table 5.4, the more polar the fiber, the more thoroughly it was washed free of oily soil.

The work of adhesion of the oil for the fiber immersed in a surfactant solution is a measure of the affinity of the oil for the fiber relative to the affinity of water for the fiber. As can be seen in Table 5.6, $\underline{W}_{\underline{FO}/\underline{W}}$ decreased as the fiber polarity increased. This is in accordance with Young's equation (5.26) because $\theta_{\underline{O}}$ increases and consequently cos $\theta_{\underline{O}}$ decreases as the polarity of the fiber surface increases. The equivalent equation

$$\underline{W}_{\underline{FO}/\underline{W}} = \gamma_{\underline{WO}} + \gamma_{\underline{FW}} - \gamma_{\underline{FO}} \qquad (5.27)$$

leads qualitatively to the same conclusion, because $\gamma_{\underline{FW}}$ decreases, and

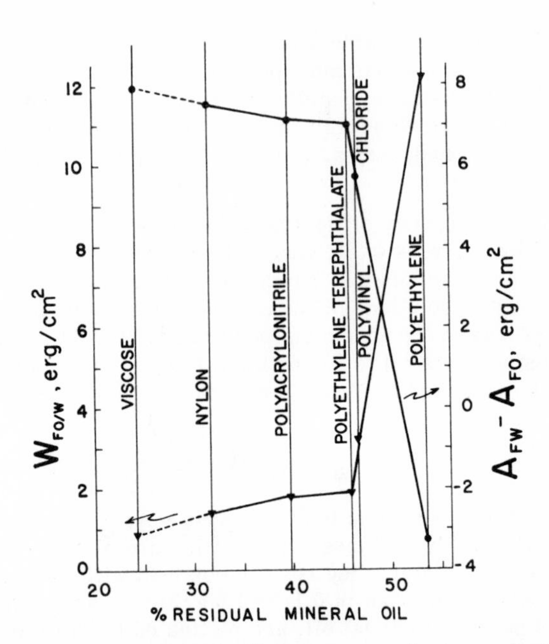

FIG. 5.8. The system mineral oil-aqueous nonionic surfactant-fiber: work of adhesion of oil to fiber in 0.1 g/liter Lissapol N and difference in adhesion tensions, as a function of residual oil after washing in 0.1 g/liter Lissapol N (30).

γ_{FO} increases for most soils, with increasing fiber polarity. However, γ_{FW} and γ_{FO} are difficult to estimate.

The difference in adhesion tensions $A_{FW} - A_{FO}$, which is the parameter governing the rolling-up process according to Eqs. (5.22) and (5.24) (1, 48, 49), was found by Stewart and Whewell to increase with increasing fiber polarity (see Table 5.6). This is in agreement with Eq. (5.21), and with the experimental observation that the more polar fibers are more thoroughly washed free of oily soil.

One seeming discrepancy in the system olive oil-0.5 g/liter Lissapol N is that a higher $A_{FW} - A_{FO}$ value and a lower $W_{FO/W}$ value were obtained for polyvinyl chloride than for the two fibers with immediately higher and lower polarity. This could be due to the presence of plasticizer in the polyvinyl chloride, or to penetration of olive oil into the polymer. The amorphous polyvinyl chloride is often plasticized, and commonly with esters resembling olive oil in polarity. Owing to their high crystallinity, polyethylene terephthalate and polyethylene are not readily plasticized nor appreciably penetrated by esters at room temperature.

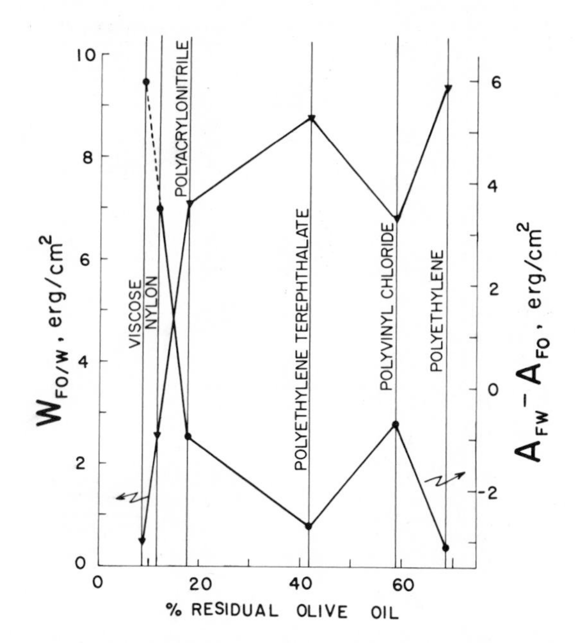

FIG. 5.9. The system olive oil-aqueous nonionic surfactant-fiber: work of adhesion of oil to fiber in 0.5 g/liter Lissapol N and difference in adhesion tensions, as a function of residual oil after washing in 0.5 g/liter Lissapol N (30).

Olive oil, which consists mainly of triglycerides of unsaturated fatty acids, is more polar than mineral oil, which consists of saturated hydrocarbons. Therefore, it adheres more strongly to a given fiber than mineral oil and is less completely removed. For most fibers, the percentage of residual oil after washing in 0.1 g/liter Lissapol N was nearly twice as great for olive oil as for mineral oil (30); the residual work of detergency calculated by Eq. (5.25) was from 10 to 100 times greater.

For a given fiber immersed in 0.1 g/liter Lissapol N, $W_{FO/W}$ of olive oil was between 10 and 20 times greater than for mineral oil except in the case of polyethylene, where it was only twice as great. In terms of Eq. (5.25) the contribution of γ_{WO} to the difference in the works of adhesion of the two oils is small. The chief factor is polymer wettability as measured by the contact angle: For all fibers except polyethylene immersed in 0.1 g/liter Lissapol N, θ_O of mineral oil was greater than 130°, making cos θ negative (see Table 5.6), whereas θ_O of olive oil was 70° or less. Comparing the works of adhesion of Figs. 5.8 and 5.9, olive oil is seen to have greater $W_{FO/W}$ values for a given fiber in 0.5 g/liter Lissapol N

TABLE 5.6

Effect of Surfactant Concentration on Contact Angles, Work of Adhesion, and Difference in Adhesion Tensions of Mineral Oil on Different Fibers[a]

Fiber	Lissapol N concentration, g/liter, and γ_{WO} (dyn/cm)								
	0 (58)			0.1 (9)			2.0 (2)		
	θ_O, deg.	$W_{FO/W}$, erg/cm²	$A_{FW} - A_{FO}$, erg/cm²	θ_O, deg.	$W_{FO/W}$, erg/cm²	$A_{FW} - A_{FO}$, erg/cm²	θ_O, deg.	$W_{FO/W}$, erg/cm²	$A_{FW} - A_{FO}$, erg/cm²
Polyethylene	52	94	-36	68	12.2	-3.3	110	1.5	0.7
Polyvinyl chloride	71	77	-19	130	3.2	5.7	> 150	< 0.3	> 1.9
Polyethylene terephthalate	66	82	-24	142	1.9	7.0	> 150	< 0.3	> 1.9
Polyhexamethylene adipamide	94	54	4	147	1.4	7.5	> 150	< 0.3	> 1.9
Polyacrylonitrile	117	32	26	143	1.8	7.1	> 150	< 0.3	> 1.9
Viscose	144	11	47	> 150	< 1.2	> 7.7	> 150	< 0.3	> 1.9

[a]From Stewart and Whewell (30).

solution than mineral oil in 0.1 g/liter Lissapol N despite the higher surfactant concentration, except for polyethylene.

For a given fiber immersed in 0.1 g/liter Lissapol N solution, the difference in adhesion tensions $\underline{A}_{FW} - \underline{A}_{FO}$ for mineral oil was higher than for the more polar olive oil, again indicating more extensive removal by rolling up of the former. In fact, as is seen in Fig. 5.9, the $\underline{A}_{FW} - \underline{A}_{FO}$ values for olive oil in 0.5 g/liter Lissapol N were negative for all fibers except viscose and nylon owing to θ_W angles greater than 90°—a situation not conducive to the rolling up of olive oil on the less polar fibers.

Just as Kling and Lange (55) showed for sodium dodecyl sulfate (Fig. 5.7), so Stewart and Whewell observed for Lissapol N that increasing surfactant concentrations reduced $\underline{W}_{FO}/\underline{W}$ and increased $\underline{A}_{FW} - \underline{A}_{FO}$ by lowering γ_{WO} and cos θ_O [see Eqs. (5.26) and (5.21), and Table 5.6]. Evidently, higher surfactant concentrations resulted in more extensive adsorption of surfactant at the water-oil and fiber-water interfaces. In many instances, the addition of nonionic surfactant increased θ_O considerably above 90°, making cos θ_O negative and large, and thus causing extensive rolling up of the oil. This is why the residual oil content of skeins was found to decrease with increasing surfactant concentration. For instance, increasing the Lissapol concentration from 0.1 to 2.0 g/liter reduced the amount of residual mineral oil from two- to fivefold.

Stewart and Whewell (30) present an elegantly self-consistent picture. They found that the removal of two oily soils from different fibers could be completely explained in terms of the polarity of fiber and oil, and that the process was related to interfacial free energies derived from contact angles. Therefore, they justly conclude that the rolling-up mechanism is entirely adequate to describe the removal of liquid oils under their experimental conditions.

In a more qualitative fashion, the soiling of polymer films immersed in solutions of different surfactants by cyclohexane or by oleic acid was studied microscopically. The variables were length of time of contact between oil droplet and film required for the droplet to displace the intervening layer of detergent solution, thereby wetting the film, and surfactant concentration (61). Maximum soil stabilization (no wetting after an "infinite" time of contact) was achieved at concentrations near the cmc, if one takes into account the fact that solubilized oils lower the cmc values. Polyamide and polyester films were not wetted by the oils except for the system oleic acid-sodium oleate solution, where acid soaps may have been formed, indicating that there is no redeposition problem in the washing machine. Polyethylene and polytetrafluoroethylene films were wetted by oleic acid in every case. The less polar cyclohexane failed to wet polytetrafluoroethylene immersed in solutions of all four surfactants and polyethylene in sodium oleate and cetyl trimethyl ammonium bromide solutions. The other two surfactants used were sodium dodecyl sulfate and nonylphenol$(EO)_{20}$.

A comparable situation, namely the displacement of water from solid surfaces by oil, has been studied extensively at the U.S. Naval Research Laboratory (62, 63).

2. Studies with Radioactive Tracers

Wagg studied the removal of oily soil by built detergents at 60° in two ways. The first was to apply one of several single, pure constituents of sebum (free fatty acid, fatty alcohol, triglyceride, mineral oil) to a variety of fibers, and to follow its removal by detergent solutions microscopically (64). The second method consisted in soiling the fibers with a complex mixture simulating sebum, to which was added one of several radiotagged components (stearic acid, tristearin, cholesterol, octadecane) at a time (65). Radioactive counting permitted quantitative study of the rate and extent of removal of each of the tagged compounds in turn under the more realistic condition where the compound was part of a sebumlike mixture.

Wagg found little specificity in the ease of removal of the four labeled compounds which were part of the artificial sebum. Octadecane was more easily removed than stearic acid and tristearin, possibly owing to its greater volatility, but there was no statistically significant difference in the ease with which cholesterol, stearic acid, and tristearin were removed from the fibers.

In the microscopic study, the molten fatty soil component was applied directly to the fiber surface. Cotton and viscose were cleaned the most readily, the less polar fibers nylon and polyethylene terephthalate less readily and less completely. In the radiotracer study, the fatty soil was applied by impregnating the chopped fiber with a chloroform solution, followed by drying and aging at 40°. The fibers were ranked in the following order of increasing difficulty for removing soil containing tagged components: viscose, nylon, polyethylene terephthalate, wool, and cotton. The difficulty of cleaning cotton in the second set of experiments was probably due to the fact that the solvent had deposited some fatty soil inside the lumen or into some of the deep surface crevices of the cotton fiber not readily accessible when simply dipping the fiber into the molten soil. During washing in water, swelling of the cotton fibers probably constricted the lumen and the fissures, rendering them inaccessible to the washing liquor. Even if the washing liquor could penetrate into the lumen, any rolled-up oil droplets formed inside it would probably be too large to leave the interior of the fiber. The fatty material deposited from solution into the lumen will be removed by subsequent solvent extraction of the fiber used to determine residual oily soil, however, because molecularly dissolved fat can diffuse through pores far too small to permit the passage of emulsified droplets. The amount of solvent-deposited oily soil trapped in the lumen is about 1% of the weight of the cotton fabric (66).

This morphological aspect is probably the solution to the current controversy regarding the difficulty of removing oily soil from cotton (67), which according to some investigators surpasses the difficulty of removing oily soil from polyamide, polyester, and acrylic fabrics (41, 68, 69). There is certainly general agreement that regenerated cellulose fibers are more readily washed free of oily soil than synthetic fibers because of their greater polarity. Therefore, the exceptional position of cotton, even when freed of noncellulosic constituents, with regard to the removal of oily soil applied from solution, cannot be due to its chemical nature. Morphological factors (lumen, fissures) are the likely culprits. It is doubtful that under natural soiling conditions, any sebum is deposited in the lumen of cotton fibers (70).

Wagg and collaborators (64, 65) found that, in general, soaps (sodium oleate and a tallow-based soap) and an ethylene oxide-based nonionic surfactant removed oily soils more efficiently from all fibers in distilled water than did the following synthetic anionic surfactants: alkylaryl sulfonate, sulfonated hydrocarbon oil, and secondary alkyl sulfate.

Qualitative experiments by other investigators also confirmed that the ease of removal of oily soil by rolling up increased with increasing fiber polarity (71, 58) and with decreasing soil polarity (71). Thus, on polyethylene and polyester fibers, the rolling-up process stopped at rather small θ_O values. Owing to the large area of contact between the spherical segments of oil and the fiber surface, the oil was not readily detached by agitation. On the other hand, the oil was rolled up into nearly spherical droplets on 6- and 66-nylon, acrylic fiber, viscose, and wool, which were shaken off with ease. Mineral oil was more extensively rolled up than triglycerides (71).

Some results of careful quantitative work by other investigators cannot be explained in terms of the rolling-up mechanism. Gordon et al. (72, 73) worked with swatches of cotton, 66-nylon, and polyethylene terephthalate fabric to which a mixture of lubricating oil (25%), tristearin (10%), stearic acid (15%), oleic acid (15%), cholesterol (7%), octadecanol (8%), and peanut oil (20%) was applied from solvents. All but the last component were available in radioactive form, the first two labeled with tritium and the next four with carbon-14. This permitted to follow the removal of two soil components at a time by using one tagged with ^{14}C and one with ^{3}H together with four inactive components. The radiochemical purity of the labeled compounds was established by gas-liquid, thin layer, or column chromatography combined with radioactive assay. This turned out to be a necessary precaution because several of the commercial labeled compounds contained surprisingly large proportions of radioactive impurities (73). After washing in the Terg-O-Tometer for 10 min at 16 or 49° in hardened water with built detergents, the swatches were immersed in a toluene-based liquid scintillation counter to measure residual radioactivity and thereby the amount of residual soil components.

Direct comparison of the amount of soil retained by the different fabrics after washing may be misleading since the fabrics probably had different surface areas. Therefore, detergency results are expressed as percentage of soil remaining or removed, based on the amount initially present in the fabric. On this basis, nylon generally retained the least soil, followed either by the polyester fabric or by cotton, depending on the soil component. The poor washability of cotton relative to the two more hydrophobic fabrics is probably due to deposition of soil by solvents in the lumen or fissures of the cotton fibers, where it is inaccessible to the detergent solution because cotton swells in water. This is corroborated by the lower selectivity in the removal of soil from cotton than from the other two fabrics. In order to determine selectivity, Gordon used a mixture of all six labeled soils, at a ratio of the $^{14}C/^{3}H$ fractions equal to 0.82. The ratio of the two soil fractions in the fabrics before washing was 0.82 for cotton, 0.77 for nylon, and 0.81 for polyester, indicating little selectivity in soil uptake. After washing at 49°, the ratios of the residual $^{14}C/^{3}H$ fractions were 0.63 and 0.65 for cotton, 0.37 and 0.52 for nylon, and 0.37 and 0.57 for polyester fabric. The first of each pair of numbers is after washing in a detergent with an ethoxylated nonionic surfactant, the second with a linear alkylaryl sodium sulfonate. Washing changed the soil ratio least for cotton, as one would expect if a portion of the whole soil were removed extensively and another portion not removed at all because of inaccessibility. The lower selectivity of the anionic surfactant was at least partly due to its lower washing efficiency; it removed less of the single and of the mixed soils than three nonionic surfactants at both temperatures.

Fractionation of the soil mixture during washing at 49° was observed with the three fabrics and the four built surfactants. The individual soil constituents present in the mixture were removed more extensively in the order of increasing polarity: namely, lubricating oil $<$ tristearin $<$ cholesterol = octadecanol $<$ oleic acid $<$ stearic acid (73). This contrasts with the removal of single soil components by rolling up, where the least polar material is rolled up and removed the most extensively (see Sec. II,B,1. It seems impossible to reconcile *any* selectivity in the removal of oily soil constituents from a mixture applied to a fabric with the rolling-up mechanism, regardless of whether the most or the least polar constituents are removed preferentially, as long as the mixture is a homogeneous liquid at the washing temperature. If the mixed soil on the fiber surface consists of a single liquid phase, it would be rolled up into droplets of uniform composition. Extensive separation of the soil constituents during drying of their common solution applied to fabrics seems unlikely (70).

Wagg found no significant preferential removal of different constituents from a sebumlike mixture (65). This may have been due to poor reproducibility, particularly because he used only ^{14}C labeling and could follow only one component at a time. This required a newly soiled batch of chopped fibers for each component, and it is not easy to achieve uniform soiling from batch to batch. Aside from reproducibility, the main differences in the techniques of Wagg (65) and Gordon (73) is that the former used chopped fibers, while the latter used fabrics and washed with strong agitation.

The effect of yarn structure was described in a recent paper (74) comparing the uptake and removal of oily soil from two polyester fabrics, one made of staple yarn and the other of continuous filament. The two fabrics had comparable weights per unit area and a relatively small difference in specific surface area. When treated with a sebumlike oily soil emulsified in water, the staple fabric picked up 20.0% oil based on the weight of the fabric, compared to only 3.6% pickup by the filament fabric. After five successive washes, the staple fabric still retained 7.5% oil (63% removal); the filament fabric retained only 0.3% oil (92% removal).

The following experiments, also based on polyester, show that the difference in soiling was due mainly to yarn structure rather than fabric structure (74). Oil uptake from the aqueous emulsion was 4.8% for a filament fabric and 13.3% for the corresponding filament yarn, compared to 21.1% for a staple fabric and 38.0% for the corresponding staple yarn. Ground-up powdered staple fabric picked up 26.8% oil and ground-up filament fabric picked up 26.7% oil. The difference in soiling between staple and filament fabrics persisted for the respective yarns but disappeared for the powders.

Scanning electron micrographs revealed heavy deposits of oil on the staple fabric, which were largely held between the crisscross, randomly spaced fibers. After washing, residual oil remained in the V-shaped areas where the fibers cross one another. The oil picked up by the filament fabric was spread fairly uniformly over the filaments themselves; very little oil was held between adjacent filaments (74).

An ingenious experimental arrangement was used by Fort et al. (75, 76) to study the removal of radioactively labeled fatty constituents from the surfaces of cellulose, 66-nylon, polyethylene terephthalate, and a tetrafluoroethylene-hexafluoropropylene copolymer. They used the polymers in the form of thin films as end windows of a Geiger counter to cover the counter tube like a drumhead. The tube and the attached film, which had been soiled on the outside with radiotagged oily soil, were immersed into the stirred surfactant solutions. By measuring the radioactivity remaining on the soiled film as a function of washing time, the soil removal process was continuously monitored and automatically recorded.

In each experiment, one of the following four compounds, labeled with ^{14}C, was applied to a film from a carbon tetrachloride solution: octadecane, octadecanol, tristearin, and stearic acid. After the solvent evaporated, each film was polished to distribute the soil evenly to a final thickness of between three and eight monolayers. The rate or extent with which the four pure compounds were removed (as percentage of the amounts initially present) increased with increasing soil polarity, in the order listed above. This was observed at 20 and 60° for all substrate-surfactant combinations, and indicates that removal was not by rolling up. The order of increasing effectiveness of the surfactants in 0.01 M solution at 20 and 60° was nonylphenol$(EO)_{9.5}$ >> sodium palmitate > sodium oleate > cetyl trimethyl ammonium bromide > sodium dodecyl sulfate. This refers to the removal of pure soil components and of their mixture; the soaps were not studied at 20°. All surfactants have cmc values below 0.01 M. The nonionic surfactant removed tristearin nearly quantitatively at 60° from all four substrates within the first 3 min. Cetyl trimethyl ammonium bromide and sodium dodecyl sulfate cleaned the films only partially, in the order cellulose > nylon > fluoropolymer > polyethylene terephthalate. The pronounced swelling of cellophane in water may have mechanically aided in removing the film of oil.

While the initial rate of soil removal was found to increase with increasing temperature, the amount of soil removed in a given wash time sometimes reached a maximum at intermediate temperatures. For instance, the percentage of octadecanol (MP = 58.5°) removed by sodium dodecyl sulfate from polyethylene terephthalate reached a constant level within about 10 min. This level increased in the order 40° > 60° > 80° > 20°. This effect is due to competition between soil removal from the film surface by the surfactant solution and soil diffusion into the interior of the film, where it is inaccessible to the surfactant solution. Substantial diffusion begins at the second-order transition temperature which, for polyethylene terephthalate, is between 68 and 71° (77), and increases exponentially with temperature. This shows that the highest wash temperatures do not necessarily produce the cleanest clothes, especially for synthetic fibers.

Finally, the removal of mixed soils was studied by applying a mixture of the four soil constituents, of which only one was labeled in each experiment, to polyethylene terephthalate. The nonionic surfactant cleaned the films almost completely at 60° but cetyl trimethyl ammonium bromide and sodium dodecyl sulfate left considerable residual soil behind, which reached a constant level within 5 min of washing. In the case of these two ionic surfactants, the amount of residual tristearin, expressed as percentage of the amount initially present, was reduced by the presence of the other soil components; that of stearic acid was increased; and that of octadecanol was not significantly affected. The presence of the other soil constituents tended to offset the effect of polarity on the extent of removal of single constituents. This could signify that the mixed soils were removed simultaneously rather than selectively (75).

It is easy to imagine why the rolling-up process would not operate under the conditions of Fort's experiment. When a smooth film of oily soil completely covers a smooth solid surface, the aqueous phase cannot establish contact with the latter. A boundary line along which the three phases meet is necessary for the oil to be rolled up (see Fig. 5.5). The three interfacial tensions operate along this boundary to decrease the resultant tension by replacing the solid-oil interface with the solid-water interface. This wedge effect of the surfactant solution, sweeping up the oil spread out on the solid surface and collecting it into droplets, cannot occur if the surfactant solution does not come in contact with the solid surface.

3. Soap Formation

Builders contained in detergent formulations render the washing liquor alkaline and transform the free fatty acids of sebum into sodium soaps. As expected, soluble soaps formed in situ greatly aid in dispersing and emulsifying the oily soil because of their strong detersive activity (64, 78). On the distaff side, hardness, particularly calcium ions, transform the soluble sodium soaps into insoluble lime soaps, which form hydrophobic, hard-to-remove deposits in the fabric (47).

Triglycerides, which are the major components of sebum, are not saponified by the washing liquor, not even when the pH is boosted above 11 by the addition of sodium metasilicate (64).

4. Solubilization

Surfactant molecules associate in dilute aqueous solution to form micelles. Critical micelle concentration values are commonly between 0.01 and 0.0001 M. Micelles are aggregates of 50-300 surfactant molecules or more, depending on the nature of the surfactant, its concentration, temperature, and added electrolytes. The hydrocarbon portions of the surfactant molecules are turned inward, away from the water (hydrophobic bonding); the hydrophilic portions (ionized groups or polyoxyethylene chains) are directed toward the water. A wide variety of water-insoluble compounds, including fatty acids and alcohols, glycerides, and hydrocarbons, are solubilized by surfactant solutions, i.e., dissolved in the hydrocarbon interior of the micelles. If the solubilized molecules have polar groups such as hydroxyls or carboxyls, these are usually located in the outer, hydrophilic layer of the micelles. Solubilization results in clear and stable solutions of the fatty material as opposed to emulsions, which are cloudy and inherently unstable. Solubilization of oily soil was first proposed by McBain as an important mechanism in detergency (79) and has been reintroduced more recently (80, 81).

The arguments against solubilization as a mechanism in the removal of oily soil is that it requires micelles. Consequently, the surfactant concentration has to exceed the cmc, and little or no detergency is to be expected below the cmc. Preston (82) has shown that the detergent activity of anionic surfactants, as their surface activity, increases with increasing surfactant concentration until the cmc is reached, and then practically levels off. The addition of surfactant to a solution at and above the cmc augments almost exclusively the number and possibly the size of the micelles. Since the concentration of nonassociated surfactant molecules is only slightly increased beyond the cmc, the wetting ability of the solution is hardly improved, and interfacial tensions are only slightly reduced. This indicates that the nonassociated surfactant molecules or ions are responsible for washing, for altering contact angles, and for lowering surface and interfacial tensions, and implies that detergency results from surface activity. This is the basis of the rolling-up mechanism. According to Eq. (5.21), the difference in adhesion tensions, which governs the extent of rolling up, remains constant if γ_{WO} and θ_W are not further lowered by additional surfactant; this occurs when the cmc is reached. Furthermore, as is seen in Fig. 5.7, the residual work of detergency, i.e., the work required to complete the displacement of oil, levels off near the cmc and does not decrease further at higher surfactant concentrations. According to Preston's viewpoint, the role of micelles, should they be present in the wash liquor, is to act as a reservoir of nonassociated surfactant molecules, dissociating to restore their concentration to the cmc if they are depleted by adsorption on fabric and soil or by foaming.

Ginn and Harris presented the following data in favor of solubilization as a mechanism contributing to the removal of oily soil. By using radiotagged triglycerides, they showed that nonionic surfactants are capable of solubilizing as well as emulsifying triglycerides, while solubilization by sodium oleate and by synthetic anionic surfactants was negligible (80). Removal of triglycerides from frosted glass by nonionic surfactants resulted in clear solutions, which indicates solubilization. For all nonionic and for some anionic surfactants, maximum soil removal occurred at concentrations well above the cmc (81).

Perhaps the surfactant solutions failed to roll up the oily soil smeared on glass surfaces for the same reason advanced for the nonoccurrence of rolling up on film (75). However, even for cotton fabric, removal of oily soil (triglyceride plus mineral oil tagged with graphite) by nonionic surfactants was found to increase markedly above the cmc (81).

These results indicate that solubilization of oily soil plays a role at higher concentrations of nonionic surfactants. They do not preclude rolling up as the primary removal mechanism, however. Solubilization was indicated by the fact that the final surfactant solutions were clear and free of oil droplets. But this would occur regardless of whether the oil was first displaced en masse by preferential wetting and the suspended droplets subsequently solubilized, or whether the oil was removed from the fabric layer by layer through direct solubilization. There is certainly ample visual evidence for the former mechanism with both nonionic and anionic surfactants (50, 56-58), and none for the latter.

From a materials balance in the washing machine, it should be possible to decide whether enough surfactant is present to solubilize a substantial part of the oily soil. Owing to variations in the cloth: bath ratio, degree of soiling of the load, amount and composition of the detergent used, water hardness, etc., these calculations represent only an estimate. It is assumed that the fabric adsorbs an equilibrium amount of surfactant, and that only the excess surfactant is available for solubilization. This ignores the possibility that the oily matter to be solubilized competes with the fabric for surfactant. On the other hand, loss of surfactant through foaming and through adsorption by particulate soil is ignored. The figures used refer to moderately soiled cotton clothes and relatively soft water.

A machine load of 9 lb of clothes per 14 gal of wash liquor corresponds to 77 g cotton/liter. A detergent level of 3.15 oz per 14 gal corresponds to approximately 219 mg/liter anionic surfactant or 169 mg/liter nonionic surfactant. Of this, approximately 154 mg anionic surfactant or 77 mg nonionic surfactant will be adsorbed by the cloth, leaving 65 mg anionic or 92 mg nonionic surfactant per liter in solution. The cmc of a representative anionic surfactant in the absence of oil is 100 mg/liter. The presence of oil may reduce the cmc to the 65 mg/liter level, but not much below. This precludes extensive micelle formation of and oil solubilization by the anionic surfactant.

A representative nonionic surfactant has a cmc of 50 mg/liter in the absence of oil; this may be reduced to 35 mg/liter by solubilized oil, exceeding the 92 mg/liter level of nonadsorbed surfactant by 57 mg/liter. This amount of micellar nonionic surfactant is capable of solubilizing about 11 mg of oily soil, or 0.014% of the weight of the clothes. If garments in intimate contact with the skin contain 1% oily soil (83) and make up 10% of the load, only a small fraction of the oily soil present could be removed by solubilization even if the remaining 90% of the load contains none. Only about one-third of the oily soil could be solubilized in the unlikely case that all of the nonionic surfactant is used up by solubilization and none is adsorbed by the fabric.

5. Formation of Mesomorphic Phases as a Mechanism for Removal of Oily Soil

Micelles formed in dilute aqueous surfactant solutions are small and approximately spherical. Radii commonly range from 20 to 40 Å. By increasing the surfactant concentration, by adding salts or solubilized materials, and, for nonionic surfactants, by raising the temperature, the micelles become larger and pronouncedly asymmetric. Finally, mesomorphic or liquid crystalline phases appear. Viscoelastic gels are sometimes formed. This has been observed for soaps (84-86) and synthetic anionic (86, 87), cationic (87, 88), and nonionic surfactants (87, 89). Mesomorphic phases are liquids with a certain degree of order resulting from the alignment and mutual organization of the anisometric micelles. They are birefringent and possess sharp X-ray diffraction patterns (86, 87).

Mesomorphic phases are also formed in fairly concentrated ternary systems comprising a surfactant, a fatty alcohol or acid or other amphiphilic compounds, and water (90-92). By contrast, solubilization occurs already in very dilute systems and is less specific: Not only amphiphiles but also hydrocarbons and other compounds without strongly polar groups are solubilized. Lawrence (93-95) applied the formation of mesomorphic phases in ternary systems to detergency, proposing a soil removal mechanism based on the penetration of aqueous surfactant solutions into oily soil to produce mesomorphic phases with the polar soil constituents. He offers ample microscopic evidence that surfactant solutions do penetrate solid and liquid fatty alcohols and acids, as well as mineral oil containing at least 10-20% of a dissolved fatty acid or alcohol, to produce mesomorphic phases. Owing to the adsorption of surfactant at the water-oil interface to form a close-packed monolayer, the local surfactant concentration is high enough for extensive penetration into the oily soil and formation of liquid crystalline phases, despite the very low overall surfactant concentration in the wash liquor. Subsequently, the mesomorphic phases are swollen and broken up by the osmotic influx of water. Their dispersion occurs sometimes through the formation of myelinic figures—sinuous tubes projecting into the aqueous phase (78)—or by gradual breaking up into fine particles or droplets from the periphery inward (58, 78). Dispersion of the mesomorphic phase in water is followed by its decomposition through dissolution of the surfactant. This liberates the oily soil portion in a finely subdivided form. The surfactant either stabilizes this emulsion or, if its concentration is high enough, solubilizes the oil.

This mechanism favors the removal of the most polar soil because surfactants form mesomorphic phases with fatty acids and alcohols but not with triglycerides and hydrocarbons. Mesomorphic phase formation may well result in preferential removal of the former two types of compounds from sebum, leaving the latter two in the fabric. It differs fundamentally from the rolling-up process because it is based on penetration of the bulk of the oily soil by aqueous surfactant, whereas rolling up brings only the surface of the oily soil in contact with the surfactant solution.

Mesomorphic phases are very viscous liquids or even gelatinous. They begin to form in the outside layer of the oily soil, which is in direct contact with the aqueous surfactant. This outer, viscous membrane delays the penetration of fresh aqueous surfactant into the interior of the oil, which is required to convert it progressively into a mesomorphic liquid. The high viscosity also delays the subsequent influx of water and the breaking up of the mesomorphic phases (78). Even though the layer of oily soil on fabrics is usually quite thin, their high viscosity may so slow down the formation and dispersion of mesomorphic phases as to render this detergency mechanism unimportant compared to other, faster mechanisms within the 10 min wash cycle of the washing machines. It is also questionable whether natural soils are polar enough for extensive mesomorphic phase formation.

6. Specific Soil-Surfactant Interaction

In addition to the nonstoichiometric liquid crystalline phases formed in the presence of water, ionic surfactants form molecular association complexes with related amphiphilic compounds in the solid state and in monolayers. Examples of these are acid soaps like the 1:1 stearic acid-sodium stearate complex (96), and the 1:2 complex between dodecanol or tetradecanol and sodium dodecyl or tetradecyl sulfate (97, 98). Whether the formation of such complexes plays a role in the removal of oily soil is not known.

Proteins, which are constituents of natural soil (41, 42), form association complexes with anionic (99) but not with nonionic surfactants (100). This was probably the reason why sodium dodecylbenzene sulfonate was superior to nonylphenol $(EO)_{9.5}$ in washing naturally soiled cloth and in removing proteinaceous soil from it, while the nonionic surfactant was superior in washing artificially soiled cloth containing no protein (42).

C. Removal of Solid Organic Soil—Cold-Water Detergency

According to Fig. 5.4, sebum is largely solid at 10° but almost completely molten at 37°. At intermediate temperatures, it is a pasty solid. Of the three mechanisms of fatty soil removal discussed so far, rolling up can operate only with liquid soil; solubilization and mesomorphic phase formation are slower for solid than for liquid soil.

There is a dearth of publications dealing with the removal of solid fatty soil. As mentioned in Sec. B,2 above, Fort et al. (75, 76), using single soil components applied to films, found substantially greater values for initial removal rates and for percentages of removal within a given wash time at higher wash temperatures. Exceptions were found for percentage removal within a given time, which sometimes reached a maximum at intermediate temperatures owing to substantial diffusion of soil into the interior of the film at the highest wash temperatures.

Scott (66) studied the effect of temperature on the removal of single, pure soil constituents from cotton fabric by a built sodium dodecylbenzene sulfonate. Octadecane, tripalmitin, octadecanol, and stearic acid, labeled with ^{14}C, were applied to the fabric from carbon tetrachloride solutions. With increasing wash temperature, the removal of hydrocarbon and triglyceride after 10 min washing was small and constant below the melting points of the two compounds, increased suddenly at the melting points, and reached a second constant level above the melting points. Similar observations were made with other hydrocarbon soils on cotton (83). The removal of mineral oils which were liquid at all wash temperatures increased little, if any, with increasing temperature. The removal of a paraffin wax showed an abrupt increase at its melting point (74). This behavior indicates that the molten soils were removed by the rolling-up mechanism.

On the other hand, Scott found that the removal of octadecanol and of stearic acid by the built alkylbenzene sulfonate increased smoothly with rising wash temperature, beginning at 38 and 48° below the respective melting points. The two plots of percent soil removed versus temperature showed no inflection at the melting points but seemed to level off at somewhat higher temperatures. The removal of the two polar solids is attributed to the formation of mesomorphic phases. This is confirmed by photomicrographs of the interaction of surfactant solutions with the soil constituents in bulk (66). Even the removal of octadecanol from the fabric by a built nonionic surfactant, which was a stepwise function of temperature like the removal of octadecane and tripalmitin by the built anionic surfactant, was shown to occur through the formation of mesomorphic phases, because the sudden increase in removal occurred not at the melting point of octadecanol but 10° below, which is the temperature at which the nonionic surfactant solution began to penetrate octadecanol crystals to form mesomorphic phases. Below that temperature, the surfaces of crystals immersed in solutions of the nonionic surfactant remained smooth, and the removal of octadecanol from cotton was low and independent of temperature.

Scott found that the time of washing was an important variable in the removal of solid octadecanol by surfactant solutions because solids are only slowly penetrated by these solutions and because the resultant mesomorphic phases are very viscous. As mentioned in Sec. B,5 above, it is doubtful that the formation and dispersal of mesomorphic phases play a significant role in the detergency of liquid fatty soil under practical laundry conditions because of their very high viscosity and the short duration of the wash cycle. This applies even more strongly to cold-water detergency because the viscosity of the liquid crystals is even higher at the lower wash temperatures and because the penetration of solid fatty soil by surfactant solutions proceeds doubtlessly even more slowly than the penetration of liquid fatty soil.

Even solid hydrocarbons and triglycerides are removed to some extent by surfactant solutions at temperatures well below their melting points (66, 75, 76, 83), without the benefit of mesomorphic phase formation. This is ascribed to breaking up of the polycrystalline aggregates which make up the solids into the individual crystallites, as the surfactant solution penetrates into cracks and boundaries between these microcrystals. Evidence for this mechanism is the retention of appreciable amounts of radioactive material when the surfactant solutions used to wash films or fabrics soiled with radiotagged solid octadecane and tristearing (75, 76) or tripalmitin (66) were filtered through membranes. Filtration of surfactant solutions used to wash fabrics soiled with radiotagged octadecanol or stearic acid at temperatures well below the melting points left no residue even on the finest membranes (66). Evidently, the removal of the polar solids, which began with the formation and osmotic dispersal of mesomorphic phases, resulted in their dissolution, possibly by solubilization. Removal of the low-polarity soils, on the other hand, took place by simple dispersal of the solids.

The breaking up of polycrystalline aggregates can be observed microscopically by placing a fragment of solid triglyceride, preferably containing some free fatty acid, in a surfactant solution. If the solution is alkaline, such as one of sodium oleate, or of sodium dodecyl sulfate containing sodium carbonate, the solid is seen to break up into a cloud of fine particles, mostly below 1 μ.

Gordon *et al.* (73), using the soil mixture described in Sec. II,B,2, found surprisingly small differences in soil removal at 16 and 49° for the less polar, tritium-tagged fraction and for the more polar, ^{14}C-tagged fraction. On nylon, the main difference was observed with two nonionic detergents, namely, a higher removal of the less polar fraction at 16 than at 49°. On cotton, the average removal values for three nonionic and one anionic detergent were 38% at 16° versus 49% at 49° for the ^{3}H-labeled fraction, and 44% at 16° versus 58% at 49° for the ^{14}C-labeled fraction. On polyester fabric, the average soil removal figures were 28% at 16° versus 27% at 49° for the ^{3}H-labeled fraction and 64% at 16° versus 74% at 49° for the ^{14}C-labeled fraction (73). The small temperature effect observed by Gordon and co-workers could be attributed to the strong agitation employed by them (100 cycles/min of the Terg-O-Tometer), which may have overshadowed the effect of increased temperature (70).

D. Conclusions

Most cellulosic fabrics can be washed relatively free of oily soil with present-day detergents and washing machines without undue difficulty. However, the detergency of some modified or resin-treated cottons and of some synthetic fabrics presents a problem, which is best solved by understanding the physical chemistry of the soil removal process as it occurs in the washing machine. While the arts of making textile fibers, of developing formulated detergents, and of designing washing machines are fairly well advanced, our understanding of the fundamentals of detergency under full-scale laundry conditions is incomplete even in the case of the simplest soil, namely, a fatty liquid. Only recently has the effect of yarn structure been investigated (74), and there is still disagreement regarding the effect of surfactant concentration (81, 82). The effect of agitation in determining the mechanism of oily soil removal is unknown (70). Another unresolved question is whether and under which operating conditions the commonly used laboratory-scale washing equipment (Terg-O-Tometer, Launder-Ometer, Deter-Meter) (101) can duplicate the level and kind of agitation prevailing in the washing machine. The often employed Terg-O-Tometer setting of 90 cycles/min (73, 74) provides unrealistically violent agitation.

The limiting factor is not a lack of physicochemical knowledge of the mechanisms involved. The four known soil removal mechanisms (rolling up, solubilization, mesomorphic phase formation, and breaking up of polycrystalline aggregates) are adequately if not completely understood, and we know how to make each mechanism become operative in the laboratory by choosing appropriate conditions. Lacking is quantitative information on the prevalence, effectiveness, and limitations of each of the soil removal mechanisms in the washing machine as a function of the important process variables, namely: fiber, yarn, and fabric geometry and surface properties; nature of the soil; chemical constitution and concentration of surfactants and builders; temperature and time of washing; and level and type of agitation. Once it is known which detergency mechanism(s) is most effective in full-scale equipment and how to make it operative there, we have the basis for the rational development of construction and surface finishes for easy-to-wash fabrics, of detergents, and of washing machines.

REFERENCES

1. N. K. Adam, The Physical Chemistry of Surfaces, 3rd ed., Oxford Univ. Press, London, 1941.

2. J. L. Moilliet, B. Collie, and W. Black, Surface Activity, 2nd ed., Spon, London, 1961.

3. F. M. Fowkes, ed., Contact Angle, Wettability, and Adhesion, Adv. Chem. Series 43, American Chemical Soc., Washington, D.C., 1964.

4. A. W. Adamson, Physical Chemistry of Surfaces, 2nd ed., Wiley, New York, 1967.

5. F. M. Fowkes and W. D. Harkins, J. Amer. Chem. Soc., 62, 3377 (1940).

6. R. Shuttleworth and G. L. J. Bailey, Discuss. Faraday Soc., 3, 16 (1948).

7. F. E. Bartell and J. W. Shepard, J. Phys. Chem., 57, 211, 455 (1953).

8. A. J. G. Allan and R. Roberts, J. Polym. Sci., 39, 1 (1959).

9. R. N. Wenzel, Ind. Eng. Chem., 28, 988 (1936).

10. A. B. D. Cassie and S. Baxter, Trans. Faraday Soc., 40, 546 (1944).

11. D. J. Crisp, in Waterproofing and Water-Repellency (J. L. Moilliet, ed.), Elsevier, Amsterdam, New York, 1963, p. 437.

12. E. W. Washburn, Phys. Rev., 17, 273 (1921).

13. N. R. S. Hollies, M. M. Kaessinger, and H. Bogaty, Text. Res. J., 26, 829 (1956).

14. H. Bogaty, N. R. S. Hollies, J. C. Hintermaier, and M. Harris, ibid., 23, 536 (1953).

15. N. R. S. Hollies, M. M. Kaessinger, B. S. Watson, and H. Bogaty, ibid., 27, 8 (1957).

16. K. Durham and M. Camp, Proc. 2nd Int. Congr. Surface Activity, London, 4, 3 (1957).

17. S. Baxter and A. B. D. Cassie, J. Text. Inst., 36, T67 (1945).

18. J. M. Preston and M. V. Nimkar, ibid., 43, T402 (1952).

19. K. Schaefer, Z. Elektrochem., 59, 273 (1955).

20. H. Lange, Kolloid-Z., 136, 136 (1954).

21. F. M. Fowkes, J. Phys. Chem., 57, 98 (1953).

22. E. Wolfram, Kolloid-Z. u. Z. Polym., 182, 75 (1962).

23. V. R. Gray, Forest Prod. J., 12, 452 (1962).

24. V. R. Gray, Chem. Ind., 1965, 969.

25. P.-J. Sell and A. W. Neumann, Angew. Chem., 78, 321 (1966); Int. Educ., 5, 301 (1966).

26. J. L. Gardon, J. Phys. Chem., 67, 1935 (1963).

27. L.-H. Lee, J. Appl. Polym. Sci., 12, 719 (1968).

28. D. A. Olsen, R. W. Moravec, and A. J. Osteraas, J. Phys. Chem., 71, 4464 (1967).

29. B. R. Ray, J. R. Anderson, and J. J. Scholz, ibid., 62, 1220 (1958).

30. J. C. Stewart and C. S. Whewell, Text. Res. J., 30, 903 (1960).

31. S. Newman, J. Colloid Interface Sci., 25, 341 (1967).

32. D. K. Owens, J. Appl. Polym. Sci., 8, 1465 (1964).

33. F. van V. Vader and H. Dekker, J. Phys. Chem., 68, 3556 (1964).

34. F. E. Bartell and B. R. Ray, J. Amer. Chem. Soc., 74, 778 (1952).

35. L.-H. Lee, J. Polym. Sci., A-2, 5, 1103 (1967).

36. H. Schonhorn and F. W. Ryan, J. Phys. Chem., 70, 3811 (1966).

37. H. Schonhorn, Macromol., 1, 145 (1968).

38. E. Wolfram, Kolloid-Z. u. Z. Polym., 211, 84 (1966).

39. F. M. Fowkes, Ind. Eng. Chem., 56, No. 12, 40 (1964).

40. J. Berch, H. Peper, and G. L. Drake, Text. Res. J., 35, 252 (1965).

41. O. Oldenroth, Fette, Seifen, Anstrichm., 61, 1142, 1220 (1959); 62, 13 (1960).

42. S. Tomiyama and M. Iimori, J. Amer. Oil Chem. Soc., 42, 449 (1965).

43. V. R. Wheatley, Soap, Perfumery, Cosmetics, 29, 181 (1956).

44. A. W. Weitkamp, A. M. Smiljanic, and S. Rothman, J. Amer. Chem. Soc., 69, 1936 (1947).

45. K. Bey, Fette, Seifen, Anstrichm., 65, 611 (1963); 66, 579 (1964).

46. C. B. Brown, Res., 1, 46 (1947).

46a. V. McLendon and F. Richardson, Am. Dyestuff Reporter, 52, 27 (1963).

47. W. C. Powe and W. L. Marple, J. Amer. Oil Chem. Soc., 37, 136 (1960).

48. N. K. Adam, J. Soc. Dyers Colour., 53, 121 (1937).

49. W. Kling, Kolloid-Z., 115, 37 (1949).

50. N. K. Adam and D. G. Stevenson, Endeavour, 12, 25 (1953).

51. H. Schott, Text. Res. J., 37, 336 (1967).

52. H. L. Rosano and M. Weill, Amer. Dyest. Rep., 42, 227 (1953).

53. H. L. Rosano and J. B. Montagne, Rev. Franc. Corps Gras, 10, 9 (1963).

54. W. Kling and H. Koppe, Melliand Textilber., 30, 23 (1949).

55. W. Kling and H. Lange, Kolloid-Z., 142, 1 (1955).

56. R. C. Palmer, J. Soc. Chem. Ind., 60, 56 (1941).

57. W. Kling, E. Langer, and I. Haussner, Melliand Textilber., 25, 198 (1944); 26, 12, 56 (1945).

58. R. P. Harker, J. Text. Inst., 50, T189 (1959).

59. D. G. Stevenson, ibid., 42, T194 (1951).

60. D. G. Stevenson, ibid., 43, T112 (1952).

60a. T. H. Grindstaff, H. T. Patterson, and H. R. Billica, Textile Res. J., 40, 35 (1970).

61. H. Sonntag and K. Strenge, Tenside, 4, 129 (1967).

62. W. D. Bascom and C. R. Singleterry, J. Phys. Chem., 66, 236 (1962).

63. H. R. Baker, P. B. Leach, C. R. Singleterry and W. A. Zisman, Ind. Eng. Chem., 59, No. 6, 29 (1967).

64. R. E. Wagg and G. D. Fairchild, J. Text. Inst., 49, T455 (1958).

65. R. E. Wagg and C. J. Britt, ibid., 53, T205 (1962).

66. B. A. Scott, J. Appl. Chem., 13, 133 (1963).

67. A. Kling and H. H. Hofstetter, Seifen, Oele, Fette, Wachse, 92, 323 (1966).

68. O. Oldenroth, ibid., 93, 59 (1967).

69. H. Harder, Fette, Seifen, Anstrichm., 63, 25 (1961).

70. H. Schott, Text. Res. J., 39, 296 (1969).

71. R. S. Hartley and F. F. Elsworth, J. Text. Inst., 49, P554 (1958).

72. B. E. Gordon, W. T. Shebs, and R. U. Bonnar, J. Amer. Oil Chem. Soc., 44, 711 (1967).

73. B. E. Gordon, J. Roddewig, and W. T. Shebs, ibid., 44, 289 (1967).

74. C. B. Brown, S. H. Thompson, and G. Stewart, Text. Res. J., 38, 735 (1968).

75. T. Fort, H. R. Billica, and T. H. Grindstaff, ibid., 36, 99 (1966).

76. T. Fort, H. R. Billica, and T. H. Grindstaff, J. Amer. Oil Chem. Soc., 45, 354 (1968).

77. D. Jeschke and H. A. Stuart, Z. Naturforsch., 16a, 37 (1961).

78. D. G. Stevenson, J. Text. Inst., 44, T12 (1953).

79. J. W. McBain, in Advances in Colloid Science, Vol. 1 (E. O. Kraemer, ed.), Wiley-Interscience, New York, 1942, pp. 99-142.

80. M. E. Ginn, E. L. Brown, and J. C. Harris, J. Amer. Oil Chem. Soc., 38, 361 (1961).

81. M. E. Ginn and J. C. Harris, ibid., 38, 605 (1961).

82. W. C. Preston, J. Phys. Colloid Sci., 52, 84 (1948).

83. W. C. Powe, J. Amer. Oil Chem. Soc., 40, 290 (1963).

84. A. Skoulios, Adv. Colloid Interface Sci., 1, 79 (1967).

85. N. Pilpel, J. Colloid Sci., 9, 285 (1954).

86. V. Luzzati, H. Mustacchi, A. Skoulios, and F. Husson, Acta Cryst., 13, 660 (1960).

87. F. Husson, H. Mustacchi, and V. Luzzati, ibid., 13, 668 (1960).

88. T. Nash, J. Appl. Chem., 6, 539 (1956).

89. K. Kenjo, Bull. Chem. Soc. Japan, 39, 685 (1966).

90. D. G. Dervichian, Proc. 2nd Int. Congr. Surface Activity, London, 1, 327 (1957).

91. P. Ekwall, I, Danielsson, and L. Mandell, Kolloid-Z., 169, 113 (1960).

92. A. S. C. Lawrence, Discuss. Faraday Soc., 25, 51 (1958).

93. A. S. C. Lawrence, Nature, 183, 1491 (1959).

94. A. S. C. Lawrence, Chem. Ind., 1961, 1764.

95. A. S. C. Lawrence, in Surface Activity and Detergency (K. Durham, ed.), MacMillan, London, 1961, Chap. 7.

96. F. V. Ryer, Oil & Soap, 23, 310 (1946).

97. M. B. Epstein, A. Wilson, C. W. Jakob, L. E. Conroy, and J. Ross, J. Phys. Chem., 58, 860 (1954).

98. H. C. Kung and E. D. Goddard, ibid., 67, 1965 (1963); 68, 3465 (1964).

99. J. A. Reynolds, S. Herbert, H. Polet, and J. Steinhardt, Biochem., 6, 937 (1967).

100. R. M. Dowben and W. R. Koehler, Arch. Biochem. Biophys., 93, 496 (1961).

101. J. C. Harris, Detergency Evaluation and Testing, Wiley Interscience, New York, 1954, Chap. 4.

Chapter 6

REMOVAL OF PARTICULATE SOIL

Hans Schott
School of Pharmacy, Temple University
Philadelphia, Pa.

REMOVAL OF PARTICULATE SOIL

I. INTRODUCTION

Organic or greasy soil is displaced from fabrics mainly by a process involving preferential wetting (see Chap. 5). The forces binding inorganic, particulate soil to fabrics are of a different nature, as discussed in Chap. 4. Therefore, their removal occurs by different processes. Because solid dirt particles are often covered with oily soil, which may act as a cement between them and the fabric substrate, their removal may involve mechanisms characteristic of both oily soil and particulate soil.

II. NATURE OF PARTICULATE SOIL

Most particulate dirt, street dust as well as vacuum cleaner dust, consists of one-half to two-thirds of inorganic matter, the main constituents of which are aluminum and calcium silicates and quartz sand (1-3). The source of silicates is largely clays (4) and other minerals. The surfaces of most of these particles are rather close-packed layers of oxygen or hydroxyl ions, if they are not contaminated by organic soil. They have negative charges due to silicic acid groups. Ashes from fossil fuels and incinerators constitute another type of inorganic particulate soil. Their composition varies from source to source.

Soot from smokestacks and exhausts of internal combustion engines may be only a minor soil constituent but, owing to its deep color, it is very evident and therefore particularly objectionable. Airborne particulate soil, collected from air-conditioner filters in heavily populated industrial areas, lost 40% by weight on ashing and its color changed from dark gray to light beige. This shows that most of the reduction in whiteness of fabrics soiled with urban particulate dirt is due to carbonaceous material like soot (5). Soot, like manufactured carbon blacks, consists largely of carbon. Carbon blacks contain varying amounts of oxygenated groups such as carboxyl, phenol, quinone, and lactone. Their surfaces range from almost purely graphitic and inert to rather polar, with negative charges due to carboxylic and phenolic groups (6-8). Graphitization (heating at about 2700°C in the absence of oxygen) removes the acidic groups as carbon monoxide, thereby practically eliminating the base-binding capacity of the carbon blacks, and induces crystallization of the carbon. Some carbon blacks contain basic ashes (8).

In view of their mode of formation, soot particles are likely to contain tar and their surfaces are probably rich in oxygenated groups. Commercial carbon blacks are commonly used as model soils. They vary widely in particle size, surface composition, ash and tar content, and porosity, depending on the manufacturing process; the corresponding properties of soot have not been reported. Therefore, different blacks give different responses in detergency experiments, and it is questionable whether any are representative of soot.

Silicate dust, ashes, and soot are largely airborne and occur in the fabric as discrete particles. By contrast, the buildup of lime soaps through the reaction of water hardness with fatty acids from sebum and with synthetic anionic surfactants, and the deposition of waterborne colloidal hydrous oxides of iron and manganese result in fairly uniform stains. Portions of fabric covered with lime soap are hydrophobic. While the lime soaps have little intrinsic color, they prevent the uptake of optical brighteners by the fabric and thereby impart a yellowish tinge to it.

The hydrous oxides of iron(III) and of manganese(II) or (III) may occur as positively charged sols in natural waters; the former may also originate from rusty pipes. Owing to their charge, they are strongly taken up by the negatively charged fabrics during the rinse cycle of the washing machine (9). No such problem exists during the wash cycle, where there are two mechanisms for changing the charge of the sols from positive to negative, thereby reducing or preventing their uptake by the fabrics through electrostatic repulsion. First, the alkalinity of the builders boosts the pH of the wash liquor above the zero points of charge of the hydrous oxides of iron(III), which is 8.5, and of manganese(III), which is about 8.1, but not above the $pH > 12$ value of manganous hydroxide, if it were present (10). Second, the adsorption of excess anionic surfactant or of condensed phosphates reverses the charge of the sols even in neutral media (9).

The hydrous oxides of iron and manganese range in color from yellow to very dark brown. Since they are strongly colored, trace amounts suffice to impart a yellowish or brownish tinge to white fabrics and to render the colors of colored fabrics dull. Several milligrams of iron per pound of fabric is a common level; several tens of milligrams usually produces visible yellowing (11). Iron and manganese oxides catalyze the oxidation and polymerization of unsaturated constituents of sebum in air, transforming them into colored compounds which are much more difficult to wash out than the original compounds. They also promote the oxidative degradation of cellulose by some bleaches. The removal of iron and of lime soaps is discussed under "Sequestration" (Chap. 11).

Calcium phosphates constitute a major type of inorganic incrustation of the fabric. They are precipitated by the reaction of water hardness with residual polyphosphates during the rinse cycle of the washing machine, where the polyphosphate concentration is too low to sequester the calcium. Ash levels of several percent of the weight of the fabric, consisting mainly of calcium phosphates, are not uncommon (12). While these incrustations cause only minor discoloration in the absence of heavy metals, they may impart a harsh feel to the fabric. Silicate incrustations are sometimes produced by the reaction between water hardness and sodium silicates contained in commercial detergent formulations.

III. EQUILIBRIA IN THE SYSTEM: FIBER-SOIL-SURFACTANT SOLUTION

James W. McBain, one of the pioneers in the application of physical chemistry to colloidal phenomena, described the detergency process (13) by the equation

$$\text{Fabric·Dirt} + \text{Surfactant} = \text{Fabric·Surfactant} + \text{Dirt·Surfactant} \tag{6.1}$$

This equilibrium can be broken down into three processes (14), namely,

$$\text{Dirt} + \text{Surfactant} = \text{Dirt·Surfactant} \tag{6.2}$$

$$\text{Fabric} + \text{Surfactant} = \text{Fabric·Surfactant} \tag{6.3}$$

$$\text{Fabric·Dirt} = \text{Fabric} + \text{Dirt} \tag{6.4}$$

Addition of these three equilibria leads to the overall detergency equation (6.1). In this chapter, the binary equilibria represented by Eqs. (6.2), (6.3), and (6.4) are discussed in that order, followed by discussion of the three-component system of Eq. (6.1). With the exception of the soiling process [inverse of (6.4)], the interactions take place in excess water. As the concentration of water remains practically unchanged, it is not shown as a component in Eqs. (6.1)-(6.3). Commercial detergents contain several ancillary components in addition to the surfactant.

IV. INTERACTION BETWEEN PARTICULATE SOIL AND SURFACTANTS

Surfactants play a dual role in soil removal. They overcome the attraction between soil and fabric by attaching themselves to both. Not only do they loosen the soil from the fabric but they deflocculate it at the same time, i.e., they break it up into colloidal particles and stabilize their aqueous dispersion. Soil which forms a fine and stable dispersion in the wash liquor is much less prone to reattach itself to the fabric during the remainder of the wash cycle than soil present as a coarse and unstable suspension. The ability of particulate soil to form stable dispersions and the role of adsorbed surfactants in this process are discussed next.

A. The Nature of Solid-Water Interfaces and Hydrophilic Dispersions

Hydrosols are colloidal dispersions of solids in water. The solids can be classified into hydrophilic and hydrophobic. Colloidal particles of the former are highly hydrated and surrounded by a shell of water of hydration resembling ice. Hydrophilic solids in bulk form disperse spontaneously in water, and their hydrosols are consequently stable. Examples are globular proteins, starch and colloidal cellulose, sodium bentonite, methylcellulose, polyethylene glycol, gum arabic, and sodium polyacrylate. The main difference between sols of hydrophilic solids and true aqueous solutions (e.g., of sucrose) is that the particle size in the former is in the colloidal range and in the latter is of molecular dimensions. There is no sharp dividing line; particle radii between 10 and 100 Å (10^{-7} and 10^{-6} cm) are generally considered the lower limit of the colloidal range.

Surfactant adsorption by sols of hydrophilic substances does not increase their stability, and may even decrease it if the hydrocarbon chains of the surfactant molecules render the surfaces less hydrophilic. Owing to the hydrophilic and polar nature of these solid particles, the transition in going from solid to interface to water is relatively small. The interfacial free energies are low; they would not be lowered appreciably by the adsorption of surfactants and could even be increased. Because of this and of the concomitant negative change in entropy, surfactants are not adsorbed at such interfaces unless specific interactions, called chemisorption, take place. Examples are the reaction of anionic surfactants with cationic $-\overset{+}{N}H_3$ groups in proteins (wool, silk), and the uptake of cationic surfactants via their reaction with carboxylic acid groups of cellulose or with silicic acid groups of glass and clays.

B. The Stability of Hydrophobic Dispersions

Hydrophobic dispersions are inherently unstable compared to large crystals or bulk solid of the same material. If the particle surfaces are inert or nonpolar, such as those of graphite and of hydrocarbon elastomers, the particles will coagulate rapidly (caducuous dispersions). The free energy of a graphite-water interface is considerable, and coagulation will lower it. However, the coagulated particles touch each other only at a few points of contact unless their surfaces are flat, or unless the materials are soft enough to coalesce slowly, such as elastomeric latexes. Otherwise, the specific interfacial area of a dispersion is not very much reduced by coagulation.

If there are ions or charged groups in the surface, electrostatic repulsion may prevent particles from approaching each other closely enough for the attractive forces to predominate and cause aggregation. The role of this electric barrier in detergency is described in Chap. 4, in Ref. 15, and mentioned below. It renders inherently unstable, hydrophobic dispersions resistant to coagulation for long periods of time. Such sols are said to be diuturnal. Many inorganic sols are thus stabilized by electric double layers comprised of ions in the particle surface and a halo consisting primarily of oppositely charged ions (counterions) surrounding the particle to establish electroneutrality.

Figure 6.1 represents the solid-water interface of a barium sulfate particle precipitated by the reaction between sodium sulfate and a very slight excess barium chloride (16). There are more Ba^{2+} than SO_4^{2-} in the surface, so that the particle is said to be positively charged. Of course the particle is surrounded by an atmosphere of ions in solution where Cl^- or OH^- predominate over Na^+ or H^+ so that the particle plus its ionic atmosphere have no net charge. Three planes define the electric double layer which surrounds the particle; each has a characteristic potential associated with it. The potential ψ_0 at the surface of the particle is the thermodynamic or Nernst potential, which operates in concentration and galvanic cells. Adjoining the surface is the compact Stern layer of counterions (Cl^-) which are so strongly attached to the surface by electrostatic and van der Waals forces that thermal agitation cannot render it diffuse. The electric potential drops linearly with distance across the Stern layer from ψ_0 to ψ_d. The ions in the Stern layer are in dynamic equilibrium with those in the contiguous diffuse Gouy layer. The potential at the boundary between the two is ψ_d. In the Gouy layer, the potential decays by a factor $1/e$ over a distance $\delta = 1/\kappa$, where κ is the Debye-Hueckel parameter; δ is taken as the "effective thickness" of the diffuse double layer, which has no sharp outer boundary. It terminates where the concentration of Cl^- is equal to the bulk concentration of Cl^-, i.e., where the composition has become identical with that of the bulk solution. At that point, $\psi_d = 0$.

A shell of water molecules thicker than the Stern layer is bound to the charged surface by ion-dipole interaction. This hydration shell moves with the particle. The potential at the plane of shear or slip, namely, the

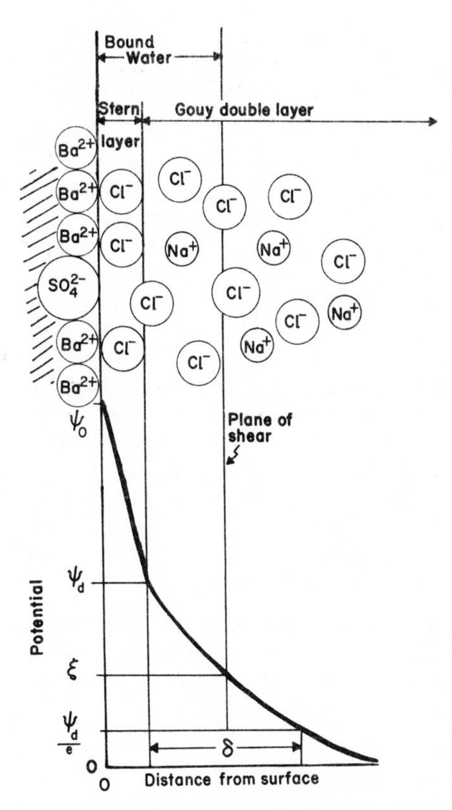

FIG. 6.1. Schematic representation of the electric double layer surrounding a barium sulfate particle, and the characteristic potentials.

plane which separates water of hydration from free water, is the electrokinetic or ζ (zeta) potential. It does not include the Stern layer but only that part of the Gouy layer which lies outside the hydration shell. The ζ potential is determined by measuring electrophoretic mobilities (see Sec. V,B,1). It is usually somewhat smaller than ψ_d. Polyvalent counterions in the Stern layer sometimes reverse the charge, so that ψ_d and ζ have opposite signs from ψ_0 (17).

Sols of highly charged, hydrophobic particles are very sensitive to electrolytes, the important factors being the valence and concentration of the counterions. Coagulation by added electrolytes occurs as a consequence of decreases in the double layer thickness and in the ζ potential. Potential-determining ions are ions capable of becoming an integral part of the suspended solid, such as Ba^{2+}, Pb^{2+} or Ca^{2+}, and SO_4^{2-} or CrO_4^{2-} for $BaSO_4$, and Al^{3+}, Fe^{3+} or Mn^{2+}, and OH^- for $Al(OH)_3$. Their concentration in solution largely determines the thermodynamic or ψ_0 potential of the particles. Therefore, the potential-determining ions are very effective in stabilizing suspensions if they are of the same charge as the particles, and in coagulating suspensions if they carry the opposite charge.

The Derjaguin-Landau-Verwey-Overbeek theory of the stability of hydrophobic dispersions is discussed in Chap. 4. It is summarized in the potential energy curve showing the interaction between two identical, charged particles as a function of their distance of separation, when they are about to collide because of Brownian motion, convection currents or because the suspension is stirred (Fig. 6.2). This curve represents the balance among the following three effects: London-van der Waals attraction (by convention, negative energy) tends to make particles stick to each other once they are fairly close together; coulombic repulsion, which results from the resistance against interpenetration and deformation of the two electric double layers as the particles approach, tends to keep them apart; Born repulsion, operative at very small distances only, prevents the interpenetration of the two particles once they are in intimate contact. Curves similar to those of Fig. 6.2 represent not only the interaction between like particles but also that between different particles of like charge such as negatively charged particles and fibers.

When two particles approach, they begin to repel each other as soon as their double layers start to overlap. The particles have to overcome this coulombic barrier B before they can come sufficiently close to one another

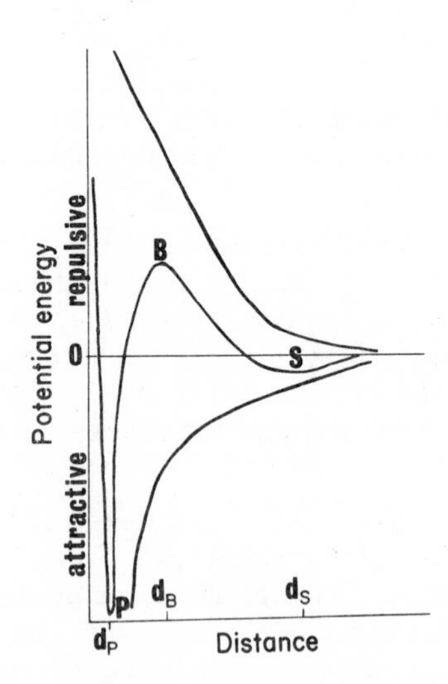

FIG. 6.2. The potential energy-distance curve representing the interaction between two identical charged particles.

to adhere through London-van der Waals attraction. Thus, higher potential barriers B or greater ζ potentials and thick double layers result in more stable dispersions. Barriers of the order of 25 kT units usually render dispersions permanently stable or diuturnal; k is the Boltzman constant and T the absolute temperature. Once thermal motion brings two particles closer together than ζ or d_B, they have surmounted the repulsive barrier; London-van der Waals attraction takes over and causes them to adhere to each other; their equilibrium distance then becomes d_P (17). The primary minimum in potential energy is deep, and a state of such low potential energy is a stable one. Particles located in this energy well are strongly attached to one another. Separating them is even more difficult because it also requires overcoming the coulombic barrier; it cannot be accomplished through stirring. When all of the particles have become attached to other particles, destruction of the dispersion or coagulation has occurred. Colloidal particles attached to fibers are in a similar situation.

The repulsive potential decreases exponentially with distance within a range of the order of ζ, while the attractive potential decreases with the second power of the distance of separation between the particles, provided this distance is small compared with their thickness. For isolated atoms the London-van der Waals attraction decays with the sixth power of the distance. The fact that the attractive energy between two particles is approximately equal to the summation of the energies of attraction for all of their atoms renders the attraction effective over much longer ranges. While repulsion prevails at intermediate distances, attraction predominates at small and large distances. Thus, there are two potential energy minima, the deep primary well at d_P and the secondary minimum at d_S, a distance several times greater then ζ. The secondary minimum is usually quite shallow, say one or less kT unit deep, and hence is without effect (18). However, for large particles or for particles which are large in one or two dimensions (rods and plates), the secondary minimum S can be deep enough to trap them at distances d_S from each other (18). This requires a depth of about six kT units. This fairly long-range attraction produces loose aggregation or flocculation, which can be reversed by agitation or by removal of flocculating electrolytes. The importance of the secondary minimum is being increasingly recognized (19), and it is likely to play a major role in the retention of particulate soil by fabrics (20).

Silicate soils such as clays, brick, and cement dust are hydrated. The foregoing theory refers to hydrophobic dispersions, and will have to be modified in order to be applicable to most actual particulate soils.

C. Adsorption of Surfactants by Particulate Soil

In order to produce any cleansing or suspending, the surfactants must be adsorbed from aqueous solution onto the soil. The qualitative and quantitative aspects of this process are described below.

1. Nature of the Surfactants

Surfactants are amphiphilic or amphipatic compounds; each molecule contains an oleophilic or hydrophobic and a hydrophilic moiety. The former is usually a hydrocarbon "tail" and consists of material readily soluble in "oils" or solvents of low polarity. Sometimes it is a fluorocarbon. The second moiety is chemically similar to compounds which are hydrated and extensively soluble in water. It consists of a negatively or positively charged ionic "headgroup" in the case of anionic or cationic surfactants plus the counterions required for electroneutrality. In the case of nonionic surfactants, the hydrophilic moiety is usually a polyoxyethylene chain produced by adding ethylene oxide to the active hydrogen of a hydroxyl, carboxyl, or sulfhydryl group attached to the hydrocarbon moiety. Sometimes, it is a sugar derivative.

The size of the hydrocarbon portion relative to the hydrophilicity of the charged headgroup or to the length of the polyoxyethylene chain determines whether the surfactants will be soluble in oils or in water. The former are of interest in dry cleaning, the latter in detergency.

2. Mechanisms of Surfactant Adsorption

Even when the surfactant is water soluble, i.e., when the hydrophilicity of the headgroup overcomes the hydrophobicity of the hydrocarbon chain, the surfactants preserve their amphiphilic character. For this reason, they accumulate at water-air, water-oil, and solid-water interfaces, with the hydrocarbon chains oriented toward the medium of lower dielectric constant (air or oil) and the hydrophilic group immersed in water.

The concentration of surfactant is much greater at these interfaces than in the bulk of the aqueous solution. For the water-air and water-oil interfaces, the extent of the adsorption of surfactants at the boundary, the so-called surface excess Γ expressed in moles per unit area, can be calculated from interfacial tension γ-concentration $\underline{C}$ data by means of the Gibbs adsorption equation (21, 22)

$$\Gamma = -\left(\frac{\underline{C}}{\underline{RT}}\right)\left(\frac{d\gamma}{\underline{dC}}\right) = -\left(\frac{1}{\underline{RT}}\right)\left(\frac{d\gamma}{\underline{d} \ln \underline{C}}\right) \qquad (6.5)$$

Accumulation of surfactants at interfaces increases Γ and thereby lowers γ.

For the solid-water interface, interfacial tension is replaced by the surface pressure π of the film of adsorbed surfactant molecules, which is the difference between the interfacial tensions at the solid-water and the solid-solution interfaces (41). From Equation (5.1) of Chapter V,

$$\pi = \gamma_{\underline{SL}} - \gamma^*_{\underline{SL}} = \gamma^*_{\underline{LA}} \cos \Theta^* - \gamma_{\underline{LA}} \cos \Theta \qquad (6.5a).$$

Asterisks refer to the surfactant solution, parameters without asterisks to pure water, and Θ is the contact angle of the liquid on the solid surface (see Chapter V). It is assumed that $\gamma_{\underline{SA}} = \gamma^*_{\underline{SA}}$.

The Gibbs equation then takes the form (41)

$$\Gamma = \left(\frac{1}{\underline{R}\,\underline{T}}\right)\left(\frac{d\pi}{\underline{d}\ \ln \underline{C}}\right) \qquad (6.5b).$$

From the value of the surface excess, one can calculate the area A occupied by an adsorbed surfactant molecule as

$$A = \left(\frac{1}{\Gamma\,\underline{N}}\right) \qquad (6.5c).$$

where $\underline{N}$ is Avogadro's number.

The Gibbs equation describes nonspecific or physical adsorption, a process observed with all surface-active substances which is a consequence of their amphiphilic nature and involves only van der Waals forces. There is a dynamic equilibrium between the surfactant molecules adsorbed at interfaces and those dissolved in the bulk solution. The adsorption is reversible. If the surfactant concentration in the solution is lowered by dilution, net desorption of surfactant from the interface proceeds until the distribution of surfactant between interface Γ and bulk solution $\underline{C}$ again reaches equilibrium.

When the surfactant concentration at the interfaces is high, i.e., when the adsorbed molecules are densely packed in the interfaces, further increases in overall surfactant concentration cause the molecules to associate in solution, forming micelles. In these aggregates, the hydrocarbon tails are oriented inward, away from the water, and in contact with each other (hydrophobic bonding), while the hydrophilic groups are oriented outward, in contact with water, where they are ionized and/or hydrated. The lowest surfactant concentration at which this aggregation occurs is called the critical micelle concentration (cmc).

Solid surfaces containing acidic groups bind cationic surfactants by ion exchange. This is called chemisorption because primary valences are involved. It is often followed by physical adsorption of a second surfactant layer. Examples of surfaces with base-exchange capacity are bentonite (23) and glass (24), which contain silicic acid groups, and cellulose (25-28) and nylon (27, 28), which contain carboxyl groups. In the case of glass and cetyltrimethyl ammonium bromide, the ion-exchange process at the glass-water interface can be represented by

$$-\overset{|}{\underset{|}{Si}} - O^-Na^+ + C_{16}H_{33}(CH_3)_3N^+Br^- \rightleftharpoons$$

$$-\overset{|}{\underset{|}{Si}} - O^-N^+(CH_3)_3C_{16}H_{33} + Na^+Br^-$$

The surface-active cation is chemisorbed with its ionic headgroup attached to the glass as cetyltrimethyl ammonium silicate. This changes the charge or ζ potential of the glass surface from negative to nearly zero. The hydrocarbon tail of the surface-active cation is oriented outward, toward the aqueous phase. The glass surface is rendered hydrophobic because its outermost layer consists of hydrocarbon chains. This process leads to the coagulation of aqueous dispersions of finely powdered glass and of sodium bentonite because the particles have lost their charge and their hydration layer. Furthermore, the free energy of the interface between the hydrocarbon-covered glass and water is high, so that any reduction of the interfacial area by coagulation of the dispersion produces a significant lowering of the interfacial free energy.

The chemisorption of the surface-active cation goes to completion at low surfactant concentrations; the solid adsorbs it very avidly from solution. As the surfactant concentration is increased, a second layer of surfactant is adsorbed by the solid, with the opposite orientation to that of the first layer: the hydrocarbon tails of the second layer are oriented toward the glass surface, where they are attracted by van der Waals forces to the hydrocarbon tails of the first surfactant layer. The ionic headgroups are oriented outward, toward the aqueous phase. Because the first layer was chemisorbed through an ion-exchange process, it contains no bromide. The physically adsorbed second layer contains the entire surfactant molecule including the bromide counterion. Adsorption of the second layer is completed near the cmc. Increase in surfactant concentration beyond that required to form a close-packed second layer on the solid surface leads to aggregation into micelles. In fact, the double layer of adsorbed surfactant molecules can be considered as a flat micelle attached to that surface. The second layer of cationic surfactant in effect covers the glass surface with an outer layer of ionized ammonium bromide; this renders the glass surface hydrophilic and wetted by water. The glass surface has acquired a positive ζ potential, and the dispersion of powdered glass in water has become deflocculated and stabilized again. However, its charge has been inverted, from negative to positive. The physical adsorption of the second surfactant layer is reversible, i.e., the second layer can be washed off with water.

Since most fibers and particulate soils are negatively charged, cationic surfactants are unsuitable as detergent actives (29). They would coagulate particles suspended in the wash liquor and redeposit them onto the fabric. If the repulsion between negatively charged particles and negatively charged fibers was not sufficient to detach the particles from the fibers, chemisorption of the first layer of cationic surfactant, which reduces the charges to near zero and renders the surfaces hydrophobic, will attach the particles more firmly to the fabric (30). At that point, removal of the particles requires adsorption of a second layer of surfactant molecules which renders fiber and particle surfaces hydrophilic and positively charged. This consumes large amounts of surfactant (29). Furthermore, the physically adsorbed second layer of surfactant molecules would be desorbed during the rinse cycles, leaving fabric and suspended soil with near zero ζ potential and covered with an outer layer of hydrocarbon chains belonging to the irreversibly chemisorbed first layer. Soil redeposition would then occur.

Many fabric softeners are quaternary ammonium salts with two long alkyl chains R_1 and R_2. Being chemisorbed by cotton, viscose, and nylon, they are added at very low levels in the last rinse cycle. They are probably stripped from the fabric in the next wash cycle by a reaction of the type:

$$\text{Cellulose - COO}^{-}\text{N}^{+}(\text{CH}_3)_2\text{R}_1\text{R}_2 + \text{R-SO}_3{}^{-}\text{Na}^{+} \rightleftharpoons$$

$$\text{Cellulose - COO}^{-}\text{Na}^{+} + \text{R - SO}_3{}^{-}\text{N}^{+}(\text{CH}_3)_2\text{R}_1\text{R}_2$$

which is driven to completion by the relatively high concentration of anionic surfactant in the wash liquor and because the tetraalkylammonium alkylarylsulfonate has low solubility. Buildup of fabric softener can cause serious problems of rewettability (31) and detergency.

Anionic surfactants are not adsorbed at all by sodium bentonite (at least not below the cmc), presumably owing to the hydration and to the negative charge of the clay surface. They are adsorbed by calcium bentonite, probably via calcium bridges (32):

$$-\text{Si - O}^{-}\text{Ca}^{2+}(\text{OH})^{-} + \text{R - SO}_3{}^{-}\text{Na}^{+}$$

$$-\text{Si - O}^{-}\text{Ca}^{2+-}\text{O}_3\text{S - R} + \text{Na}^{+}\text{OH}^{-}$$

The bivalent calcium ion is anchored by one valence to the clay surface, while the second valence binds the surface-active anion.

Cuming and Schulman (16) observed a two-step adsorption process when a positively charged barium sulfate suspension was treated with sodium dodecyl sulfate. The first step was the chemisorption of the dodecyl sulfate anion via anion exchange, with the hydrocarbon portion oriented outward (structure I). This rendered the suspended solid poorly wetted by water and reduced its ζ potential to near zero. Coagulation of the suspension was indicated by its large sedimentation volume. Physical adsorption of a second layer of complete sodium dodecyl sulfate molecules with the ionic groups oriented toward the water was completed near the cmc. This imparted a hydrophilic character and a negative charge to the surface (structure II). The solid became readily wetted by water, and deflocculation of the suspension was shown by its decreased sedimentation volume (16).

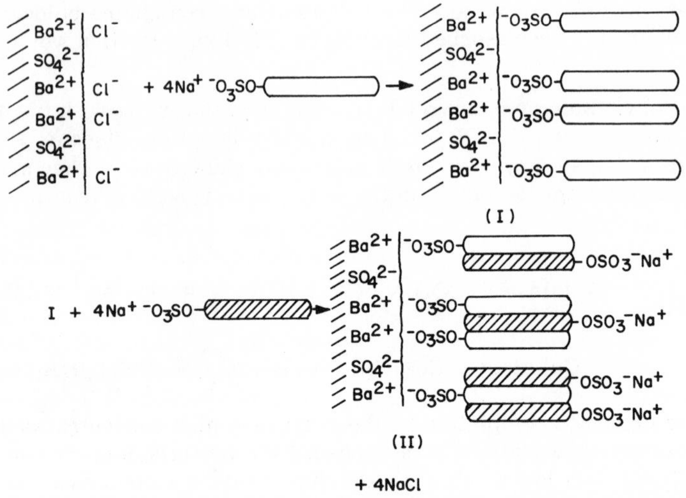

Ba^{2+} | Cl^-
SO_4^{2-} |
Ba^{2+} | Cl^- + $4Na^+ \, ^-O_3SO-$ → Ba^{2+} | $^-O_3SO-$
Ba^{2+} | Cl^- SO_4^{2-} |
SO_4^{2-} | Ba^{2+} | $^-O_3SO-$
Ba^{2+} | Cl^- Ba^{2+} | $^-O_3SO-$
SO_4^{2-} |
Ba^{2+} | $^-O_3SO-$

(I)

I + $4Na^+ \, ^-O_3SO-$ → Ba^{2+} | $^-O_3SO-$ $-OSO_3^-Na^+$
SO_4^{2-} |
Ba^{2+} | $^-O_3SO-$ $-OSO_3^-Na^+$
Ba^{2+} | $^-O_3SO-$
SO_4^{2-} | $-OSO_3^-Na^+$
Ba^{2+} | $^-O_3SO-$ $-OSO_3^-Na^+$

(II)

+ 4NaCl

3. Configuration of Adsorbed Surfactant Molecules

In order to determine the manner in which the adsorbed surfactant molecules are oriented at the solid-liquid interface, or the cross-sectional area per adsorbed molecule, one must know the amount of surfactant adsorbed per gram of solid and the specific surface area of the solid. The former is readily obtained from analysis of the treated solid (after correcting for the amount of surfactant contained in the solution retained by capillarity and dried onto the solid), or from analysis of the concentration of residual surfactant in solution and a balance of materials. The specific surface area of the solid is often determined by nitrogen or krypton adsorption using the BET technique, or by adsorption of stearic acid from benzene solution. Both methods require that the solid be dry. Therefore, they give low area estimates for sorption from aqueous solution if the solid swells in water. For solids with pores large enough to accommodate nitrogen molecules but too small to admit surfactant molecules, the area obtained by nitrogen adsorption would be too large (33, 34). There is evidence (35) that these two factors cancel each other out approximately in the case of cotton, so that the value for the specific surface area obtained by nitrogen adsorption can also be used for adsorption from aqueous solution.

Sodium bentonite or montmorillonite is a solid for which the specific surface area is accurately known, and which acts as a molecular caliper, permitting direct observation of the orientation of the adsorbed molecules by X-ray powder diffraction. Like many clays, bentonite has a layer structure. A single lattice layer consists of a sheet of boehmite, AlO(OH), sandwiched between two sheets of tetrahedral silica, and is 9.4 Å thick. Isomorphous replacement of silicon by aluminum and of aluminum by magnesium creates an excess negative charge, which is neutralized by countercations located between the lattice layers. Natural countercations are sodium and calcium. The cation-exchange capacity of bentonite, at about 0.9 mequiv/g, is of the same order of magnitude as that of commercial ion-exchange resins. Polar molecules like water or ammonia penetrate between adjacent lattice layers and push them apart. This produces intracrystalline swelling, i.e., lattice expansion along the c axis (see Fig. 6.3). The repeat distance $d_{(00l)}$ in the c direction between successive basal or a-b planes (the planes of the clay layers) is increased by Δ. In dry, unswollen clay, $d_{(00l)}$ = 9.4 Å is equal to the thickness of one lattice layer, indicating that adjacent layers practically touch each other: Δ = 0. For the expanded clay lattice, the difference Δ between $d_{(00l)}$ and 9.4 Å is equal to the thickness of the layer of adsorbed molecules inserted between adjacent lattice layers. The value of Δ gives information about the orientation of the adsorbed and intercalated molecules.

When Jordan treated sodium bentonite with n-alkylamine hydrochlorides, Δ was 3.9 ± 0.1 Å for propyl- through decylamine and 7.9 ± 0.1 Å for dodecyl- through octadecylamine hydrochloride (23). The first value is equal to the cross-sectional diameter of a normal paraffin chain, indicating that the surfactant molecules were adsorbed lying flat in the plane of the clay lamellas with their zigzag hydrocarbon chains parallel to them, i.e., in the a-b plane. The second Δ value equals twice the cross-sectional

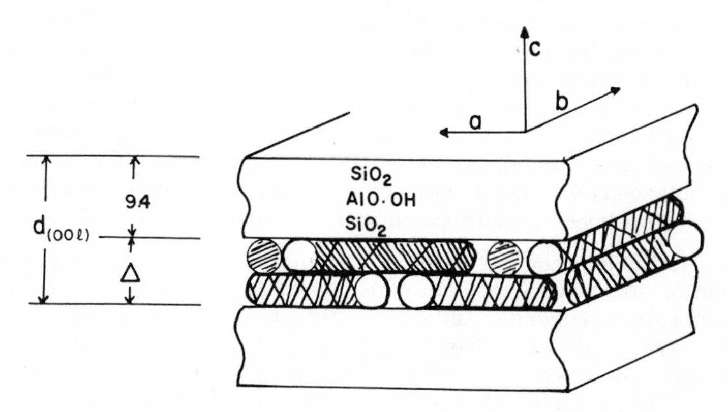

FIG. 6.3. Cross section through bentonite lattice, showing inserted surfactant molecules.

diameter of a paraffin chain, indicating that two layers of adsorbed surfactant molecules were inserted between adjacent lattice layers of clay lying flat in the a-b plane. Both outer surfaces of a lattice layer of clay consist of a sheet of oxygen ions which attract the hydrocarbon chain of the alkylammonium ion through secondary valence forces, while the head-group is attached to the cation-exchange site by an electrovalence.

The density of dry sodium bentonite is 2.83 g/cm^3, and the specific surface area of the fully swollen clay is

$$\frac{2}{(2.83\ g/cm^3)(9.4 \times 10^{-8}\ cm)} = 7.50 \times 10^6\ cm^2/g = 750\ m^2/g$$

The factor 2 arises from the fact that the top and bottom surface of each lattice layer must be counted separately. This area represents the basal area or the area of the a-b planes; the lateral or edge area is negligible. The calculated value is in excellent agreement with experimental values obtained from the adsorption of polar molecules.

A cation-exchange capacity of 0.90 mequiv/g corresponds to

$$\frac{7.50 \times 10^6\ cm^2/g}{0.90 \times 10^{-3} \times 6.023 \times 10^{23}\ sites/g} = 1.38 \times 10^{-14}\ cm^2$$

or 138 $Å^2$ per site. The basal areas occupied by a methylene group and a $-NH_2^+$ ion are 6 and 22 $Å^2$, respectively. When the exchange capacity of the clay is saturated with butylammonium, 4 6 + 22 = 46 $Å^2$ of each 138 $Å^2$ of basal area, or 33%, are covered with amine. Therefore, the butylammonium ions attached to the bottom surface of one clay layer fit into the gaps between the butylammonium ions attached to the top surface of the lattice layer immediately below it: A single layer of adsorbed butylammonium ions is capable of saturating the cation-exchange sites in two adjacent basal clay surfaces. As the hydrocarbon length increases, that layer becomes increasingly crowded. Each dodecylammonium ion occupies $12 \times 6 + 22 = 94$ $Å^2$ per 138 $Å^2$, or 68% of the basal area. Since this exceeds 50%, the dodecylammonium ions required to saturate the cation-exchange sites of two adjoining clay surfaces cannot fit into a single, close-packed layer one molecular diameter high. Therefore, they pack into two layers, corresponding to a Δ spacing of 8 Å (23).

The equilibrium concentration of cationic surfactants in the presence of sodium bentonite with residual cation-exchange capacity is very low: The chemisorption process results in nearly quantitative removal of the surfactants from solution. Even though nonionic surfactants are not that completely adsorbed from solution by bentonite, the process is quite extensive for physical adsorption (36, 37). Single-layer sorbates are formed at low concentrations, with the chains lying flat between adjacent lamellas: the Δ value of the dried sorbates was 4.2 Å for ethylene oxide adducts of n-alkanols (37) and 4.6 Å for adducts of i-nonylphenol (36). The close-packed single layer of adsorbed surfactant was completed just below the cmc. The double-layer sorbates, formed at higher concentrations of n-alkanol adducts, gave a Δ value of 8.4 Å. Polyethylene glycols (the hydrophilic portion of nonionic surfactants) as well as a polypropylene glycol also formed double-layer sorbates, with $\Delta = 8.2 \pm 0.1$ Å.

Single-layer sorbates increased their Δ value by 3.8 Å when exposed to a relative humidity of 85%. This increased in Δ and the associated moisture regain correspond to insertion of a single layer of water molecules between adjacent clay lamellas. The Δ value of the double-layer sorbates remained constant at 8.4 Å for water contents of over 50% (corresponding to the wet clay cakes separated by centrifugation from suspensions treated with surfactant), 5-10% (after conditioning at 85% relative humidity), and 0% (after drying over P_2O_5) (37). The highly hydrated nonionic surfactants (38), upon adsorption by the water-swollen clay, lost their own water of hydration and displaced that of the clay, a process requiring considerable energy.

Anionic surfactants were adsorbed by calcium bentonite in a fully extended configuration, with the chains perpendicular to the plane of the clay lamellas, i.e., parallel to the c axis. The Δ values were 17.5 Å for sodium decane sulfonate and 20.3 Å for sodium dodecyl sulfate, which is nearly identical to the lengths of fully extended $C_{10}H_{21}SO_3^-$ (17.2 Å) and $C_{12}H_{25}$-OSO_3^- (20.5 Å), respectively (32).

Knowing the weight and volume of surfactants adsorbed on bentonite permits one to calculate their density. For a double-layer sorbate of $C_{12}(EO)_{14}$, the weight was found to be 0.310 g surfactant/g clay. The volume is one-half of the basal area of the clay times Δ, namely,

$$(1/2)(7.5 \times 10^6\ \text{cm}^2/\text{g clay})(8.4 \times 10^{-8}\ \text{cm}) = 0.315\ \text{cm}^3\ \text{surfactant/g clay}$$

The quotient, 0.98 g surfactant/cm^3 surfactant, is 10% below the density of the solid surfactant in bulk.

4. Adsorption Isotherms

At constant temperature, quantitative relationships between the concentration of surfactant adsorbed at the solid-water interface and the concentration of surfactant in aqueous solution are represented by adsorption isotherms. These are usually plotted as gram or mole of adsorbed surfactant x per gram of solid adsorbent m (or, when the area of the solid is known, as gram or mole of surfactant per square centimeter of solid), versus the equilibrium surfactant concentration C, i.e., the concentration of residual or nonadsorbed surfactant. The Langmuir adsorption isotherm [Eq. (6.6) or (6.7)] often describes the adsorption of surfactants:

$$\frac{\underline{x}}{\underline{m}} = \frac{\underline{abC}}{1 + \underline{bC}} \tag{6.6}$$

$$\frac{\underline{C}}{(\underline{x}/\underline{m})} = \frac{1}{\underline{ab}} + \frac{\underline{C}}{\underline{a}} \tag{6.7}$$

From Eq. (6.6), it is seen that $\underline{a}$ is related to the surface area of the solid because the limiting value for $\underline{x}/\underline{m}$ at high $\underline{C}$ values is $\underline{a}$. The constant $\underline{b}$ is related to the heat of adsorption $\Delta\underline{H}$ by

$$\underline{b} = \underline{k} \exp(-\Delta\underline{H}/\underline{RT}) \tag{6.8}$$

Typical type I or Langmuir adsorption isotherms are shown in Figs. 6.4 and 6.5 for the adsorption of the homogeneous nonionic surfactant

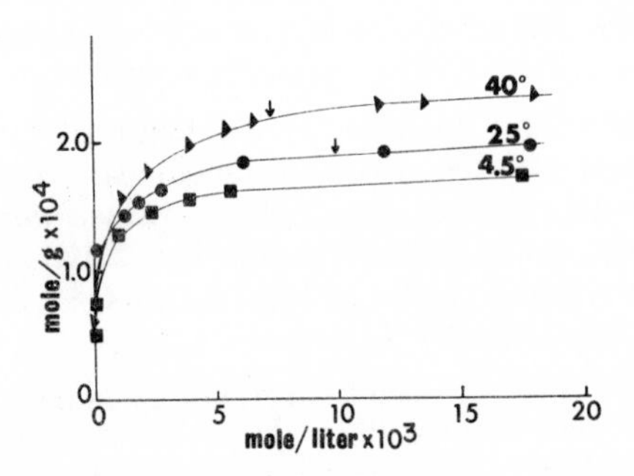

FIG. 6.4. Adsorption isotherms for $C_8(EO)_6$ on Graphon. Arrows indicate cmc. Reprinted from Ref. 39, p. 981, by courtesy of The Faraday Society.

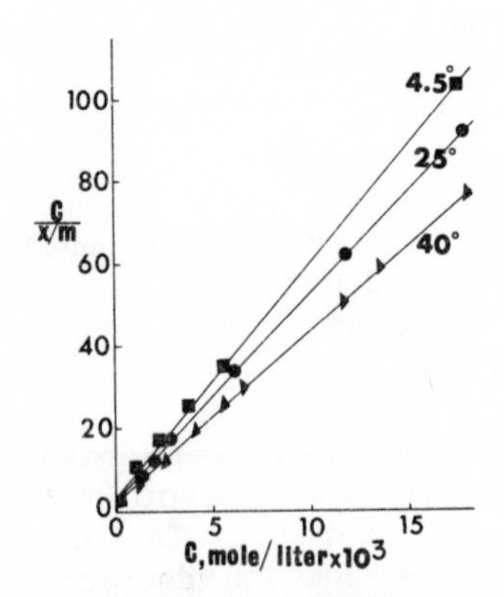

FIG. 6.5. Adsorption isotherms for $C_8(EO)_6$ on Graphon (39), plotted according to Eq. (6.7).

$C_8(EO)_6$ on Graphon graphitized carbon black (39). The values of $\underline{a} \times 10^4$ calculated as the reciprocal slopes in Fig. 6.5 are 1.75, 2.02, and 2.435 mole surfactant/g carbon at 4.5, 25, and 40°C, respectively. If the adsorbed surfactant molecules lie flat with both the hydrophobic and hydrophilic portions in contact with the surface, the area per adsorbed molecule $\underline{A}$, for unit density, is

$$\underline{A} = \underline{M}/[2(4.2 \times 10^{-8})\underline{N}] = 7.8 \times 10^{-15}\ \text{cm}^2 = 78\ \text{Å}^2$$

where $\underline{M}$ is the molecular weight (395) and $\underline{N}$ is Avogadro's number. The specific surface area of the solid is

$$\underline{S} = \underline{N}\underline{A}\underline{a} \tag{6.9}$$

The values are $\underline{S}$ = 82, 95, and 114 m^2/g at 4.5, 25, and 40°C, respectively. The value of $\underline{S}$ determined by nitrogen adsorption was 91 m^2/g, and Graphon is known to have little or no porosity. This good agreement indicates that the entire surfactant molecule is adsorbed in contact with the solid substrate.

The leveling off of the Langmuir isotherms (Fig. 6.4) often occurs in the vicinity of the cmc. The Langmuir adsorption isotherm was derived for the physical or chemical adsorption of up to a monolayer of adsorbed molecules, and precludes lateral interaction between those molecules. It does not apply to multilayer adsorption. Therefore, the adsorption of surfactants does not always follow a Langmuir-type isotherm. For instance, there is a step or an intermediate plateau beginning at the cmc in the adsorption isotherms of nonionic surfactants on calcium and sodium bentonite (37) and on silver iodide (40). In the case of the clay, it has been shown by X-ray diffraction that the beginning of the intermediate plateau coincides with the completion of a close-packed monolayer of adsorbed surfactant molecules. At higher concentrations, a second surfactant layer is inserted between adjacent lattice layers. This produces a second leveling off in the isotherms.

5. Thermodynamics of Adsorption

Studies of the energy changes accompanying adsorption from aqueous solution are complicated by heterogeneities in the solid surface and by competition between surfactant and water for that surface. Changes in free energy $\Delta\underline{G}$, in heat content or enthalpy $\Delta\underline{H}$, and in entropy $\Delta\underline{S}$ associated with the adsorption of surfactants can be calculated from the equilibrium constant $\underline{K}$ for the distribution of surfactant between solid surface and solution

$$\underline{K} = \frac{\underline{C}_{solid}}{\underline{C}_{soln}} = \frac{\Gamma}{\underline{h}\underline{C}_{soln}} = \frac{\underline{x}/\underline{m}}{\underline{C}_{soln}} \tag{6.10}$$

Activities have been replaced by concentrations $\underline{C}$; subscripts "solid" and "soln" refer to the adsorbed and dissolved surfactant, respectively; and $\underline{h}$ is the thickness of the adsorbed film. Moreover,

$$\Delta \underline{G} = -\underline{R}\underline{T} \ln \underline{K} \tag{6.11}$$

For positive adsorption, $\underline{C}_{solid} > \underline{C}_{soln}$ so that adsorption results in a decrease in free energy (41). In principle, one can determine heat and entropy of adsorption from the temperature coefficient of $\Delta\underline{G}$ through the Gibbs-Helmholtz equation

$$\Delta \underline{H} = \Delta \underline{G} + \underline{T}\,\Delta \underline{S} = \Delta \underline{G} - \underline{T}\,\frac{d\Delta \underline{G}}{d\underline{T}} \tag{6.12}$$

An equivalent treatment is the use of the van't Hoff equation

$$\log \frac{\underline{K}_2}{\underline{K}_1} = \frac{\Delta \underline{H}}{2.303\ \underline{R}} \left(\frac{\underline{T}_2 - \underline{T}_1}{\underline{T}_1 \underline{T}_2} \right) \tag{6.13}$$

The heat of adsorption of $C_8(EO)_6$ on Graphon estimated from the values of $\underline{b}$ by means of Eq. (6.8) was of the order of 3000 cal/mole. The values of $\underline{b}$ were obtained from the intercepts in Fig. 6.5. Equation (6.13) gave a range of 1400 to 2200 cal/mole. Calorimetric measurements of the heats of immersion of Graphon in water and in solutions of $C_8(EO)_6$ at different concentrations showed that the heat of adsorption as a function of $\underline{x}/\underline{m}$ became practically constant near the point corresponding to coverage by a close-packed, flat surfactant monolayer. Therefore, the differential heat of adsorption per mole of surfactant did not differ measurably from zero. The range of values for the (integral) heat of adsorption was about three times higher than that calculated above (39).

The exothermic elimination of the hydrocarbon-water interface was apparently offset by the endothermic desolvation of the surfactant molecules accompanying their adsorption. The negative $\Delta\underline{G}$ of the adsorption process must have been caused by a net increase in entropy resulting from the difference between the entropy increase due to the release of water of hydration of the surfactant and due to the removal of the structure-promoting hydrocarbon tails from water, and the entropy decrease associated with the adsorption of the surfactant on the solid. A complicating factor is the decrease in hydration of the dissolved nonionic surfactant molecules with increasing temperature, which tends to shift the distribution of the surfactant between solid and solution increasingly toward the solid surface as the cloud point is approached. The adsorption behavior of nonionic surfactants is strongly influenced by the interaction between surfactant and water in addition to that between surfactant and solid (39).

The heat of adsorption of sodium dodecyl sulfate and of sodium dodecylbenzenesulfonate on Graphon, measured from heats of immersion, was between -7000 and -9000 cal/mole (42). Measurements with sodium n-alkyl sulfates ranging from C_{12} through C_{18} on a channel black also resulted in negative heats of adsorption, i.e., the adsorption process was exothermic and proceeded more extensively at lower temperatures (43). This is expected because only the headgroup of the anionic surfactant is hydrated, and its degree of hydration is unlikely to be reduced by adsorption.

D. Stabilization of Dispersions and Deflocculation by Surfactants

Particles which are not made up of smaller ones are designated primary particles. When they aggregate, they usually preserve their identity because, being solid and rigid, they touch only at a few points unless they are flat. The aggregates or secondary particles usually contain some trapped suspending medium. The process by which secondary particles are formed from primary ones is called coagulation or flocculation, the reverse process deflocculation or peptization. Flocculation forms weaker aggregates than coagulation, i.e., aggregates which can be broken up or redispersed more easily, often by stirring (44). The secondary minimum in the potential energy-distance curve usually leads to flocculation, the primary minimum to coagulation.

Surfactants stabilize dispersions of largely hydrophobic colloids, rendering them more resistant to coagulation by salts, solvents, shear, heating, etc. If dispersions have been coagulated, surfactants often redisperse the aggregates. The mechanisms are the same in both cases, and depend on whether the surfactant is ionic or nonionic.

1. Stabilization by Anionic Surfactants

Ionic surfactants stabilize or peptize dispersions of those particles which have charges of the same sign as the surfactant. Particles of the opposite charge adsorb surfactant molecules with the hydrocarbon chains oriented toward the water. This reduces the magnitude of the charge of the particles and usually decreases their hydration and promotes coagulation.

Adsorption of anionic surfactants by negatively charged particles increases the charge density as shown by greater electrophoretic mobility and increased negative ζ potential. Surfactant adsorption and increase in electrophoretic mobility are greatest for hydrophobic particles, i.e., particles of low polarity, low charge density, and low hydration, such as carbon blacks, paraffin, and triglycerides (45-51), and produce the greatest deflocculation and increase in stability of their dispersions. The adsorption isotherms of anionic surfactants on carbon blacks and the plots of electrophoretic mobility versus surfactant concentration are similar in shape. Adsorption and mobility increase until the cmc is reached, and then level off (49, 51). Dispersions of more hydrophilic and ionic metal oxides and of clays which are not seriously flocculated and have appreciable negative ζ potentials in water alone undergo correspondingly smaller increases in the absolute values of their ζ potentials and in stability upon the addition of anionic surfactants (45, 48, 52). Powdered glass (48) and sodium bentonite (32) adsorb little if any anionic surfactant and have nearly identical ζ potentials in solutions of anionic surfactants and of sodium chloride of comparable ionic strength.

Aqueous dispersions of paraffin and soot are negatively charged, the former in part by preferentially adsorbing hydroxyl ions from water, and both by the presence of carboxyl groups. These charges stabilize the dispersions. According to the theory of the stability of hydrophobic sols (Sec. IV, B), the addition of enough of a salt such as sodium chloride compresses the diffuse part of the double layer surrounding the particles and reduces the height of the potential barrier to near zero. Particles can therefore approach close enough to adhere to one another, and the sol coagulates. The electrophoretic mobility of the particles at that point is quite small (53). Adsorption of anionic surfactants added to the dispersions prior to sodium chloride increases the charge density and the thickness of the double layer surrounding the particles and raises the potential barrier. A much greater concentration of sodium chloride is now required to reduce the thickness of the double layer and the height of the potential barrier enough to suppress the coulombic repulsion.

However, even when the electrophoretic mobility is quite small, the adsorbed anionic surfactant molecules exert a stabilizing action because of the water of hydration associated with their ionic headgroups (54), which forms a protective shell around each particle (53). Theoretical calculations predict that the interaction energy between particles surrounded by a layer of water of hydration is several times smaller than the interaction energy between unsolvated particles (55). Causing hydrophobic dispersions to become hydrated increases their stability considerably. The ability of anionic surfactants to stabilize hydrosols even in the presence of salts is probably the reason why many detergents based on anionic actives are reasonably effective in sea water, which contains 0.5 *M* NaCl, particularly if they are built with sequestering agents for magnesium.

2. Stabilization by Nonionic Surfactants

Nonionic surfactants were found to reduce the magnitude of the negative ζ potential of powdered glass (48), carbon black (47), a clay (52), and silver iodide sols (56, 57), although measurements on other carbon blacks showed only minor changes in ζ potential (58). Apparently, adsorption of nonionic surfactant molecules shifted the plane of shear further into the solution. They may also have displaced some potential-determining ions from the surface or the Stern layer.

Despite this effect, the adsorption of nonionic surfactants increases considerably the resistance of hydrophobic sols against coagulation by electrolytes. Flocculation values, i.e., the minimum electrolyte concentrations producing rapid and thorough flocculation or coagulation of sols, are increased by one or two orders of magnitude in the presence of nonionic polyoxyethylated surfactants (53, 56, 57). The reason for this stabilization is that the hydrophilic polyoxyethylene moiety represents two-thirds to three-fourths of the entire molecule, so that the adsorption of a nonionic surfactant by a hydrophobic particle renders its surface quite hydrophilic, in effect coating a large portion of it with polyoxyethylene (59). The flocculation by electrolytes of hydrophilic dispersions, and of hydrophobic dispersions rendered hydrophilic through the adsorption of nonionic surfactants, is a salting-out process which requires very high electrolyte concentrations, several moles per liter. Unlike the coagulation of hydrophobic sols with or without adsorbed ionic surfactants, the flocculation values in the presence of nonionic surfactants do not depend on the valence of the counterions. The flocculation value of a given electrolyte was found to increase with increasing concentration of the nonionic surfactants; the largest increase occurred just below the cmc. Dilution of these salt-flocculated sols with water caused almost complete peptization (57).

$C_8(EO)_6$ was found to be adsorbed on Graphon in a flat configuration, with both the hydrophilic and the hydrophobic moiety in contact with the surface (39). However, analogy with the adsorption of polymers at solid-liquid interfaces suggests that commercial polyoxyethylated surfactants, which consist of larger molecules, are adsorbed by hydrophobic sols with the entire hydrocarbon moiety in direct contact with the solid surface but with only part of the polyoxyethylene moiety so adsorbed. The remaining part of the polyoxyethylene moiety is likely to extend into the aqueous phase in the form of loops and loose chain ends, which exert a dual protective action. They act as a steric barrier preventing a close approach of the particles. They also increase the magnitude of the hydration layers surrounding the particles because the entanglement of neighboring loose chain segments traps additional amounts of water, similarly to what occurs in micelles of nonionic surfactants (38).

Even polyethylene glycols, i.e., compounds consisting entirely of the hydrophilic moiety of nonionic surfactants, stabilized paraffin sols against coagulation by salts. Their effectiveness increased with their molecular weight (53), possibly because the larger molecules were adsorbed more extensively. Polyethylene glycol also deflocculated clay dispersions (see the next section).

3. Suspendability

Surfactants not only stabilize dispersions against coagulation by salts, shear, and other factors. They are also instrumental in dispersing, suspending, and deflocculating powdered solids in water. This process consists in breaking up aggregates into smaller aggregates or into primary particles. Because of their pronounced tendency toward adsorption at solid-water interfaces, one can imagine that surfactant molecules act as wedges on the areas of contact between primary particles, prying them apart in order to become adsorbed at the resultant, newly created solid-water interfacial areas. This process is assisted by shear stresses produced by stirring. Shear alone is usually insufficient to produce stable dispersions of hydrophobic solids. Once aggregates are broken up into smaller particles, adsorption of surfactant molecules generally prevents their reagglomeration.

Agitation of aqueous suspensions of carbon black and of ferric oxide produced some deflocculation even in the absence of surfactants. The average floc size was found to decrease with decreasing solids concentration and increasing level of agitation. The floc size remained nearly constant for hours when standing at rest after a period of agitation. This was a pseudo stable state, however, because gentle agitation produced extensive flocculation (60). Even ultrasonic irradiation, which produces very concentrated stresses and is capable of breaking down aggregates into primary particles, was insufficient to produce stable dispersions of carbon blacks in water without added surfactants (61): Surfactants were required to prevent renewed flocculation of the peptized particles. The size of the aggregates in an aqueous suspension of carbon black stirred with a Waring Blender was found to decrease with increasing sodium dodecyl sulfate concentration up to the cmc (62).

Description of the experimental techniques for assessing the deflocculation of dispersions by surfactants is beyond the scope of this book. Pertinent methods include measurements of the following: rheological properties, sedimentation rate and volume, filtration rate, and water permeability of the filter cake. These methods were largely developed for pigments by the paint and printing-ink industry (63) and for coating formulations by the paper industry (64).

Two techniques are useful for following the deflocculation of dilute dispersions by surfactants. Light transmission is very sensitive to particle size and hence to flocculation and deflocculation. The optical density or turbidity T of a dispersion is the logarithm of the ratio of the intensity of the incident beam I_0 to the intensity of the transmitted beam I. For relatively large and opaque particles.

$$T = \log(I_0/I) = KCLS \tag{6.14}$$

where C is the concentration (grams of particles per cubic centimeter), L the length of the light path, and S the shadow projected area of the particles in the beam (square centimeters per gram of particles). The total scattering coefficient or opacity factor K varies for a given material as a function of particle size, going through a maximum often located at about

5 μ (65). A further complication is the effect of the solvent included inside the aggregates on their refractive index and on K (66).

The second technique, developed by R. D. Vold, is to determine suspension isotherms. This involves the following steps: shaking the powder with the surfactant solution, undisturbed settling in a cylindrical container at constant temperature for a constant period of time, and determining the amount of solid remaining in suspension down to a constant depth of supernatant liquid (67). The concentration of powder remaining in suspension as a function of increasing surfactant concentration usually went through a sharp maximum. Increased suspendability was due to peptization of the powders by the surfactants. This was shown by measuring particle size and sedimentation volume (67, 68). Decreased suspendability in the presence of excess ionic surfactant was caused by flocculation. Apparently, the excess ionic surfactant reduced the ζ potential of the particles just as other electrolytes would, by increasing the ionic strength of the solution (69). The maxima in the suspension isotherms obtained with nonionic surfactants were considerably less pronounced. In particular, the descending branch was nearly horizontal (58, 68). The fact that higher concentrations of nonionic surfactants produced a small amount of flocculation might be ascribed to cross-linking and concomitant agglomeration of particles by micelles. A polyoxyethylated octylphenol exerted no suspending action on manganese dioxide (67) but was very effective with two carbon blacks (68). A polyoxyethylated nonylphenol deflocculated quartz sand. The maxima in the suspension isotherms of sand and of carbon blacks by nonionic surfactants were located at low surfactant concentrations (68, 70).

Since nonionic polyoxyethylated surfactants are strongly adsorbed by sodium bentonite, it is not surprising that they deflocculated dispersions of this clay: They increased the turbidity and reduced the viscosities and sedimentation volumes. Polyethylene glycols, which constitute the hydrophilic portion of these surfactants and are also strongly adsorbed by bentonite, likewise deflocculated its aqueous dispersions. The glycol concentrations producing maximum deflocculation were over an order of magnitude higher than those observed with nonionic surfactants, because the latter have a very strong tendency to concentrate at solid-water interfaces. Some flocculation of bentonite was observed at still higher concentrations of polyethylene glycol. Anionic surfactants, which are not adsorbed by sodium bentonite, produced no measurable deflocculation (71).

4. Importance of Deflocculation and Stabilization in the Washing Process

Once the particulate soil is detached from the fabric and suspended in the wash liquor, stabilization of its aqueous dispersions will reduce or eliminate soil redeposition (72). The suspendability of particulate soil determined by suspension isotherms or sedimentation rates is not to be taken literally as a reduced tendency to settle and redeposit onto the fabric under the influence of gravity (73). Rather, it is a measure of the degree of deflocculation of the soil, which should parallel the stability of its dispersions. Deflocculated particles in stable dispersions will not usually become attached to the fabric when they collide with it under the influence of the agitation in the washing machine or of Brownian motion. On the other hand, the attachment of particles to fabrics becomes stronger as their size decreases (see Sec. VI, B, 3).

During the washing process, the aggregated soil particles on the fabric are either detached and suspended in the aggregated state and then deflocculated, or the aggregates are broken up into primary particles by the deflocculating action of the detergent while still attached to the fabric: The secondary particles are removed from the fabric either as aggregates or piecemeal. It is difficult to determine which mechanism predominates, because the forces of adhesion between particles or aggregates and fiber surfaces, and the forces of adhesion holding together primary particles in aggregates are generally similar (see Sec. VII,A,4).

Flocculation of particulate soils is often caused by polyvalent cations, chiefly Ca^{2+} and Mg^{2+}. Therefore, builders which sequester these cations also deflocculate particulate soil. This effect is discussed in Chap. 11.

V. INTERACTION BETWEEN FIBERS AND SURFACTANTS

Wetting of fabrics by surfactant solutions, which is accompanied by the adsorption of surfactant molecules at the fiber-water interface, has been discussed in Chap. 5. The adsorption process is discussed in greater detail below, particularly as it affects the ζ potential of the fibers and the removal of particulate soil.

A. Adsorption of Surfactants by Fibers

Surfactant adsorption of the fiber-water interface is a necessary step in detergency. In fact, the extent of adsorption of various surfactants by different fibers was found to correlate with their ability to wash the corresponding fabrics (74).

1. Mechanism of Adsorption

Fibers adsorb surfactants from aqueous solution by the same processes as described for particulate soil. Physical adsorption occurs because of the tendency of the surfactant molecules to concentrate at interfaces, thereby lowering the interfacial tensions. The process predominates for nonionic surfactants. Ion exchange results in chemisorption via electrovalences. For instance, the uptake of alkyltrimethylammonium halides by different samples of purified cellulose occurred in two stages. Cation exchange according to

$$\text{Cellulose} - COO^{-}Na^{+} + R(CH_3)_3N^{+}Br^{-} \rightleftharpoons \text{Cellulose} - COO^{-}N^{+}(CH_3)_3R + Na^{+}Br^{-}$$

proceeded until the cation-exchange capacity of the cellulose, resulting from the presence of carboxyl groups, was exhausted. This process was completed below the cmc. Ion-pair adsorption, i.e., physical adsorption of the surface-active cation plus its halide counterion, began in the vicinity of the cmc (26-28).

The protein fibers wool and silk contain carboxyl groups and amino and other basic groups in side chains. Amino and imino groups chemisorb anionic surfactants. According to the equation (75)

$$\text{Polypeptide} - NH_2 + R - SO_3^{-}Na^{+} + H_2O \rightleftharpoons \text{Polypeptide} - NH_3^{+}{}^{-}O_3S - R + Na^{+}OH^{-}$$

it is seen that binding occurs most extensively in acid media. Polyamide fibers contain carboxyl and amino and groups. The former interact with cationic surfactants according to

$$\text{Polyamide} - COOH + R(CH_3)_3N^{+}Br^{-} \rightleftharpoons \text{Polyamide} - COO^{-}N^{+}(CH_3)_3R + H^{+}Br^{-}$$

Thus, binding is more extensive in alkaline media.

In general, the adsorption of anionic surfactants by cotton, viscose, acetate, nylon, and wool was found to decrease with increasing pH while the adsorption of cationic surfactants increased (76). Several polyamide fibers adsorbed n-alkylsulfuric acids of different chain lengths in amounts about ten times greater than the corresponding sodium salts (77). Chemisorption by amino end groups cannot account completely for the adsorption of n-alkylbenzenesulfonic acids by wool and polyamide fibers, because hydrochloric acid of the same strength was adsorbed in lesser amounts (77). In both instances, physical adsorption of the surface-active acids may have played a role in their uptake by these fibers. Alternatively, the amide groups may have acted as proton acceptors for the free acids.

In the case of 6-nylon, the amino end groups caused the chemisorption of sodium dodecyl sulfate which, as expected, decreased when going from a pH of 3 to 7 to 11. The carboxyl end groups chemisorbed cetyltrimethylammonium bromide; the extent of the adsorption decreased markedly when going from a pH of 11 to 7 to 3. The saturation uptake of sodium dodecyl sulfate at pH 7 was the same as that of cetyltrimethylammonium bromide at a pH of 11-12, namely, 1×10^{-4} mole/g (27, 28). This agrees with the customary polymer molecular weight of 10,000 as well as with the fact that 6-nylon (poly ϵ -aminocarproic acid) must contain the same number of amino and carboxyl end groups.

Another mechanism of chemisorption is the formation of calcium bridges, for instance to bind anionic surfactants to cellulose:

$$\text{Cellulose—COO}^- \ Ca^{2+} \ {}^-O_3SO\text{—R}$$

However, precipitation of $Ca(O_3SO\text{—}R)_2$ onto cellulose can give similar results (see below).

Uptake of surfactants by fibers that takes place in the absence of specific chemical reactions is due to physical adsorption.

2. Adsorption Isotherms

The early work on the adsorption of surfactants by fibers has been summarized by Harris (78). The following sources of error were often overlooked and render much of the results useless. Soaps, the alkali salts of fatty acids, are decomposed to free fatty acids by hydrolysis and by the action of the ubiquitous carbon dioxide. The liberated fatty acids form poorly soluble acid soaps, 1:1 association complexes between the free acid and the soap (79). Alkyl sulfates, the earliest synthetic surfactants studied, undergo hydrolysis, which is especially fast in acid and alkaline media and above the cmc. This reduces the equilibrium surfactant concentration by means other than adsorption. Consequently, the apparent surfactant uptake of the fibers, determined by measuring the residual surfactant concentration, is too high.

Most of the early studies of surfactant adsorption were conducted on cotton. In many instances, insufficient purification left some of the waxes, oils, proteins, pectins, and inorganic constitutuents on the fabric. This led to two erroneous results: (i) anionic surfactants were reportedly adsorbed by cotton from distilled water below the cmc (62, 76, 80-86); and (ii) maxima or points of inflection were observed in the adsorption isotherms of these surfactants on cotton (62, 80, 82, 84).

The following results were obtained when the role of noncellulosic impurities of cotton was examined. The adsorption of sodium alkyl sulfates was shown to be directly proportional to the nitrogen content (amount of protein) of raw cotton (86). In the case of cotton fabric, wax was found to be responsible for the adsorption of sodium alkylbenzenesulfonates, as well as for the maxima in the adsorption isotherms near the cmc. Extraction of wax reduced the adsorption near the cmc to zero (87). Evidently, the anionic surfactants were adsorbed by the waxy impurities of cotton rather than by the cellulose below the cmc. Near the cmc, the wax was solubilized by the surfactant and removed from the fabric. This reduced the amount of surfactant adsorbed by the fabric and produced a maximum in the adsorption isotherm. This artifact would have become apparent to the earlier investigators had they studied desorption as well as adsorption. Acid treatment of cotton fabric eliminated the adsorption of sodium dodecylbenzenesulfonate (DDBS) at concentrations below the cmc by removing polyvalent cations (88, 89) which could have retained the surfactant via salt bridges. When purified fabric was treated with this surfactant in water hardened with calcium chloride, some DDBS was retained. By using radioactively labeled DDBS and/or calcium, Schwarz et al. (90) showed that the mechanism was precipitation of calcium DDBS onto the fabric, for the following reasons: (i) The molar ratio of retained calcium to retained DDBS was 1:2, which corresponds to $Ca(DDBS)_2$, whereas a calcium bridge would have given a 1:1 ratio. (ii) The DDBS could be removed by extraction with ethanol, which is a solvent for $Ca(DDBS)_2$ but which would not have broken calcium bridges between cellulose carboxylate and the DDBS anion. (iii) Repeated washing in hard water with sodium DDBS caused a continuous increase in the amounts of calcium and DDBS retained by the fabric (90). Similarly, the presence of calcium ions was required to produce measurable adsorption of dodecyl sulfate by microcrystalline cellulose (27, 28).

Treatment of cotton fabric consisting of the following steps is recommended to remove noncellulosic impurities: alternate, exhaustive extractions with toluene and with ethanol to remove natural waxes and lubricating oils; boiling under nitrogen with 1% sodium hydroxide to remove pectins and proteins; the caustic solution is preferably changed a few times after it has turned yellow; washing with distilled water, treatment with cold 0.01 N hydrochloric acid to remove metal ions, followed by washing with distilled water to remove the acid. This procedure reduced the ash content of the fabric to 0.015%, leaving behind mainly silica (25, 35). Cotton thus purified adsorbed only a few milligrams of sodium decane sulfonate or sodium dodecyl sulfate per 100 g below the cmc (91). Substantial adsorption of anionic surfactants occurred only above the cmc. For instance, sodium tridecylbenzenesulfonate adsorption at twice the cmc was 30 mg/100 g (87).

Maxima in the adsorption isotherms may occur if the fibers swell extensively in water and less extensively in the surfactant solutions (84). Most adsorption isotherms are of the Langmuir type, leveling off in the vicinity of the cmc (27, 28, 35, 92, 93). This proves that the surfactants are adsorbed as single molecules or ions rather than as micelles.

Few studies of the adsorption of surfactants include desorption data. Gordon et al. (94) at the Shell Development Co. used radioactive sodium alkyl sulfates and linear alkylbenzenesulfonates in built detergents with cotton fabrics in a commercial washing machine in hardened water. They found that a nearly constant surfactant buildup was reached after the first wash-rinse cycle. Subsequent cycles produced only small increments. Rinsing with water removed small amounts of the retained surfactant; this may have resulted from shaking off some of the calcium alkyl sulfate or sulfonate which had been precipitated onto the fabric. On the other hand, the radioactive surfactant on the fabric was readily removed by exchange with unlabeled surfactant in the bath (94).

A comparable study with a detergent based on a polyoxethylated nonionic active showed a gradual buildup of adsorbed surfactant on the fabric during the first five or six successive wash-rinse cycles; a limiting surfactant uptake was reached after the sixth cycle. The adsorbed radioactive surfactant was removed efficiently by rinsing with water and by exchange with unlabeled surfactant in solution (92). The adsorption of $C_{12}(EO)_{14}$ on thoroughly purified cotton was found to be completely or nearly completely reversible (35). The adsorption of polyoxyethylated alkylphenols on quartz was also reversible (95).

3. Affinities of Different Fibers for Surfactants. Effect of Surfactant Type and Structure

The effects of specific chemical interactions and of artifacts such as the precipitation of calcium alkyl sulfates and sulfonates are discussed above. Other complications, which make it difficult to compare quantitatively the results of different investigators, are as follows: (i) Equilibrium adsorption is not always attained. (ii) The molecular weight of the fiber polymer is rarely quoted. Where chemisorption through end groups occurs, as in the case of anionic surfactants and the amino groups and cationic surfactants and the carboxyl groups of polyamides, the surfactant uptake increases with decreasing molecular weight. (iii) Uptake data are usually expressed as mole or gram of surfactant per gram of fiber rather than per square centimeter of fiber, making it difficult to compare different data unless the fiber denier is specified. (iv) Adsorption of a series of surfactants on different fibers was often determined at only one surfactant concentration, which varied from investigator to investigator. (v) Fabric finishes were not always completely removed. For these reasons, only qualitative comparisons are made below.

a. Anionic Surfactants. For sodium *n*-alkyl sulfates, Weatherburn and Bayley (National Research Council of Canada) found the following order of increasing uptake: cotton < viscose < nylon < acetate < wool (76). For sodium soaps, the order was similar, but depended on whether the fatty acid portion or the alkali component was being considered. Cotton and viscose adsorbed the alkali component in preference to the fatty acid component. Acetate, nylon, and wool showed preferential adsorption of the fatty acid component for lower molecular weight soaps and preferential adsorption of the alkali for higher molecular weight soaps (79). Other measurements (77) showed only small differences between the adsorption of sodium *n*-alkyl sulfates by acetate, cuprammonium rayon, and 66-nylon, but different polyamides showed increasing uptake for a given surfactant in the order: 6,10-nylon < 11-nylon < 66-nylon < 6-nylon.

For a given fiber, the uptake generally increased with increasing chain length of the alkyl group. This order is to be expected since it parallels the order of increasing interfacial activity. In other instances, maximum adsorption was found for the C_{14} or C_{16} soap or alkyl sulfate (76, 77, 79).

This may have been an artifact caused by insufficient times of contact. Surfactant uptake by polyamide fibers was found to increase over a period of many hours (77), much longer than the times of contact commonly employed (76, 77). If the adsorption process did not proceed to equilibrium, surfactants with the greatest hydrocarbon moiety (C_{18} and C_{20}) would be the most affected because they are the slowest to diffuse and to reach equilibrium uptake. Thus, an apparent maximum would be observed for the adsorption of the C_{14} or C_{16} surfactants at intermediate contact times, where the C_{18} and C_{20} surfactants are farthest from reaching final uptake. When equilibrium is reached for all surfactants, the uptake probably increases monotonically with chain length.

The long times required to reach equilibrium uptake in the case of the polyamide fibers indicate that surfactant ions or molecules are adsorbed in the bulk of the fibers as well as at the fiber-water interface, because the latter process is fast while diffusion into the polymer is slow (94).

b. Nonionic Surfactants. The following orders of increasing uptake of nonionics by different fabrics were recorded: for three ethylene oxide adducts of diisobutylcresol, wool $<$ nylon $<$ viscose $<$ acetate $<$ cotton (76); for an ethylene oxide adduct of *n*-dodecanol, polyester $<$ cotton $<$ nylon (92); for dimethyldodecylphosphine oxide, polyester $<$ cotton $<$ viscose $<$ nylon (93). Mast and Benjamin of the Procter & Gamble Co. found that nylon fabric adsorbed nearly 50 times more dimethyldodecylphosphine oxide than a polyester fabric of comparable geometry, indicating that the surfactant diffused extensively into the nylon fibers (93). The area per adsorbed surfactant molecule was the same for the polyester fabric as for a carbon black, indicating that the surfactant was adsorbed only at the polyester-water interface and did not penetrate into the bulk of the fibers. In the case of the hydrophilic cellulose substrates, extensive adsorption of dimethyldodecylphosphine oxide occurred only above the cmc (93). Polyoxyethylated surfactants are adsorbed extensively by purified cellulose even below the cmc (35).

In a series of polyoxethylated surfactants having the same hydrocarbon moiety, adsorption by different fibers increased with decreasing polyoxyethylene content, i.e., with increasing interfacial activity (76). Similar results had been obtained for adsorption on carbon black (39). Surfactants prepared by the addition of ethylene oxide to fatty alcohols have a broad molecular weight distribution. In adsorption experiments with cotton (35, 92), nylon, and polyester (92), these surfactants were shown to undergo fractionation. Molecules of lower degree of polyoxyethylation were adsorbed preferentially. The residual surfactant in solution contained the fractions of higher molecular weight and lower interfacial activity.

Adsorption and desorption experiments on cotton were performed with two series of nonionic surfactants, one consisting of ethylene oxide adducts of normal primary C_{12} through C_{20} alkanols with a constant polyoxyethylene content of 60%, the second consisting of five n-dodecanol adducts with polyoxyethylene contents ranging from 55 to 75%. In both series, the adsorption was found to decrease and the desorption to increase with increasing molecular weight (96). Unfortunately, many of the adsorption experiments were performed at temperatures above the cloud points of the surfactants, so that solubility in water may have been a dominant factor in controlling the extent of their adsorption on cotton.

B. The Electrokinetic Properties of Fibers and the Effect of Surfactants

Adsorption of ionic surfactants greatly affects charge and electrokinetic properties of the fibers. This is discussed below in terms of the electrokinetic or ζ potential, which can be measured readily.

1. Measurements of Zeta Potential

Details of the three experimental techniques can be found in Refs. 97 and 98. Zeta potential is the potential difference between the plane of slip or shear (boundary of the hydration layer attached to a solid surface) and the bulk of the aqueous medium. The plane of slip is located inside the diffuse part of the electric double layer. Therefore, when the liquid moves relative to the surface, the ions in the mobile part of the double layer, outside the plane of slip, are entrained and move along the surface. This produces a potential difference proportional to the ζ potential.

If the fibers are formed into a porous pad or plug through which solution is forced, the resulting interfacial potential is called "streaming potential." If an outside conductor prevents the buildup of a potential gradient between the two end of the pad, a "streaming current" flow through it, the intensity of which is a function of the ζ potential (99, 100). If an electric field is applied between the two ends of the fiber pad, motion of the ions in the mobile part of the double layer along the fiber surface causes the associated water of hydration to flow (electroosmosis) and produces an electroosmotic pressure. If the fibers are chopped up into small pieces and suspended in the liquid, they will migrate under the influence of an electric field toward the pole of opposite charge (electrophoresis) with a velocity proportional to the ζ potential. The bound part of the double layer, extending to the plane of shear, is carried along. The electrokinetic potentials of a given fiber derived from the magnitudes of the streaming potential or current, of the electroosmotic flow rate or pressure, and of the electrophoretic mobility are identical (97, 98). This indicates that the hydrodynamic plane of slip is the same when the fibers are stationary and the liquid moves past them as when the liquid is stationary and the fibrils move through it. Particulate soil is most conveniently studied by electrophoresis, but it can be formed into a porous plug and studied by the other two methods as well.

2. The Zeta Potential of Fibers

The electrokinetic potential does not depend on the method of measurement; however, it does depend on added electrolytes and on pH. Fabric finishes and dyes (99) are common sources, of variation, as are adsorbed anions and cations, particularly polyvalent ones. Different concentrations of ionic end groups can result from a slight excess diacid or diamine respectively dialcohol used in the condensation polymerization of polyamides or polyesters. Various initiators used in vinyl polymerization, e.g., persulfate, perborate, or sulfite, introduce different end groups. These affect the ζ potential of the fibers markedly, as do fiber impurities. For instance, the ζ potentials of dewaxed cotton containing pectins and of the pectic substance extracted from cotton were approximately 6 and 9 times greater, respectively, than that of dewaxed and depectinized cotton (71). Electrokinetic potentials of various textile fibers in neutral aqueous media are listed in Table 6.1.

The ζ potential of cotton was found to decrease monotonically with increasing concentrations of NaCl in neutral solution (102). However, maxima in the ζ potentials of cotton and six other fibers between 10^{-4} and 10^{-3} M NaCl and KCl have also been reported, with a steady reduction at higher concentrations (99,106). Electrolytes compress the diffuse part of the double layer, so that increasing concentrations should produce lower ζ potentials.

The sign and magnitude of the charge of fiber surfaces depend very much on the pH (see Fig. 6.6). At high pH values, the carboxyl groups of protein, polyamide, polyester, and cellulosic fibers are ionized, conferring a large negative charge to the fibers. The magnitude of this charge and of the resulting ζ potential decrease with decreasing pH. At low pH values, the charge of amphoteric fibers is large and positive owing to protonation of amino or imino groups, while the carboxyl groups are not ionized. This charge decreases with increasing pH. At an intermediate pH value called the isoelectric point or zero point of charge, the contributions of the charge arising from the acidic groups are equal and opposite to those from the basic groups. The net charge of the fiber surface and the ζ potential are zero at that pH value, provided that no other ions are adsorbed at the fiber surface. The isoelectric point depends somewhat on the concentration of added electrolytes and on the temperature. Characteristic values at room temperature and low salt concentration are listed in Table 6.1.

3. Effect of Surfactants on the Zeta Potential of Fibers

Ionic surfactants are often chemisorbed by fibers via reactions with functional groups of the opposite charge. Adsorption of increasing amounts gradually reduces the charge and the magnitude of the ζ potential of the fiber surface to near zero, and usually renders the fiber surface more

TABLE 6.1

Electrokinetic Potentials of Textile Fibers in Neutral Aqueous Media, and Isoelectric Points at Low Electrolyte Concentrations

Fiber	ζ potential, mV	Isoelectric point, pH
Wool	-33 (101); -34 (102); -40 (103); -48 (104); -54 (105); -60 (106)	3.4 (102); 3.5 (101); 4.2 (103)[a]; 4.5 (103)[b]
Silk		
raw		4.3 (107)
degummed	-1 (104); -18 (107); -24 (108); -34 (105); -42 (106)	3.3 (105); 3.7 (107)
Polyamides		
66-nylon	-22 (109)	2.6 (102); 2.7 (109)[c]; 3.9(109)[d]
6-nylon	-34 (105); -42 (106); -50 (109a)	
Cotton		
raw[e]	-30 (52, 108); -33 (50, 106); -38 (104); -39 (58)	
mercerized[f]	-15 (104); -24 (108)	
dewaxed[g]	-28 (109a)	1.7 (110)
purified[h]	-9 (110)	2.5 (110)
Viscose	-4 (104); -3, -9 (108); -14 (105); -30 (106)	
Secondary cellulose acetate	-36 (104); -38, -50 (108); -57 (106)	
Acrylic fibers	-38 (108)[i]; -5. -6 (108)[j]; -47 (108)[k]	5.9, 5.7 (108)[j]
Polyester	-15 (111); -64, -74 (108)	
Polyvinyl chloride	-51 (108); -53 (106)	

[a] Powdered.
[b] Scales and cortical cells.
[c] Drawn and oriented fibers.
[d] Unstretched fibers of lower crystallinity.
[e] Contains waxes, pectins, and proteins.
[f] Probably lost some wax and pectin.
[g] Waxes removed by solvent extraction.
[h] Dewaxed by solvent extraction; pectins and proteins removed with boiling caustic.
[i] Acrylonitrile homopolymer.
[j] Polyacrylonitrile containing small amounts of vinylpyridine comonomer.
[k] Polyacrylonitrile containing small amounts of a vinyl ester.

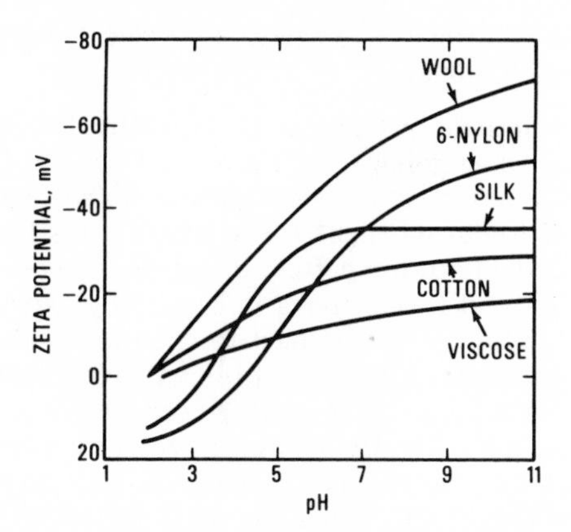

FIG. 6.6. Effect of pH on the ζ potentials of commercial fibers. Reprinted from Ref. 105, p. 312, by courtesy of Verlag Chemie, GmbH.

hydrophobic. Subsequent physical adsorption of additional surfactant increases the charge and the ζ potential, but changes their sign to that of the surface-active ion (see Sec. IV,C,2). For instance, the negative ζ potential of cotton fibers is due to carboxyl groups; most of these are part of the pectic impurities; a few belong to the cellulose proper. In the presence of increasing concentrations of cationic surfactants, which are chemisorbed through ion exchange by cotton, the ζ potential is increased to zero. Subsequent physical adsorption changes the ζ potential to increasingly positive values (58, 102, 106). The common textile fibers have negative ζ potentials in water. The effect of increasing concentrations of dodecylpyridinium chloride on their ζ potentials is shown in Fig. 6.7.

If anionic surfactants are to be adsorbed at the fiber surfaces, their interfacial activity must overcome the repulsion between their negatively charged headgroups and the negative charges of the fiber surfaces. The more hydrophobic the fibers, the greater the free energy at their interface with water and the lower the concentration at which anionic surfactants are adsorbed appreciably. Hydrophilic fibers have low fiber-water interfacial free energies, hence substantial adsorption of anionic surfactants occurs only at relatively high concentrations, in the vicinity of the cmc for viscose and pure cotton.

Zeta potentials are sensitive indicators for the adsorption of ionic surfactants. For instance, the ζ potential of purified cotton was not affected by sodium dodecylbenzenesulfonate until the surfactant concentration was increased to 90% of the cmc. This was also the concentration at which surfactant adsorption became significant. Likewise, the ζ potential of dewaxed cotton decreased by a mere 3 millivolts when going from water to sodium dodecyl sulfate solutions of concentrations up to twice the cmc, and by an additional 3 mV only, to - 35 mV, at concentrations up to four times the cmc (109a).

The following reductions in ζ potential, in millivolts, were observed for different fibers when going from water to a neutral 0.001 M sodium tetradecyl sulfate solution [this concentration corresponds to one-half of the cmc (106)]: 37 for 6-nylon; 31 for wool; 21 for silk; 14 for acetate; 10 for polyvinyl chloride; 6 for cotton; and 3 for viscose. The large drop in ζ potential of the first three fibers is caused by extensive uptake of the anionic surfactant, which is due in part to chemisorption through amino groups, and the concomitant shifting of the balance between free amino and carboxyl groups in favor of the latter (see Sec. V,A,1). The reduction of the ζ potential of all seven fibers by equivalent concentrations of anionic surfactants decreased when going from sodium tetradecyl to dodecyl to decyl sulfate (106). This parallels the order of decreasing

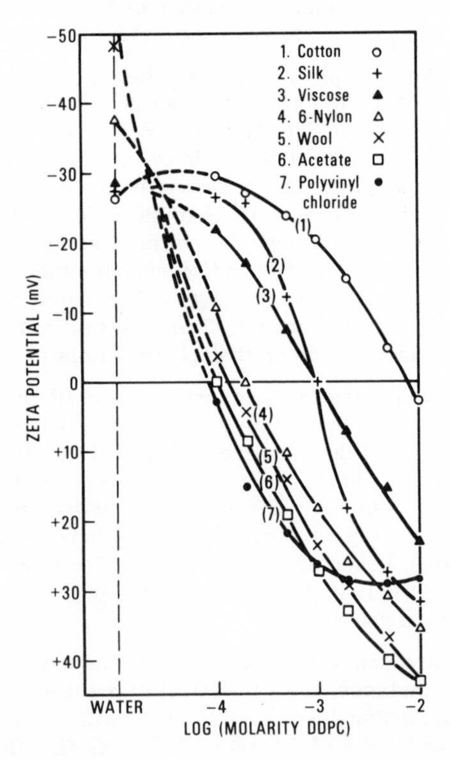

FIG. 6.7. The ζ potentials of textile fibers in water and in solutions of dodecylpyridinium chloride (DDPC) of different concentrations. Reprinted from Ref. 106, p. 76, by courtesy of Dr. Dietrich Steinkopff Verlag.

interfacial activity, i.e., the tendency of the surfactants to accumulate at interfaces, and hence the order of decreasing adsorption at the fiber-water interface.

In comparable experiments, the ζ potentials of 6-nylon and polypropylene fibers were measured in solutions of sodium n-alkyl sulfates as a function of surfactant concentration. Initially, the ζ potentials of the fibers decreased fast with increasing surfactant concentration, but they levelled off at or just below the cmc's. Further increases in the surfactant concentrations produced no additional decreases in the ζ potentials but, rather, small increases, i.e., their absolute values became slightly smaller (111a).

The reductions in ζ potential from the values measured in water to those measured in the surfactant solutions at the respective cmc's were as follows: For sodium dodecyl sulfate, 45 mV for both fibers; for sodium tetradecyl sulfate, 60 mV for polypropylene and 64 mV for 6-nylon; for sodium hexadecyl sulfate, 72 mV for polypropylene and 82 mV for 6-nylon. The longest-chain, most surface-active homolog caused the greatest reductions in ζ potential even though the measurements were made at the cmc values, which are in the molarity ratios of 11 : 3.2 : 1 for the C_{12}, C_{14}, and C_{16} sulfates. The ζ potentials of the fibers determined in mixtures of pairs of these surfactants in various proportions indicate that the higher homolog was more extensively adsorbed by the fibers, just as mixed micelles of pairs of these surfactants were enriched in the higher homolog (111a).

The amphoteric fibers nylon and wool adsorb more cationic surfactant and less anionic surfactant with increasing pH. However, the following changes in ζ potential were observed for 66-nylon between pH 3 and pH 11: from -65 to -70 mV in sodium dodecyl sulfate; and from +40 to -5 mV in dodecylpyridinium bromide, both at one-tenth of their cmc (102). Apparently, the change in ζ potential was determined by ionization of the end-group species not tied up through chemisorption with the ionic surfactant. Dodecyl sulfate ions combine with NH_2 end groups, and the reason that the ζ potential was more negative at pH = 11 than at pH = 3, despite the more extensive chemisorption of the surface-active anion at pH 3, is the ionization of the carboxyl end groups in the alkaline region. Likewise, dodecylpyridinium ions combine with carboxyl groups, and even though the surfactant uptake at pH = 3 was probably less extensive than at pH = 11, protonation of the NH_2 end groups at pH = 3 resulted in a higher ζ potential in the acid region. Similarly, adsorption of dodecylpyridinium bromide increased the isoelectric point of wool from pH 3.4 to 7 and to 10 at concentrations equal to 11% and to 39% of the cmc, respectively (102). Evidently, the alkylpyridinium ions neutralized many of the COO^- ions on the fiber surface, thereby shifting the balance between amino and carboxyl groups in favor of the former.

Nonionic surfactants reduced the absolute value of the ζ potential of most fibers by 5 to 15 mV (106) or more (102). The effect was nonspecific; plots of ζ potential versus surfactant concentration for eight fibers had similar shapes. The curves were shifted along the ζ potential axis instead of coinciding because the fibers had different ζ potentials in pure water (106). Adsorption of the nonionic surfactants may have moved the plane of slip farther away from the fiber surface toward the bulk of the solution, thereby lowering the magnitude of the ζ potential.

VI. INTERACTION BETWEEN PARTICULATE SOIL AND FABRIC

This section deals mainly with processes described by Eq. (6.4).

A. Methodology

Considerable experimental data exist on the interaction between model solid soils and fabrics. Conclusions of different authors are often at variance. This is not surprising in view of the considerable differences in the following factors: surface properties, shape, and size of the particulate soil used; the presence of admixed oily soil; the variety of fabric constructions and of type of fibers used; the diversity in surface and bulk morphology of different fibers; the presence of resins and other finishes which impart permanent press, crease resistance, soil resistance, antistatic properties, lubrication during weaving, etc.; and the different methods used to apply the soil to the fabric and to estimate the amount of soil on the fabric.

1. Model Soils

One may question whether such pigments as the oxides of iron, chromium, or manganese, ilmenite ($FeTiO_3$), rutile (TiO_2), graphite, and carbon blacks are representative of the particulate soil found in city or country. Even pure clays such as bentonite and kaolinite, which have been found on soiled garments (4, 112), are likely to have different uptake and removal properties in the presence of fatty soil originating from sebum. Vacuum cleaner dust and the residual sludge from the distillation of dry-cleaning liquids are much more realistic soils, but their widely varying compositions make it difficult to obtain reproducible data and preclude quantitative comparisons of results obtained in different laboratories.

2. Assessment of Soil Level on Fabrics

Different methods for determining the amounts of particulate soil picked up by the fabrics or retained after washing often give different results. Chemical analysis and the use of radioactive particulate soils (113-117) are generally reliable. The most convenient and widely used method, optical measurements of fabric whiteness or reflectance, frequently leads to erroneous results. The data are often interpreted with the uncorrected Kubelka-Munk equation (118, 119)

$$K/S = \frac{(1 - R)^2}{2R} \tag{6.15}$$

where K is the light absorption or reflectivity coefficient, S the light-scattering coefficient, and R the observed reflectance, for monochromatic light. In the case of a black pigment on white fabrics, where the reflectance is the same at all wavelengths, R can be measured with white light. It is generally assumed that K increases proportionally with the amount of light-absorbing pigment while S remains essentially constant, so that K/S is a linear function of the amount of pigment on the fabric (120-122). The percentage of soil removed, D, is then estimated as

$$D = 100 \left[\frac{(K/S)_{sf} - (K/S)_{wf}}{(K/S)_{sf} - (K/S)_{of}}\right] \tag{6.16}$$

where the subscripts "sf," "wf," and "of" indicate soiled fabrics, washed fabric, and original fabric (i.e., fabric before soiling), respectively.

Even corrected forms of these equations (119, 120, 123-125) may give erroneous conversions of reflectance into weight of retained soil (113, 114) because the particle size or state of flocculation of the soil on the fabric is usually influenced by surfactants and builders (62, 113). At a given fabric loading, finer dispersions of the pigment result in lower reflectance (122). It is possible to wash a soiled cloth in a surfactant solution, remove some soil from it, and at the same time have the cloth appear dirtier, i.e., of lower reflectance: The soil remaining on the cloth, while reduced in amount, is deflocculated and spread out, which results in an increased cross-sectional or projected area. No uniform distribution and/or orientation of particles on the fabric likewise lead to erroneous results (120, 123).

Even more popular is a simplified approach, namely, the linear method of calculating soil removal from reflectance values (121, 124, 126, 127):

$$D = 100 \left(\frac{R_{wf} - R_{sf}}{R_{of} - R_{sf}}\right) \tag{6.17}$$

Utermohlen and Ryan (126) determined the removal of a black iron oxide pigment from cotton fabric during washing with sodium dodecyl sulfate solutions by chemical analysis, and by reflectance measurements interpreted according to Eqs. (6.16) and (6.17). The three methods are compared in Figs. 6.8 and 6.9. Optical results calculated according to the Kubelka-Munk equation [Eq. (6.16)] are in reasonably good agreement with the correct results obtained by chemical analysis in the case of rather thoroughly washed fabrics. Results calculated by the linear method [Eq. (6.17)] are consistently low (126).

Another series of tests involved a radioactive carbon black and a group of 12 detergent compositions containing different actives and builders. Ranking of the detergents by radioactive assay differed significantly from ranking by reflectance measurements (114).

Reflectance readings generally correlate well with visual ratings of the level of soiling. However, at a given soil content, fabrics consisting of bright fibers appear dirtier than comparable fabrics with fibers containing added delustrant.

Goette, Kling and Mahl (128, 129) of Henkel & Cie. found that particle counts made on electron micrographs of surface replicas of fibers from soiled fabric swatches were in good agreement with the degree of soiling estimated from the reflectance of the swatches (128). This points to a relatively even distribution of the particulate soil on the fabric surface (129). At the stage where particulate soil could no longer be detected by electron microscopy in a series of successive wash cycles, the fabric also appeared clean to the naked eye and in the reflectance meter (128).

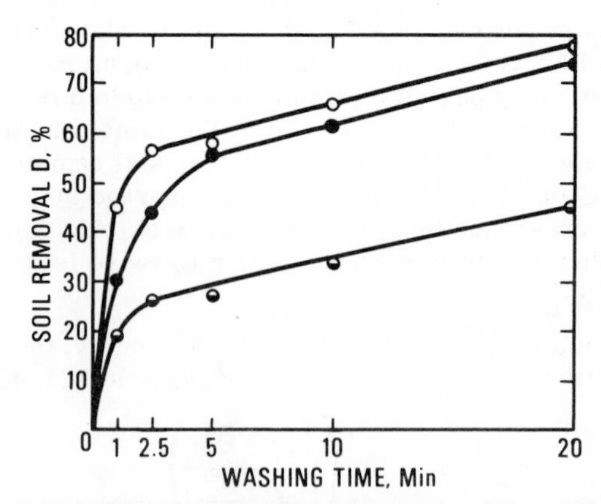

FIG. 6.8. Removal of iron oxide pigment from cotton fabric by a 0.25% sodium dodecyl sulfate solution, as a function of washing time: O, using Eq. (6.16); ◒, using Eq. (6.17); ●, by chemical analysis. Reprinted form Ref. 126, p. 2884, by courtesy of the American Chemical Society.

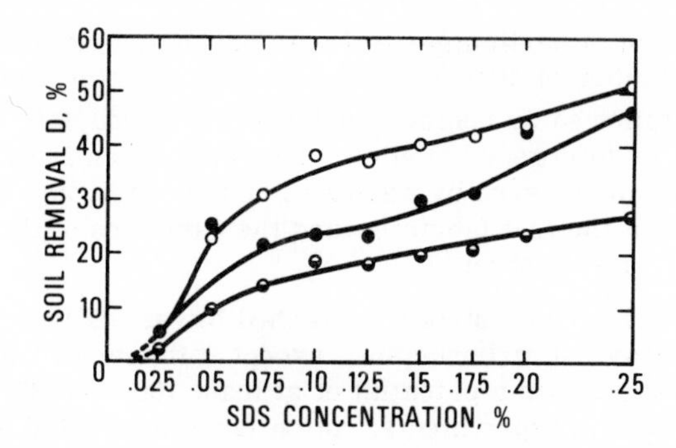

FIG. 6.9. Removal of iron oxide pigment from cotton fabric by washing for 10 minutes in sodium dodecyl sulfate (SDS) solutions of different concentrations: O, using Eq. (6.16), ◒, using Eq. (6.17); ●, by chemical analysis. Reprinted from Ref. 126, p. 2885, by courtesy of the American Chemical Society.

3. Methods of Soiling

The natural soiling of fabrics takes place either while they are being worn or otherwise put to functional uses, or during the washing process as the soil is redistributed and redeposited from dirty onto cleaner fabrics. Artificial soiling for detergency studies duplicates either the primary deposition or the redeposition process. Obviously, the manner in which a given soil is applied to a fabric exerts a profound influence on the amount taken up, the degree of dispersion, and the ease of removal. Artificially soiled test fabrics are usually much more heavily soiled than even the dirtiest pieces in a domestic wash load.

Fabrics are often soiled by agitation with suspensions of particulate soil in hardened water containing detergent, followed by rinsing and drying. Fatty soil can be added as an emulsion. Redeposition studies are also carried out by washing clean and soiled fabrics together (130). Direct application of soil onto fabrics is accomplished through treatment with suspensions of particulate soil in organic solvents which usually contain dissolved fatty matter (127). A continuous process involves spraying a moving fabric belt with two nozzles. One nozzle dispenses carbon tetrachloride containing dissolved wool fat and suspended pigments (kaolin, carbon black, and iron oxides), the other dispenses a much smaller volume of aqueous sodium chloride solution to simulate the composition of perspiration (131). Textile printing techniques are sometimes used to apply pastes of particulate soil dispersed in oily soil or in volatile or water-miscible liquids to fabrics (127).

Dry application of particulate soil with or without fatty soil is accomplished in a number of different ways simulating natural soiling processes. This is often followed by removal of loosely attached soil with a vacuum cleaner prior to detergency studies. This precaution is desirable because loose soil is removed even by washing in pure water, and hence reduces the sensitivity of the test fabric toward the variables of the detergency process under investigation.

Hardly any soil became strongly attached to vertically held fabrics on exposure to sprays of airborne soil, except if the metal hoop holding the fabric was charged with a potential of at least 1000 V (132). In other electrostatic soiling experiments, cotton fabrics mounted on metal hoops and suspended vertically were exposed to city air for several weeks. The rate of soiling increased with the potential applied to the fabrics. Positively charged fabrics were soiled more heavily than negatively charged fabrics because the atmosphere contains more negatively than positively charged particles (133). On the other hand, the soiling of carpets made from different synthetic and cellulosic fibers and wool by foot traffic was related to fiber smoothness and surface area rather than to the electrostatic properties of the fibers, specifically, to their ability to dissipate charges by conduction (134).

Rees (120) deposited particles by gravity settling from dry or wet clouds onto horizontally held cotton fabrics. In other experiments, the clouds were drawn through vertically held fabrics by means of an aspirator; the particles were retained by filtration. The former process distributed the particles randomly over the fabric surface. Filtration caused preferential soil deposition in areas of the fabric already soiled. The clouds were produced by atomizing aqueous soil suspensions by means of a jet fed with compressed air into a fine mist. Heating caused evaporation of the water and produced dry clouds of particles (120).

Contact soiling can be accomplished by mangling the fabric between soiled paper (120), or by rubbing a roughened glass plate bearing the dry or pasty soil against the fabric, followed by brushing (114). Alternatively, the soil was first applied to dense felt cubes, which were then tumbled with fabric swatches (135). Frequently, the soil is directly applied by tumbling with fabric swatches or cut yarns and glass beads or steel balls in a rotating jar (122, 132, 136).

Storage conditions of the soiled fabrics (exposure to light, moisture, heat, and air) affect soil removal, particularly if unsaturated oily soil capable of undergoing polymerization or oxidation is present (133). There is an almost complete lack of information on the effect of the different methods of soil application on washability. This makes it difficult to generalize, compare, or extrapolate detergency results, even those obtained with comparable soils and fabrics.

B. Mechanisms of Soil Retention

This section deals with the adhesion of particulate soil to fibers in the absence of oily soil. The effect of "oil bonding" is discussed in Sec.VI,D.

1. Retention by Sorptive Forces

Several attempts have been made to interpret the retention and the detachment of soil particles by and from fibers in terms of molecular forces through application of the Derjaguin-Landau-Verwey-Overbeek theory (see Chap. 4 and Refs. 15 and 137). There are two limitations. First, the theory refers to hydrophobic materials. It is not applicable to hydrophilic fibers like those of cellulose nor to hydrophilic soils like clays and silica without introducing a major correction for their hydration layers. Second, the theory refers to surfaces possessing potentials of equal sign and magnitude. There are two theories of heterocoagulation, which is the mutual coagulation of particles with different surface potentials (138, 139). They have been applied to experimental data after simplifications (20, 140).

The secondary minimum in the potential energy versus distance curve (Fig. 6.2) plays a major role in the interaction between particulate soil and fibers in aqueous media. In experiments with cellulose dispersions (20, 141), the minimum electrolyte concentrations required for flocculation were found to be surprisingly low, even lower than the flocculation concentrations for hydrophobic dispersions. This was attributed to flocculation of the cellulose particles in the secondary minimum, i.e., at comparatively large particle-particle distances.

Additional considerations of solid-solid adhesion, taking into account deformations undergone by soft or elastic substrates, are given in a comprehensive review paper by Krupp (142), and by textbooks on adhesion listed therein.

In addition to the London-van der Waals forces of attraction between solids which extend over relatively large distances owing to their additivity, there are specific interactions between the atoms or ions located in the two adhering surfaces, such as electrovalences, calcium bridges, or hydrogen bonds. These are effective over a few angstrom units only, and contribute significantly to the adhesion only if particle and fiber are in very intimate contact and if the particle is quite small, i.e., has a very high surface-to-volume ratio.

A nearly ideal example of such a situation is provided by the adhesion of sodium bentonite to cellulose. The primary clay particles are lamellas averaging 30-40 Å in thickness. Their outermost surfaces consist of a rather close-packed layer of oxygen ions, which form hydrogen bonds with the hydroxyl groups in an adjoining cellulose surface, besides inducing dipoles in the organic material which provide additional attraction.

In the adhesion of these very thin bentonite lamellas to cellulose, the attraction between the surface atoms or ions by hydrogen bonding and dipole-induced dipole interaction is at least an order of magnitude greater than the attraction of the bulk solids by London-van der Waals forces.

Because of their great flexibility and strong attraction to the fiber substrate, these flat and extremely thin clay particles covered the cellulose surface with a continuous film about 40 Å thick, and replicated all its topographic details (143): Cotton surfaces covered with this thin clay layer could not be distinguished from clay-free surfaces by electron-microscopic examination of their surface replicas, according to research at the Lever Brothers Co. The ashes of the clay-treated cotton fabric and of paper made from clay-treated wood or cotton pulp preserved the microscopic appearance of the original webs but on a shrunken scale. The quantity of bentonite retained amounted to only 8.4×10^{-7} g clay/cm^2 cellulose, which corresponds to 0.13% by weight of cotton fabric or 0.6% by weight of paper. This small amount of clay adhered so tenaciously to the cellulose that it could not be removed by stirring fabric or pulp with water (143, 144). In fact, the adhesion of bentonite clay to cellulose was found experimentally to be stronger than the autohesion of cellulose (144).

By contrast, kaolin clay taken up by cotton fabric was gradually removed by washing with water. This occurred because kaolinite particles are platelets 10 to 50 times thicker than sodium bentonite lamellas, and are therefore too inflexible to mold themselves intimately to the fiber substrate. Furthermore, the basal area of kaolinite particles is between 10 and 600 times smaller than that of bentonite particles, so that the area per unit weight of clay available to make contact with the cellulose surface is correspondingly smaller (143).

2. Mechanical Retention

In addition to "colloidal" or sorptive forces (145), particles can be retained by fabrics through mechanical entrapment. Compton and Hart (146) distinguished between macroocclusion, or entanglement of particles in the spaces between the fibers or filaments of a yarn or between yarns, and microocclusion, or entrapment of particles in the irregularities of the fiber surfaces.

The presence of surfactants and electrolytes and changes in pH of aqueous dispersions of carbon blacks did not affect their uptake by cotton, provided the particles had diameters of 0.1 mm or larger. The uptake of small carbon black particles was strongly affected by these chemicals (146). This indicates that mechanical occlusion and enmeshment are mainly responsible for the retention of large particles or aggregates, while sorptive forces are mainly responsible for the retention of the small particles, because the added chemicals are known to affect the zeta potentials of carbon particles and of cotton strongly, thereby altering the balance between van der Waals attraction and electrostatic repulsion which controls the sorptive forces.

The number of macrooccluded soil particles increased as the weight and complexity of the fabric structure increased (123). With increasing twist of the yarn, the retention of particulate soil increased at first but reached a maximum and then decreased. At low twist, the soil particles penetrated into the yarn and became trapped in the spaces between the filaments throughout the whole interior of the yarn. The interior of highly twisted yarns was inaccessible to the soil particles, which restricted them to the external yarn surface (147). Macroocclusion was prevented by chopping up skeins and fabrics in a Wiley mill to fibers 0.1 to 1 mm in length (146, 148).

Microscopic and electron-microscopic evidence has shown the following crevices, rugosities, and ridges of fiber surfaces to hold soil particles by microocclusion: the microfibrillar structure of the secondary wall of cotton (2, 129, 149-151), which is loosened by washing and bleaching (149); and the edges of the scales on wool fibers (2, 129) and on human hair (129). The primary wall of the cotton fibers is removed by kiering, bleaching, and washing (129). Electron-microscopical work on the surface replicas of cotton soiled with fine carbon black particles and with natural dust showed microocclusion; the particles were lined up in the natural rugosities of the fibers (129, 151-153). Other work gave no evidence for microocclusion (154). Most electron microscopists observed that in addition to particles trapped in surface irregularities, quite a few other particles adhered to smooth fiber surfaces or to surfaces with rugosities far smaller than the particle size (4, 129, 151, 152, 154). Even melt-spun fibers such as those from polyester and polyamide polymers, which are round and have smooth surfaces when new, develop submicroscopic cracks and crevices during washing, bleaching, or under the influence of light (149) capable of trapping small particles.

External specific surface area was among the fiber characteristics found to affect soil retention; the latter increased with decreasing filament denier. For a given type of fiber, e.g., viscose or acetate rayon, serrated cross sections and striations or channels resulted in greater soil retention than smooth cross sections (136).

Electron-microscopical examinations uncovered no evidence for the penetration of soil particles into the interior of fibers (128).

3. Effect of Particle Size on Soil Uptake and Retention

Small particles adhere to fabrics mainly through sorptive forces, i.e., London-van der Waals attraction plus specific bonds between surface atoms or ions. Increasing particulate surface-to-volume ratios promote more extensive and closer particle-substrate contact and hence increased adhesion. This was illustrated by comparing the adhesion of bentonite and of kaolinite to cotton (Sec. VI,B,1). Theoretical considerations indicate that the surface potential ψ_0 necessary to stabilize particles against deposition on fabrics (and against flocculation) increases with decreasing particle size (15). The necessary increase in ψ_0 becomes quite large for particles below 0.1 μ in diameter (137, 154). This is in good agreement with the experimental observation that, when the diameters of particles fall below 0.2 μ, their removal becomes considerably more difficult (128). The effect of particle size on the potential energy of fiber-particle interaction is particularly pronounced for the double-layer repulsive energy component. Its effect on the London-van der Waals attractive energy component is much less important (15).

It is well known in laundry practice and from laboratory experiments (128, 146, 154) that regardless of the range of particle sizes, and hence regardless of the mechanism of retention, smaller particles are deposited more readily onto fabrics and are more difficult to remove than larger particles.

Hart and Compton (150) treated chopped cotton fabric with 13 carbon blacks ranging in primary particle size from 13 to 120 mμ. According to Fig. 6.10, the reflectance of the fiber pad, which is a measure of pigment retention, increased rapidly with increasing primary particle size up to 50 mμ. For larger sizes, the increase in reflectance was much slower. This indicates that soil deposition was much more severe for particles below 50 mμ than for larger ones. Soiling by particles larger than 50 mμ was lighter and relatively independent of particle size (150). Since Compton and Hart used a dispersing agent and a colloid mill to prepare their carbon black dispersions (146), a large fraction of the aggregates was probably broken down into primary particles.

There are several explanations for the sharp break in the reflectance versus particle size curves of Fig. 6.10 occurring at 50 mμ:

(i) Compton and Hart thought that it was related to the size of the rugosities or of the spaces between fibrils in the secondary wall of cotton. These crevices are usually up to about 50 mμ wide and hence can only trap particles of lesser size.

There are two objections to this explanation. First, the size of the rugosities refers to dry cotton fibers. A more pertinent and smaller value would be their size in the surface of wet and swollen fibers, since the fibers were soiled in aqueous suspensions. Second, extension of Hart and Compton's work (148) to other fibers (Fig. 6.11) showed that similar breaks in the reflectance versus particle size curves occurred at 50 mμ for wool, silk, linen, and viscose as well, even though these fibers have surface irregularities of different sizes.

Other explanations for the greater uptake of particles with diameters below 50 mμ are the following:

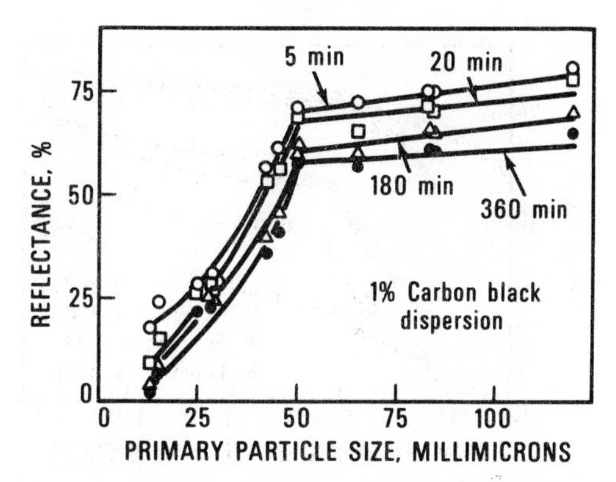

FIG. 6.10. Effect of primary particle size on soil retention by chopped cotton fibers. Reprinted from Ref. 150, p. 1138, by courtesy of the American Chemical Society.

(ii) Sorptive forces become the dominant factor in adhesion below that size while mechanical occlusion predominates for larger particles (155).

(iii) The effect of particle size on the potential energy of interaction between fiber and particle according to the Derjaguin-Landau-Verwey-Overbeek theory becomes important for diameters of 0.1 μ and less (see above and Ref. 15 and 137).

(iv) The effect is an optical artifact. Smaller particles have larger specific projected areas or cross-sectional areas per gram than larger particles and hence greater hiding power at equal weight percentages of adsorbed pigment. Furthermore, the extinction coefficient is not a monotonic function of the particle size, and there is a transition in the range where the diameter reaches 1/15 of the wavelength of light.

(v) Larger particles adhering to fabrics protrude more beyond the fiber surfaces than smaller particles and offer a bigger target. The hydrodynamic stresses exerted on the larger particles in the washing machine are therefore greater, and they are more readily removed than smaller particles.

(vi) The linear velocity of water flowing through the pores of fabrics under constant pressure is proportional to the square of the pore radius. Smaller particles are trapped by smaller pores than are larger particles. Moreover, the void space of pores partly clogged by trapped particles will normally be greater for larger pores and hence for larger particles. Therefore, liquid flowing through soiled fabrics will flow with greater velocity past large particles than past small ones, and exert a greater drag on the former. This will dislodge the largest particles most readily.

None of these explanations, except (ii), accounts very well for the abruptness of the change in uptake at the primary particle size of 50 m μ.

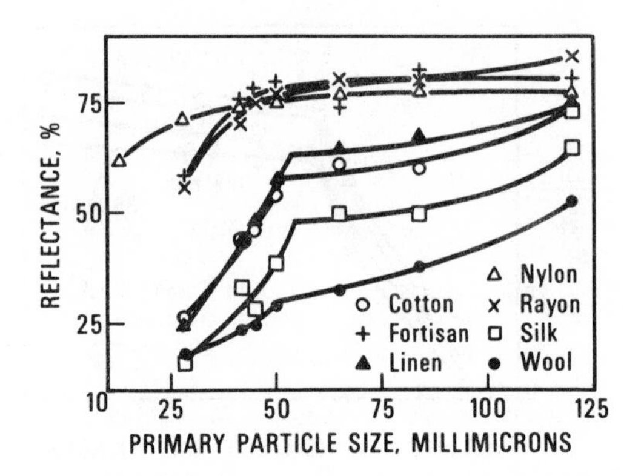

FIG. 6.11. Effect of primary particle size on soil retention by different textile fibers. Reprinted from Ref. 148, p. 165, by courtesy of the Textile Research Institute.

4. Effect of Rate of Agitation on Uptake and Retention of Particulate Soil

High levels of agitation tend to deflocculate suspended soil (60). Since smaller particles are taken up more readily by fabrics than larger particles or aggregates and are retained more tenaciously, strong agitation should lead to more severe soiling. On the other hand, soil removal is increased by increasing the level of agitation (118). Through a combination of these two opposing factors, the equilibrium amount of soil on the fabric is likely to go through a maximum at some intermediate level of agitation.

Such a maximum in soil uptake and a corresponding minimum in soil removal were found during a study of the interaction of aqueous kaolinite suspensions and cotton fabric as a function of the speed of agitation of the Terg-O-Tometer (156). The process of clay deposition and removal was found to parallel the process of clay agglomeration and dispersion. There was one stirring speed below which the rate of aggregation of the suspended clay particles increased rapidly with decreasing speed and above which it was always low. Suspensions stirred at that speed produced the most severe soiling of the fabric and the most tenacious soil retention. Evidently, lower stirring speeds left intact large clay aggregates which, owing to their size, were taken up more slowly and removed more readily than primary particles. Higher stirring speeds produced little additional deflocculation of the suspended kaolinite but enhanced the removal of the adsorbed clay from the fabric.

C. Effect of the Chemical Composition of Fibers and of Textile Finishes on Soiling and Soil Retention

The effect of the chemical nature and of the surface characteristics of the fiber-forming polymers and of wash-and-wear, permanent press, and soil-repellent finishes on uptake and removal of oily soils has been discussed in Chap. 5. Similar considerations apply to particulate soil.

1. Dry Soiling

This section deals with model studies of dry soil uptake, which are intended to duplicate the soiling of upholstery, drapery, and carpeting. The role of electrostatic charges is mentioned in Sec. VI,A,3.

According to Utermohlen et al. (157), partial acetylation was the most effective treatment to impart resistance against dry soiling to cotton fabrics, as measured by reflectance after tumbling with carbon black or vacuum sweeper dirt. Mercerization and partial carboxymethylation also reduced the degree of soiling, as did most of the anionic and cationic fabric softeners tried. Treatment with various vinyl resins and hydrophobic finishes, on the other hand, rendered the fabric less resistant to soiling (157). Cotton finishes based on silicones or on fluorocarbons increased the tendency toward dry soiling, and also affected the washability adversely (3).

Porter et al. (132) at the Southern Regional Research Laboratory of the U.S. Department of Agriculture studied the dry soiling of fabrics by tumbling swatches with a variety of carbon blacks or with a radioactive clay, and also by foot traffic, with the following results: The soiling tendency of cotton increased with the specific surface area—fabrics of coarse fibers were soiled less than fabrics of fine fibers. Untreated gray cotton, which still had the primary wall, was slightly more soil resistant than bleached cotton, which had a rougher surface owing to partial removal of the primary wall and exposure of the fibrillar structure of the secondary wall.

Of the finishes applied to retard the uptake of particulate soil by cotton fabrics, colloidal silica and alumina were the most effective. Particles of these white pigments evidently were retained on the fiber surface, blocking crevices and rugosities to penetration by the deeply colored soil particles (cf. also Ref. 2). Mercerization, which increased the smoothness of the fiber surface and obliterated some of the rugosities, did not always improve the resistance to dry soiling.

Soil uptake of yarns increased in the order: nylon < mercerized cotton < polyester < acetate < viscose rayon < glass. Since no effort was made to compare yarns of similar structure nor to remove surface finishes, this merely indicates that the smoother fiber surfaces are not necessarily soiled less extensively than the rougher ones (132).

Weatherburn and Bayley (122) soiled yarns of different polymers by tumbling with vacuum cleaner dirt. The yarns had been purified by extraction with solvents and rinsing with water. The order of increasing soiling was: cotton < acetate < viscose < nylon (variable, depending on soiling condition) < wool. All yarns showed increased soil retention with decreasing moisture content, but the order of soiling was essentially unchanged (122).

2. Wet Soiling

Experiments on the soiling and washing of fabrics in water in the absence of detergent duplicate soil deposition or removal conditions during the rinse cycle of the washing machine (158). Extensive work by Berch and Peper (158-160) at the Harris Research Laboratory of the Gillette Co. dealt with the wet soiling of cotton fabrics modified by different finishes with an acicular iron oxide (representing a "polar" or "hydrophilic" soil), a furnace black (representing a "nonpolar" or "hydrophobic" soil), and vacuum cleaner dirt.

Hydrophobic or water-repellent surface finishes (silicones and polysiloxanes, the chromium complex of perfluoro-octanoic acid, other perfluorinated finishes, and acrylic polymers) caused the fabrics to be soiled more heavily than untreated cotton by all three soils, but especially by the hydrophobic carbon black (158). In work done at the Southern Regional Laboratory, another carbon black was found to be adsorbed in increasing amounts in the following order: cotton < cotton treated with an acrylic resin < cotton treated with a silicone resin (153). Again, greater hydrophobicity of the substrate produced increased wet soiling.

This can be explained in terms of surface energetics (159): When particles (P) become attached to fibers (F) in water (W), the particle-water and the fiber-water interfaces at the site of contact are replaced by an equal area of particle-fiber interface. The change in surface free energy accompanying soil deposition is therefore

$$\Delta G = \gamma_{PF} - \gamma_{PW} - \gamma_{FW} \tag{6.18}$$

where γ represents the free energy values of the respective interfaces. The soiling process occurs spontaneously if there is a net decrease in the surface free energy G, i.e., if $\Delta G < 0$, and hence if

$$\gamma_{PW} + \gamma_{FW} > \gamma_{PF} \tag{6.19}$$

Soil deposition is favored by large γ_{PW} and γ_{FW} values and small γ_{PF}. The ideal combination for soiling is that of a hydrophobic fiber (γ_{FW} large) and a hydrophobic particle such as carbon black (γ_{PW} large); both surfaces being hydrophobic makes γ_{PF} small. Detergency is the reverse of soiling. Therefore, the change in surface free energy for soil removal is equal to the free energy change of soiling represented by Eq. (6.18) but of opposite sign.

The value of γ_{FW} of cotton treated with different finishes was estimated by casting the finishes into smooth films on glass plates and measuring the contact angle of oleic acid on these films immersed in water. The cosine of this contact angle θ_O is a measure of γ_{FW} and of the hydrophobicity of the finished fabric. For the hydrophobic carbon black, Berch and Peper (159) found a linear increase of the degree of soiling of the various cotton fabrics with increasing cos θ_O of the finishes. For the more hydrophilic iron oxide, the degree of soiling was nearly independent of cos θ_O (159).

Increasing the hydrophilicity of the fabric by hydroxyethylation reduced the soiling by the hydrophilic iron oxide pigment but not the soiling by carbon black (158). Increased hydrophilicity of the fiber means lower γ_{FW} in Eqs. (6.18) and (6.19). It should be noted, however, that pure cellulose (cotton fiber minus the primary cell wall) has a relatively low γ_{FW} value already.

Since the particulate soils were negatively charged at the prevailing conditions, anionic groups, introduced by carboxymethylation and phosphonomethylation, improved the soiling resistance of cotton through enhanced electrostatic fiber-particle repulsion. Amino groups, which confer a positive charge to cotton, caused heavy soiling of the fabric through electrostatic fiber-particle attraction (158).

Treatment with three polyacrylic ester emulsions whose films ranged from soft and pliable, to moderately firm and flexible, to hard and brittle resulted in increased soiling with decreasing hardness of the polymer (158). Higher temperatures increased the soiling of fabrics treated with these acrylics and with other thermoplastic resins; otherwise, temperature had a small and inconsistent effect on soiling. The softer finishes were more "sticky," and pigments became embedded in them (160).

D. Effect of Fatty Soil on Retention of Particulate Soil

Particulate soil in household laundry is often coated with a layer of fatty soil originating from the sebum in perspiration. This changes the surface characteristics of the particles and renders hydrophilic soils like clays and cement dust more hydrophobic. The layer of solid or liquid fatty soil intervening between particle and fiber can increase or decrease uptake, retention, and washability, depending on the polarity of fiber and particle, particle size, and the relative amounts of fatty to particulate soil.

1. Dry Soiling

Soiling of fabrics by airborne dirt is substantially more severe if the fabrics contain even small amounts of oily material. For instance, the lubricating oil of the jute backing of acetate carpets was found to increase greatly the dry soiling of these carpets. The oil migrated to the pile, where it tended to fix a large proportion of the impinging soil particles to the fibers by capillarity. Staple lubricant, carding and spinning oils in the pile also enhanced the dry soiling of the acetate carpets (161).

Another demonstration of the effect of oils is provided by comparing the dry soiling of viscose and wool yarns before and after extraction with solvents, which removed lubricating oils and/or lanolin. The presence of such oils or grease on the yarns increased the retention of vacuum cleaner dirt between 70 and 200%. On the other hand, removal of the 8% extractable fatty soil from the vacuum cleaner dirt by ether extraction had practically no effect on its retention by extracted nylon and acetate yarns (122). In the case of the yarns, the oil was probably located on the surface, forming a film, while it was distributed more unevenly and in the interior of the dirt particles. Additional examples of "oil bonding" in dry soiling are given in Getchell's review (2). The large effect of "oil bonding" does not mean that airborne soil cannot adhere strongly to fabrics in the abscence of an intervening oil film (133).

2. Wet Soiling

Soiling of shirt collars and cuffs by particulate soil plus perspiration is severe because the particles are wetted by perspiration and then dispersed and rubbed into the fabric by friction with the skin. Soil redeposition during the wash and rinse cycles of the washing machine also plays a large role in wet soiling.

Fort and Billica of the Du Pont Co. made electron-microscopical studies of cotton and polyester shirt cuffs which had been soiled either by wearing or by 20 laundry cycles with a white load in a commercial laundry (162). Soil deposited on the fabric by wearing was seen to concentrate on the outer filaments of the portion of the fabric surface in contact with the skin. These filaments were completely coated with a relatively thick, continuous sheath of fatty soil which covered up the morphology of the fiber surface. Surface replicas of the outer fibers of use-soiled cotton hardly showed the characteristic fibrillar structure of the secondary wall. In fabrics soiled by the repeated laundry cycles, the oily soil was found to be more uniformly distributed over all filaments and concentrated at fiber-fiber junction points. The laundry-soil fatty film surrounding most of the filaments was thinner than the use-soil film around the outer filaments in contact with the skin. In all cases, the particulate material appeared embedded in these fatty films (162). This does not preclude tenacious adhesion of particulate soil to fibers in the absence of fatty soil (163, 164).

The effect of an oily film around hydrophilic particles on wet soiling was studied by Berch and Peper (158) who soiled cotton fabrics treated with different finishing agents by means of aqueous suspensions of an acicular iron oxide pigment in the presence and absence of oleic acid (Ref. 158; compare also Sec. VI,C,2). Rendering the surface of the particles hydrophobic by adsorption of oleic acid increased γ_{PW} in Eqs. (6.18) and (6.19). The free energy of the interface of the particles with a hydrophilic fiber such as untreated cotton (of low γ_{FW}) was also increased, so that the tendency toward deposition, represented by the extent to which γ_{PW} plus γ_{FW} exceeds γ_{PF}, became less on balance. In the case of a hydrophobic fabric (of high γ_{FW}) such as cotton finished with silicones, fluorochemicals, or polyacrylates, or polyester fabric, the increased hydrophobicity of the soil rendered γ_{PF} smaller in addition to increasing γ_{PW}, so that the soiling tendency was enhanced (158, 159).

VII. INTERACTIONS IN THE SYSTEM: FIBER-PARTICULATE SOIL-SURFACTANT

Up to this point, only binary systems (not counting water) have been considered, with the topics indicated by Eqs. (6.2), (6.3), and (6.4). This section deals with the ternary system represented by Eq. (6.1), considering particulate soil first in the absence and then in the presence of fatty soil.

It is important to distinguish between the washing action of a detergent or its ability to remove soil from fabrics, and its ability to prevent redeposition of the soil onto fabrics or its suspending power (165). The term "suspending power" is misleading because it suggests dispersion and deflocculation of the particulate soil rather than preventing its uptake by the fabric in the washing machine; antiredeposition power is more appropriate.

A. Fat-Free Particulate Soil

1. Mechanisms of Soil Removal and of Redeposition Prevention by Surfactants

The attachment of particles to fabrics by sorptive forces is a case of heterocoagulation. Their removal by surfactants is a deflocculation process analogous to the deflocculation of aggregates into primary particles as discussed in Sec. IV,D. To be effective in removing soil from fibers and in preventing its redeposition, a surfactant must be adsorbed by the fibers or by the soil, but preferably by both. In soil removal, the two layers of surfactant molecules adsorbed at the two adhering surfaces form a wedge between fiber and particle and pry them apart, with the aid of hydrodynamic shear forces. The adsorption of anionic surfactants increases the magnitude of the negative ζ potential of fiber and/or of particle. This increases the electrostatic repulsion between the two to the point where it overcomes the London-van der Waals attraction and thus detaches the particle from the fiber.

The same mechanism prevails in the prevention of redeposition. This is illustrated by the experiments of Johnson and Lewis (20) of the Unilever Co. involving the system cotton-carbon. The electrokinetic potentials in the presence and absence of a built detergent are given in Table 6.2. Application of the Derjaguin Landau-Verwey-Overbeek theory to the system in a 0.4% solution of a built detergent based on sodium dodecylbenzenesulfonate gave the potential energy-distance curves of Fig. 6.12. The smallest particles (curve 1) will deposit on the cotton despite the presence of the surfactant because the potential energy barrier is only three kT units high. They cannot be removed from the fabric. This case is analogous to the irreversible coagulation of colloidal dispersions which occurs in the primary potential energy minimum. It explains the well-known accumulation of small particles on fabrics, which produces progressive graying. The largest particles (curve 3 in Fig. 6.12) are flocculated onto the fabric in the secondary minimum. They are desorbed easily, either by agitation of the washing machine or by dilution in the rinse cycle.

In the case of nonionic surfactants, the adsorbed molecules plus the associated water of hydration from a steric barrier between fiber and

TABLE 6.2

Zeta Potentials of Cotton and Carbon Black in Water and in a 0.4% Detergent Solution[a,b]

	ζ potential mV	
	Water	Detergent solution
Cotton	ca. - 24	-28
Carbon	- 16	-46

[a]Heavy-duty detergent based on sodium n-dodecylbenzenesulfonate and sodium tripolyphosphate, but containing no carboxymethylcellulose.

[b]From Johnson and Lewis (20).

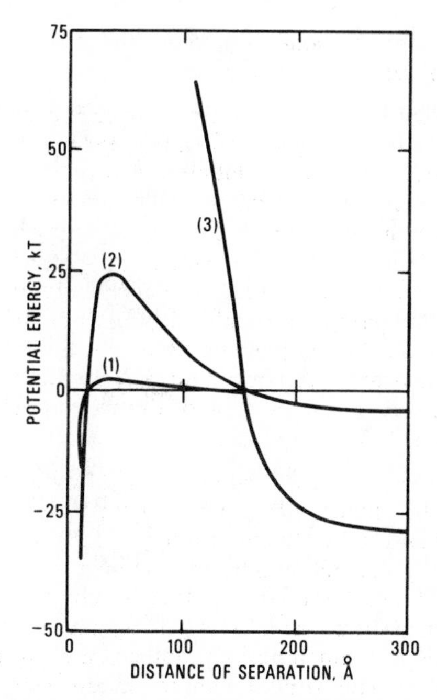

FIG. 6.12. Potential energy-distance curves for cotton-carbon in 0.4% detergent solution. Diameters of carbon particles: curve (1), 100 Å; curve (2), 1000 Å; curve (3), 1 μ. The primary potential energy minima and the increases in potential due to Born repulsion are not shown. Reprinted from Ref. 20, p. 286, by courtesy of the Society of Chemical Industry.

particle surfaces which promotes desorption of the particles from the fibers and prevents their redeposition.

In terms of surface energetics, surfactant adsorption usually reduces γ_{PW} and/or γ_{FW}. For the detergency process,

$$\Delta G = \gamma_{PW} + \gamma_{FW} - \gamma_{PF} \tag{6.20}$$

which is the reverse of the soiling process represented by Eq. (6.18). When $\gamma_{PF} > \gamma_{PW} + \gamma_{FW}$, $\Delta G < 0$, and the replacement of the particle-fiber interface by equal areas of particle-water and fiber-water interfaces reduces the surface free energy of the system, so that the separation of the particles from the fibers is the thermodynamically favored process.

Once the particles are detached from the fabric, redeposition is less likely to occur because of the presence of the adsorbed surfactant molecules at the fiber-water and/or particle-water interfaces. However, particles detached from hydrophilic fibers, of low γ_{FW}, often redeposit onto hydrophobic fibers whose γ_{FW} is so high that Eq. (6.19) applies even in the presence of adsorbed surfactant.

Particles attached to fabrics mainly through mechanical entrapment are dislodged by hydrodynamic shear of water flowing through or past fabrics, by twisting and stretching of the fabrics, and by fabric-fabric friction resulting from the agitation of the washing machine. Swelling of cellulosic and to a lesser extent of other fibers in water may aid this removal process in the case of microocclusion by evening out surface rugosities, or may render it more difficult in the case of macroocclusion by packing fibers more tightly around particles caught in interfiber spaces. The major role of the surfactant in this process is to wet the fabric. If the trapped particle is an aggregate, the surfactant and/or the builder may deflocculate it, which would result in the release of the much smaller primary particles from the fabric.

Evidence that mechanical entrapment played a dominant role in the removal of a carbon black with a 0.2 - 10 μ particle size range from fabrics was discovered by Grindstaff et al. (166) of the Du Pont Co. The ease of removal by surfactant solutions increased in the order: cotton fabric < polyester staple fabric < polyester continuous filament fabric (or nylon continuous filament fabric). The same pigment attached to films of these polymers was much more readily removed from cellophane than from nylon and polyester films. Fabrics made from continuous filaments are smoother than fabrics made from staple, and the cotton fibers have rougher surfaces than the smooth polyester fibers. Increasing fabric and fiber roughness evidently increased the mechanical entrapment and retention of the pigment. Furthermore, a nonionic and an anionic built detergent removed about equal amounts of the carbon black from a given fabric, while considerable surfactant specificity was observed in the removal of the pigment from the respective films (166). Therefore, while the pigment was attached to films through sorptive forces, its attachment to fabrics resulted primarily from mechanical entrapment.

2. Quantitative Relationships

Strauss (167) demonstrated that the interaction between carbon black and cotton fabric in the presence of sodium dodecyl sulfate is an equilibrium phenomenon and reversible (see Fig. 6. 13): At each surfactant concentration, the same limiting pigment concentration on the fabric was reached, after several hours, in adsorption and desorption experiments. The adsorption of carbon black and rutile (TiO_2) pigments on cotton followed approximately a Langmuir isotherm. The pigments were partitioned between cloth and bath (151, 167). As is seen in Fig. 6.14) increasing pigment concentrations in the bath resulted in higher pigment levels on the fabric until the uptake leveled off (167, 168).

Wagner (Waeschereiforschung Krefeld) treated cotton fabric either simultaneously (from the same bath) or sequentially with one of the

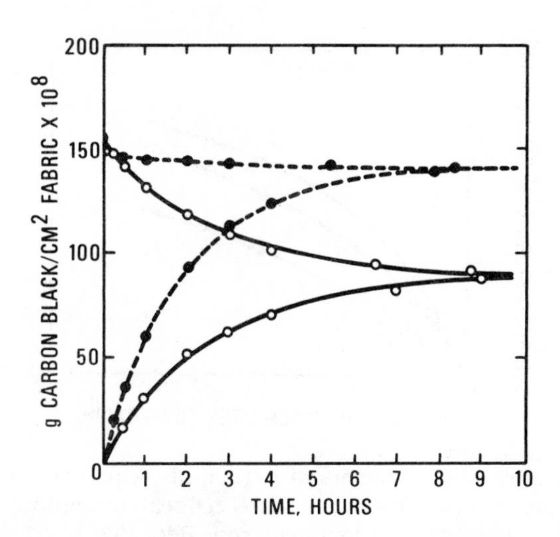

FIG. 6.13. Adsorption and desorption of carbon black on cotton: 0.1 g/liter carbon black, 95°C, Launder-Ometer. ○, 2.4 g/liter sodium dodecyl sulfate; ●, 0.8 g/liter sodium dodecyl sulfate + 1.26 g/liter NaCl. Reprinted from Ref. 167, p. 31 by courtesy of Dr. Dietrich Steinkopff Verlag.

following pairs of pigments (169),: carbon black and kaolinite, carbon black and Fe_2O_3, or Cr_2O_3 and Fe_2O_3, in aqueous suspensions containing sodium dodecyl sulfate. In every case, the uptake of one of the two pigments reduced the uptake of the second, indicating that the two pigments competed for the available fabric surface. In most desorption experiments, the retention of one pigment was reduced by the coadsorbed second pigment. The exception was carbon black which had been simultaneously adsorbed with kaolin or with Fe_2O_3. The latter two pigments were desorbed to a lesser extent in the presence of carbon black than if each had been adsorbed and desorbed by itself, while the carbon black was desorbed more extensively in their presence. Wagner (169) ascribed this behavior to the particle shape. The platelike kaolinite and the cubic Fe_2O_3, because of their flat surfaces, had greater areas of contact with the fiber surface than the spherical carbon black, and hence adhered more strongly to it (169).

The adsorption of pigments by fabrics at constant pigment concentration in the bath generally decreased with increasing surfactant concentration, up to a point (see Fig. 6.14 and Ref. 153), and leveled off as the surfactant concentration increased further (151). The relation was sometimes complex (62, 137, 167, 168). By the same token, the suspending or antiredeposition power of surfactants generally increased initially with increasing surfactant concentration and leveled off in the vicinity of the cmc (165, 170).

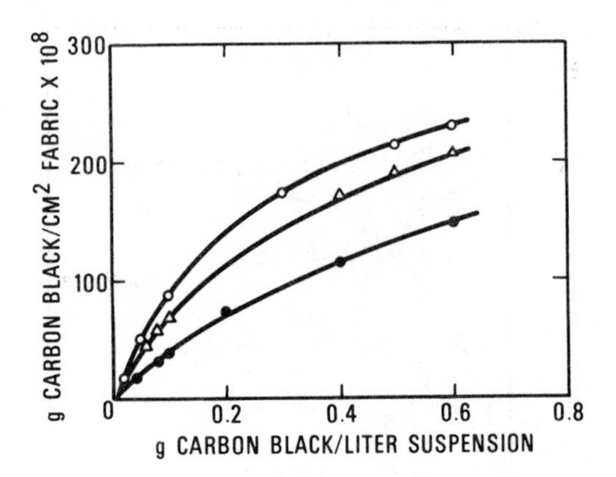

FIG. 6.14. Adsorption isotherms of carbon black on cotton: 95°C, Launder-Ometer. Sodium dodecyl sulfate concentrations: ○, 1 g/liter; Δ, 2 g/liter; ●, 2.5 g/liter. Reprinted from Ref. 167, p. 32, by courtesy of Dr. Dietrich Steinkopff Verlag.

At still higher concentrations, the antiredeposition activity often decreased (171). This effect was less pronounced at lower levels of agitation (137, 172). The initial decrease in soil adsorption or increase in suspending activity with increasing surfactant concentration is due to increasing adsorption of the surfactant at fiber and particle surfaces. This increases the magnitude of the two ζ potentials in the case of anionic surfactants and the extension of the steric barrier in the case of nonionic surfactants. Surfactant adsorption at the solid-water interfaces levels off as the layer of adsorbed surfactant molecules becomes increasingly more densely packed, causing the antiredeposition activity to level off as well. The decrease in antiredeposition activity at still higher concentrations of anionic surfactants results from the fact that the surfactant ions in solution, like ordinary electrolytes, reduce the magnitude of the ζ potentials and bring the dispersed dirt particles closer to flocculation, i.e., to deposition on the fabric. The effect of surfactant concentration on the removal or desorption of pigments from soiled fabrics, i.e., on the washing process, is similar to the effect on the redeposition process: With increasing surfactant concentration, the soil removal increases initially and then levels off.

As predicted by the Derjaguin-Landau-Verwey-Overbeek theory (see Sec. IV,B), the suspending power of anionic surfactants for particles dispersed in the bath, as well as the ability to remove adsorbed particles from fabrics, was impaired by the presence of salts. The addition of non-surface-active electrolytes increased the deposition on cotton of carbon blacks, of Fe_2O_3, and of CuO particles suspended by anionic surfactants (137, 172, 173). This effect was not due to the precipitation of alkane carboxylates or alkyl sulfates of polyvalent cations since even sodium and potassium salts caused increased deposition. It resulted from compression of the electric double layers surrounding particles and fiber surfaces by the salts and from reduction of the magnitude of the ζ potentials. This lowered the height of the potential energy barrier and led to heterocoagulation between particles and fibers in the primary potential energy minimum (137). The redeposition effectiveness of added electrolytes in the presence of anionic surfactants increased markedly with the valence of their cations. This is in agreement with the Schulze-Hardy rule and the Derjaguin-Landau-Verwey-Overbeek theory, both of which predict that the minimum electrolyte concentration required to coagulate negative dispersions is proportional to the inverse sixth power of the valence of the cation (172).

The soil removal and antiredeposition activities of nonionic surfactants are much less sensitive to added electrolytes than those of anionic surfactants, because their stabilizing action is based on surrounding particles and fiber surfaces with a steric barrier including a hydration layer rather than on increasing the ζ potentials. Electrolyte concentrations required to reduce hydration layers appreciably are an order of magnitude higher than those required to compress electric double layers (see Sec. IV,D,1 and 2).

3. Adhesion Experiments

The technique of von Buzagh (18) was applied to carborundum and calcium carbonate particles of about 20 μ diameter immersed in solutions of anionic and cationic surfactants. The suspensions were placed in quartz cells; the particles settled on the bottom and were counted. The percentage of the particles which remained attached to that quartz surface after the cell was inverted, called the adhesion number, served as an estimate of the particle-quartz adhesion. When anionic surfactants were added in increasing concentrations, the adhesion numbers remained low until the cmc was reached, and then increased sharply. When the cell and the particles were coated with $C_{32}H_{66}$, the adhesion numbers went through a minimum at intermediate surfactant concentrations (174).

The use of high-speed centrifugation to measure adhesion is not new (175). Krupp and associates (142) at the Battelle Institute, Frankfurt, recently fitted a commercial ultracentrifuge with cells capable of holding liquids, and measured the adhesion of small gold spheres to quartz and to cellophane, polyamide, and polyester films in air, in water, and in surfactant solutions (142). Distribution curves for adhesional forces were obtained by measuring the percentage of the particles remaining attached at various speeds. Particles brought into contact with the dry substrate adhered more strongly on subsequent immersion of the system in aqueous solutions than particles immersed in the solutions to make contact with the wet substrate. This is to be expected because immersion prior to particle-substrate contact results in an intervening film of water and in the adsorption of surfactant at particle and/or substrate surfaces. Typical curves, shown in Fig. 6.15, indicate how the adhesion was lowered by water, and even more by surfactant solutions (142, 176). It is difficult to compare different substrates by this technique, because the magnitude of the particle-substrate attraction depends, among other factors, on the deformability and hardness of the substrate, which determine the extent to which the particle indents the substrate and the resultant area of contact.

Visser (177) recently developed another apparatus for measuring particle-substrate adhesion in liquid media, which consists of two concentric cylinders. The substrate to be studied, in this case a cellophane film, was wrapped around the inner cylinder, and a suspension of carbon black in water was poured into the gap between the inner and outer cylinders. The number of particles per square centimeter which became attached to the inner cylinder after a period of slow rotation was counted microscopically. The carbon suspension was then replaced by a surfactant solution, and the number of particles per square centimeter which remained adhering to the inner cylinder after periods of rotation at various high speeds was again counted. There was good agreement between the tangential hydrodynamic force required to remove a given percentage of the particles measured by this technique and the perpendicular force required to remove the same percentage of particles of the same carbon black from cellophane immersed in the same solution in Krupp's ultracentrifuge (177). The

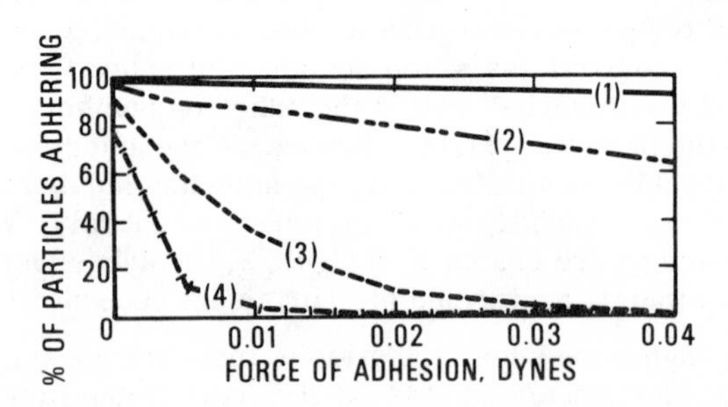

FIG. 6.15. Adhesion of gold spheres to cellophane. Particle-substrate contact prior to immersion. Medium: (1) air; (2) water; (3) 0.5 g/liter dodecanol$(EO)_8$; (4) 4 g/liter sodium dodecyl sulfate. Reprinted from Ref. 176, p. 443, by courtesy of Gordon & Breach Science Publishers, Inc.

advantage of Visser's apparatus over the ultracentrifuge is that it can be used to measure the adhesion of particles which are in the submicron range and have low densities.

4. The Effect of Particle, Fiber, and Surfactant Characteristics on Soil Removal

The affinity of the surfactant for fiber and soil has a profound effect on detergency. This is illustrated by the removal of sodium bentonite from cotton. Cotton and sodium bentonite are hydrophilic, negatively charged, and do not adsorb anionic surfactants below the cmc. Hence sodium dodecylbenzenesulfonate and sodium dodecyl sulfate removed no clay from cotton (see Table 6.3). Polyoxyethylated nonionic surfactants, which are adsorbed by both substrates, removed most of the clay. Chemisorption of cationic surfactants neutralizes the negative charges of cotton and bentonite and renders their surfaces hydrophobic (see Sec. IV,C,2). This reinforces the attachment between clay and cotton and the cohesion of the clay lamellas. Therefore, cationic surfactants removed no clay and also prevented the clay removal by nonionic surfactants. Anionic surfactants did not affect the latter process (30).

It is interesting to note in Table 6.3 that the nonionic surfactant left 0.04% clay on the fabric. This is equal to the amount of bentonite required to completely cover the cotton surface with a sheet of single lattice layers 9.4 Å thick. The initial amount of sodium bentonite, 0.13%, corresponds to a coverage of the cotton surface by stacks of three lattice layers. This results from the fact that the primary particles of aqueous sodium bentonite dispersions are stacks averaging three-four lattice layers (143). The nonionic surfactant molecules penetrated between adjacent lattice layers of the adsorbed clay particles (see Sec. IV,C,3) and pried them apart, removing all but the bottom layer from the fiber. The surfactant molecules could not penetrate between the bottom clay layer and the cellulose surface. This is an example of removal of particulate soil by deflocculation, i.e., in a piecemeal fashion (see Sec. IV,D,4).

When sodium bentonite and cotton were brought into contact in the presence of the nonionic surfactant nonylphenol $(EO)_{30}$, the fabric retained no clay at all, because both cellulose and clay particles were covered by a layer of adsorbed surfactant molecules (178). Anionic surfactants did not affect the retention of bentonite by cotton: They had no suspending or antiredeposition power.

Grindstaff et al. (166) mounted films of different polymers drumhead fashion over the end of a Geiger tube, and soiled the outside of the films with radiotagged carbon black. The end of the Geiger tube with the attached film was then immersed in stirred surfactant solutions, and the

TABLE 6.3

Removal of Sodium Bentonite from Cotton by Surfactants[a]

Surfactant	Concentration, %	pH	Removal,[b] %	Residual clay, %
Sodium dodecylbenzenesulfonate	0.30	6.8	0	0.120
Dicocodimethylammonium chloride	0.30	6.3	8	0.128
Nonylphenol$(EO)_{30}$[c]	0.025	5.9	62	0.037
	0.10	6.8	57	0.044
	0.75	6.9	63	0.037
	0.10[d]	5.9	53	0.047
Sodium dodecylbenzenesulfonate	0.15	6.7	53	0.043
Nonylphenol$(EO)_{30}$	0.15			
Dicocodimethylammonium chloride	0.15	6.5	0	0.119
Sodium dodecylbenzenesulfonate	0.15			

[a]From Schott (30).

[b]Initial amount 0.11% of weight of fabric; values reproducible within ±6%.

[c]Critical micelle concentration is 0.042%.

[d]At 75°C; all other experiments at 25°C.

desorption of the carbon black was monitored directly by measuring the decrease in radioactivity. The following was the order of increasing soil removal from nylon film by 0.01 M surfactant solutions: nonylphenol $(EO)_{9.5}$ = water < cetyltrimethylammonium bromide < sodium dodecyl sulfate. For polyester film, the order was: water < cetyltrimethylammonium bromide < sodium dodecyl sulfate < nonylphenol$(EO)_{9.5}$ (166). The reversal of the washing efficiency of the ionic and nonionic surfactants on polyester and polyamide substrates is probably due to a greater affinity of the ionic surfactants for the polyamide film (see Sec. V,A,1).

When the radioactive carbon black was rendered hydrophilic by surface oxidation with wet ozone, it was much more extensively removed from polyester film by water, sodium dodecyl sulfate, and the nonionic surfactant than was the original hydrophobic pigment. Cetyltrimethyl ammonium bromide removed the hydrophobic and the oxidized carbon black to about the same extent. Therefore, the order of increasing washing efficiency for the hydrophilic carbon black on polyester film was: cetyltrimethylammonium bromide < water < sodium dodecyl sulfate < nonylphenol$(EO)_{9.5}$. Evidently, the cationic surfactant was chemisorbed by the carboxylic groups of the surface-oxidized carbon black, with the hydrocarbon chains oriented outward (see Sec. IV,C,2), rendering the pigment surface hydrophobic again (166).

The wet soiling in the absence of surfactants of cotton fabrics treated with different finishes was described in Sec. VI,C,2 (158-160). When Berch et al. (179) washed these fabrics with a built anionic detergent, they found a consistent relation between soil deposition from water and washability. The fabrics finished with hydrophobic chemicals (silicones, acrylic resins, fluorochemicals), which suffered the heaviest uptake of particulate soil, also had the greatest resistance to soil removal. The only finish which substantially improved the removal of carbon black and of vacuum cleaner dirt from cotton was carboxymethylcellulose. This was also the only one of the finishes tested for soil removal which had promoted resistance to wet soiling. The hydrophilic iron oxide pigment was removed to the same extent from untreated and carboxymethylcellulose-treated cotton (179).

5. Effect of Surfactant and Fiber Characteristics on Soil Redeposition

As Strauss (167) has shown, the soiling of fabrics by particulate soil in the presence of surfactants is reversible. Therefore, the characteristics of fiber, soil, and surfactant should influence soil removal and soil redeposition in an analogous but opposite fashion. Increasing the hydrophobicity of the fabric was seen in Sec. VII,A,1 and 4 to decrease the extent of soil removal by surfactant solutions. Therefore, it should also increase the uptake or redeposition of soil in the presence of the surfactants. Indeed,

Oldenroth (Waeschereiforschung Krefeld) found that the natural waxes of cotton fibers (180) contained in the primary wall, which tend to render its surface hydrophobic, promoted the uptake of carbon black suspended in an anionic detergent. As increasing amounts of wax were removed from the fiber surface by solvent extraction, the deposition of carbon black was progressively reduced (180).

Another example of increased soil deposition resulting from increased hydrophobicity of the fabric is the adsorption of carbon black on cotton treated with different finishes. Adsorption experiments at the Southern Regional Research Laboratory were conducted from water and from solutions of soap, of sodium N-methyl-N-oleoyl taurate, and of a detergent based on sodium dodecylbenzenesulfonate. For each soiling medium, the uptake increased in the order: untreated cotton < cotton treated with acrylic resin < cotton treated with silicone resin, which is the order of increasing γ_{FW} (153).

According to the reversibility of soiling, the washing efficiency of surfactants should parallel their suspending or antiredeposition power. Wagner (181) examined the adsorption and desorption of carbon black on cotton in the presence of sodium n-alkyl sulfates containing 12, 14, 16, and 18 carbon atoms. Working at constant concentrations of pigment and surfactants, he found that increasing chain length of the alkyl group resulted in lower pigment deposition and in higher removal. Sodium octadecyl sulfate had the highest washing efficiency and the highest antiredeposition power, sodium dodecyl sulfate the lowest (181). Likewise, Strauss (167) found that the equilibrium concentration of carbon black on cotton fabrics was highest in the presence of sodium octyl sulfate, medium for dodecyl sulfate, and lowest for hexadecyl sulfate. This held true for the entire range of surfactant concentrations examined (0 - 3 g/liter). Naturally, the three curves representing the amount of carbon adsorbed on the fabric versus surfactant concentration in the bath converged for zero surfactant concentration (167).

Polyoxyethylated nonionic surfactants were found to prevent the uptake of a carbon black by cotton. Deposition of this black was observed in the presence of all anionic surfactants tested including sodium dibutylnaphthalenesulfonate, which produced the finest and most deflocculated suspensions of the pigment in water (182). Evidently, only the nonionic surfactants were adsorbed extensively by the fabric as well as by the carbon black, whereas the anionic surfactants were adsorbed mainly by the carbon black and thus could not prevent its deposition on the cellulose (see Sec. V,A,2).

The washing and antiredeposition powers of polyoxyethylated nonionic surfactants for various carbon blacks on cotton were found to be superior to those of anionic surfactants. The effectiveness of different anionic surfactants depended on the specific properties of the carbon black, but in general soaps were superior to alkyl sulfates and alkylbenzenesulfonates (114, 167, 170, 171). In other tests, nonylphenol$(EO)_7$ washed more carbon black from cotton fabric than sodium dodecyl sulfate (183). The statement that polyoxyethylated nonionic surfactants have superior antiredeposition power for carbon black on cotton but average to poor washing power (171) is incompatible with the reversibility of the pigment adsorption-desorption process. The antiredeposition power of surfactants for ilmenite ($FeO.TiO_2$) on cotton was found to be greatest for sodium oleate, intermediate for alkyl sulfates and an alkylarylsulfonate, and lowest for a polyoxyethylated nonionic surfactant (165, 184).

6. Effect of Temperature on Soil Removal and Redeposition

Higher temperatures favor increased removal and redeposition of particulate soil by increasing the kinetic energy kT of the particles and their Brownian motion. Since the temperature is expressed in degrees Kelvin, the increase in kinetic energy when going from 25°C (298°K) to 95°C (368°K) is only 100(368 - 298)/298 = 23%. The 66% reduction in the viscosity of water over this temperature range should further intensify the Brownian motion somewhat, but the overall increase remains small. In view of this small primary effect of temperature on detergency, secondary effects or artifacts due to temperature increases probably play dominant roles in altering the interaction among soil, fabric, and surfactant.

If the fabric or its finish tends to soften at higher wash temperatures, like thermoplastic fibers or acrylic resins on cotton, soil uptake increases and soil removability decreases with increasing temperature (see Sec. VI,C,2 and Refs. 158, 160). If fatty soil is present, exceeding its melt temperature greatly enhances the removal of particulate soil.

Another secondary effect is the influence of temperature on the solubility of the surfactant. In the case of soaps and other anionic surfactants, the solubility increases very suddenly with rising temperature at the Krafft point, and so does the washing power. Nonionic surfactants are less hydrated at higher temperatures and hence more surface active and more prone to adsorption on soil and fiber. This enchances their washing and antiredeposition power. Once the cloud point is exceeded, however, their solubility in water is very small, which may decrease their cleaning power. This is illustrated by the removal of carbon black from polyester film by nonylphenol$(EO)_{9.5}$ (166). The washing efficiency was greatest at 60°, which about coincides with the cloud point (185), smallest at 20°, and intermediate at 80°. The washing efficiency of sodium dodecyl sulfate in the same system increased monotonically from 20 to 60 to 80° (166). This occurred despite the fact that the adsorption of anionic surfactants at solid-water interfaces is generally exothermic and hence more extensive at lower temperatures (see Sec. IV,C,5). For another nonionic surfactant, dodecanol$(EO)_{13}$, the removal of carbon black from cotton fabric was found to increase monotonically from 40 to 60 to 90° (167) because its cloud point is above 100° (185). However, the removal of this carbon black by sodium dodecyl sulfate was most extensive at 40°, least at 60°, and intermediate at 90° (167). The suspending power of a soap, of sodium N-methyl-N-oleoyl taurate, and of sodium alkylarylsulfonate in the system carbon black-chopped cotton was essentially the same at 20° and at 60° (170). In the removal of bentonite from cotton by nonylphenol$(EO)_{30}$, a temperature increase of 50° had no significant effect (see Table 6.3). The cloud point of this surfactant is well above 100° (185), and its affinity for the clay was apparently quite high at all temperatures.

Wagner (151) found a greater equilibrium concentration of carbon black on cotton in pure water at 60° than at 90° at the various concentrations of carbon black in suspension. He ascribed the lower retention at the higher temperature to the greater kinetic energy of the adsorbed carbon particles (151). Where mechanical entrapment is the major factor in the retention of particulate soil by fabrics, temperature changes had no noticeable effect on soil removal (150).

B. Particulate Plus Fatty Soil

The effect of fatty soil on the removal of particulate soil depends on many factors, including the relative amounts of the two and the order of their application to the fabric. These plus differences in experimental conditions account for the diversity of conclusions.

In a photomicrographic study, Goebell and Kling (186) applied a conglomerate of pigment and oil to fabric surfaces and observed its removal upon immersion in stirred surfactant solutions. They noticed a phenomenon analogous to the rolling-up mechanism discussed in Chap. 5. The oil collected in pigment-free droplets which floated away. The solid particles which came off the fiber surface appeared largely free of oil (186).

Surfactant solutions displace oily soil from the surface of particulate soil as well as from fiber surfaces. If the oily soil had cemented solid particles to fibers, surfactant molecules become adsorbed at the surfaces of fibers and particles as the oily matter is being displaced from these solid surfaces by surfactant solutions. These layers of adsorbed surfactant molecules make redeposition of the solid particles more difficult. This is contingent on positive adsorption of surfactants on fibers and on particulate soil, and on the proper orientation of the adsorbed molecules, namely, with the hydrophilic moiety toward the aqueous phase. Two other possibilities exist. Anionic surfactants which are not adsorbed by cellulose nor by clays will not affect redeposition. If adsorption by cellulose fibers and by clays does occur, however, it is mainly through calcium bridges or by precipitation of the calcium carboxylates, alkyl sulfates, or alkylarylsulfonates, so that the hydrocarbon moieties are oriented outward (see Secs. IV,C,2 and V,A,2). This will render the solid surfaces less hydrophilic and thus promote redeposition. Calcium sequestrants prevent this adverse effect. If redeposition of particulate soil is an important factor in soiling after the oily soil which cemented it to fibers has been removed by the rolling-up mechanism from fibers and particles, the overall washing efficiency for particulate soil should either be independent of the presence of oily soil or be improved by it, provided that the detergent is properly built or that the water contians no hardness. This applies especially to cellulosic fibers because they are washed relatively free of oily soil.

Utermohlen et al., (187) using cotton and anionic surfactants, concluded that the removal of particulate soil was independent of that of oily soil. They applied lampblack or magnetic ferric oxide to the fabric together with a mixture of equal parts of mineral oil and hydrogenated cotton seed oil. The weight ratio of oily soil to particulate soil was kept constant at 9:1 for the lampblack and 4:1 for the ferric oxide, but progressive amounts of the water-insoluble oily mixture were replaced by the water-soluble butyl cellosolve, so that the ratios of oily soil to particulate soil on the fabric went as low as 1:1.6 and 1:4, and finally down to zero. The mixtures of particulate and oily soil were applied to the fabric from carbon tetrachloride, from a mixture of equal volumes of butanol and dioxane, or from water. The soiled fabric was then washed with sodium dodecyl sulfate or with soap. In no instance was the removal of particulate soil significantly affected by the presence of water-insoluble oily soil. Therefore, removal of the two types of soil were separate and unrelated phenomena (187), and redeposition played a dominant role in the attachment of particulate soil to the fabric during the washing process.

It must be stressed that the soiling levels used in many of the laboratory experiments described in this chapter were extremely high. Some of the artificial soil applied so plentifully to test fabrics is of necessity only loosely attached and is washed off the fabrics readily. This produces relatively concentrated soil suspensions in the wash liquors. Thus, many laboratory experiments place a major emphasis on the soil redeposition process, probably much more so than actual home or commerical laundering.

Redeposition was found to be important for a clay as well as for carbon soil. Gordon and Bastin (116) applied kaolin rendered radiocative by neutron irradiation to different fabrics from benzene containing dissolved, radiotagged artificial sebum. This was followed by washing with a built polyoxyethylated nonionic surfactant. Nylon retained considerably more kaolinite than cotton or polyester fabrics, and also showed the greatest redeposition. A major proportion of the residual clay on all three washed fabrics was present because of redeposition rather than because of nonremoval.

Hart and Compton (188) found that redeposition of oleophilic carbon blacks onto cellulose occurred not only from water but also from such hydrocarbons as benzene and Stoddard solvent, which solvate the surface of the carbon. They added aqueous dispersions of carbon blacks to slurried-up chopped cotton fiber which had been pretreated with mineral oil, petroleum jelly, or hydrogenated vegetable oil. The suspended fibers picked up considerable amounts of pigment. They were then filtered, dried, and slurried with different solvents which removed the fatty coat from the fibers. However, a high proportion of the pigment which had been attached to the greasy fibers remained attached after removal of the grease (188).

In their experiments with radioactive carbon black and cotton fabric, Hensley et al. (114) studied the effect of wetting the black with mineral oil prior to rubbing it into the fabric. Washing experiments conducted with the dry and with the oil-treated carbon black arranged a series of pure and built surfactants in the same order of effectiveness. The differences in the absolute removal of the oily and of the dry soil by a given surfactant may have been due to greater fabric abrasion during application of the dry soil rather than to any intrinsic differences in removability (114).

In some of the experiments on the removal of radiotagged carbon black from films stretched over the end of a Geiger tube as described in Sec. VII,A,1, Grindstaff et al. (166) deposited a sebumlike fatty mixture on the films before applying the radiotagged carbon black from a carbon tetrachloride suspension. In washing experiments with cellophane, nylon, and polyester films using either sodium dodecyl sulfate or nonylphenol$(EO)_{9.5}$, carbon black deposited on a film presoiled with fat was removed much more extensively than carbon black deposited on the clean film (166). The fat seemed to function as a weak boundary layer (189) between the particle and the substrate which was responsible for the poor adhesion.

Other experiments showed a parallel relation between the removability of oily soil and of particulate soil. Kashiwagi and Tamura (183) applied a soiling mixture of carbon black with or without dodecane or dodecanol or lauric acid to cotton fabric from carbon tetrachloride, and washed at 30° with sodium dodecyl sulfate, with nonylphenol$(EO)_{7.5}$, or with a commercial detergent. The removal of carbon black by the two pure surfactants increased in the order: lauric acid present < dodecanol present < dodecane present < no added fatty soil. Lauric acid was the best cement because it remained largely solid throughout the washing process, its melting point of 44° being above the wash temperature. The built detergent removed carbon black best in the presence of lauric acid (even better than in the absence of fatty soil), evidently owing to the formation of laurate soap under the influence of the alkaline builders (183).

Bowers and Chantrey (5) studied the uptake and removal of airborne particulate soil collected from air-conditioner filters by cotton and polyester fabrics to which natural sebum had previously been applied from carbon tetrachloride solution. The amount of particulate soil remaining on the fabric after washing was found to be proportional to the amount of residual sebum. Since polyester fabrics retained considerably more sebum than cotton, they also retained more particulate soil and suffered a far greater loss of whiteness. From their finding that sebum retention was a major factor in controlling the retention of particulate soil by polyester fabric, the authors concluded that the polyester fibers were surrounded by a layer of sebum in which particulate soil, originating from aqueous suspension in the washing machine, from the body, or from the atmosphere, became embedded (5).

A nonionic surfactant was superior to both an anionic and a cationic surfactant in removing particulate soil from polyester fabric in the presence of sebum (5). Similar results are generally observed in the absence of sebum as well (see Sec. VII,A,4). The ease of removal of graphite, applied with a complex mixture simulating sebum from chloroform to a variety of chopped fibers, increased in the order: polyester (or nylon) < cotton < cellulose acetate < viscose. The differences in graphite removal from the cellulosic fibers are probably due to differences in smoothness and in specific area of their surfaces. Soap and a polyoxyethylated nonionic surfactant were superior to an alkylarysulfonate and a secondary alkyl sulfate (190).

VIII. KINETICS AND ENERGETICS OF REMOVAL OF PARTICULATE SOIL

Wash cycles of commercial and home washing machines last only about 15 minutes. This places considerable importance on the kinetics of soil removal and redeposition, yet the information available is scant and often contradictory. The kind, level, and duration of agitation, the temperature, and the type of soil, fabric, and surfactant all affect the results. Soil redeposition is the reverse process of soil removal. At the beginning of a washing experiment, redeposition is negligible. When soil removal builds up the soil concentration in the bath, redeposition becomes important, causing an apparent reduction in the rate of the washing process. Eventually, an equilibrium might be reached. The relation between removal and redeposition depends therefore on the level of soiling, on the bath-to-cloth ratio, and on the length of time of the experiment. This makes quantitative comparisons of results derived from different experimental procedures nearly impossible. Extrapolation of kinetic data obtained with bench-scale equipment to full-size washing machines is extremely risky.

Several researchers found that the removal of particulate soil from cotton fabric followed first-order kinetics. Mathematically, if c_0 represents the total concentration of soil that can be removed from the fabric at the selected experimental conditions, and c_t the amount of soil removed in time t, $c_0 - c_t$ represents the residual removable soil content of the fabric at time t, and

$$-\frac{d(c_0 - c_t)}{dt} = \frac{dc_t}{dt} = k(c_0 - c_t) \qquad (6.21)$$

The c values refer to soil concentration on the fabric expressed in units such as grams of soil per square meter of fabric. Integrating,

$$\log\frac{c_0}{c_0 - c_t} = -\log\frac{c_0 - c_t}{c_0} = \frac{kt}{2.303} \qquad (6.22)$$

Therefore, a straight line is obtained when the logarithm of the residual soil on the fabric is plotted against time. Knowing the values of c_t at a given time t and of c_0, or two c_t values or two values of the residual soil concentration on the fabric at two different times, one can calculate the specific rate constant for soil removal k.

The fourth and sixth columns in Table 6.4 have as their basis the optimum conditions for soil removal; i.e., those conditions of surfactant type and concentration, temperature, and pH producing the fastest washing. This basis for comparison was used whenever the initial soiling levels were comparable. The exception is the level of agitation: The data used to compile columns No. 4-6 refer to the actual stirring speeds rather than to the highest speeds applied. High values for the 5-minute removal percentage in Tables 6.4-6.6 correspond to low values for the 20:5 ratio, indicating that a substantial proportion of the removable soil is removed within the first 5 minutes.

$\mathrm{Time}_{0.9}$ refers to the time required to remove 90% of the removable soil, i.e., for c_t to reach $0.9c_0$. Therefore, the specific rate constant, in min^{-1}, can be calculated by

$$k = 2.303/\mathrm{time}_{0.9} \qquad (6.23)$$

Any change in washing conditions results in a different value for k.

Experiments which followed kinetics other than first order are listed in Table 6.5. A linear relationship between soil removal, calculated according to Eq. (6.16), and the logarithm of the washing time was observed with three commercial soiled test fabrics and a Terg-O-Tometer (195). The classic paper in detergency kinetics is that of Bacon and Smith (118), who studied the percentage of soil removal S as a function of time t, surfactant concentration C, and applied force F. The latter was controlled by the number and size of the steel balls used in the Launder-Ometer and the speed of rotation. The relation found with four different surfactants in the region where increased C produced increased S values was

$$S = A(CFt)^n \qquad (6.24)$$

TABLE 6.4

Washing Experiments with Artificially Soiled Cotton Fabrics Obeying First-Order Kinetics

Soil	Equipment	Detergent	5-minute Removal, %[a]	Ratio (20/5)[b]	$Time_{0.9}$,[c] min.	Ref.
Carbon black + lubricating oil + shortening	Launder-Ometer	Soap + $Na_2CO_3/NaHCO_3$	17	1.32	64	191
Lampblack + mineral oil + shortening	Launder-Ometer	Nonionic	58	1.33	36	192
		Soap + Na_2SiO_3	58	1.49	24	"
Magnetic iron oxide + mineral oil + shortening	Launder-Ometer	Soap	52	1.76	13	193
India ink + mineral oil	Rotating disk	Na n-alkyl sulfate + Na_2CO_3 + $(NaPO_3)_x$	100	—	2	194
Kaolinite	Terg-O-Tometer	None	—	—	450	143

[a]Amount of soil removed during first 5-minute wash period, as percent of removable soil.

[b]Lowest ratio of soil removal after initial 20 minutes to soil removal after initial 5 minutes.

[c]Shortest time required to remove 90% of removable soil.

TABLE 6.5

Desorption and Adsorption Experiments of Artificial Particulate Soil on Cotton Fabrics Following Other than First-Order Kinetics

Soil	Equipment	Detergent	5-minute removal, %[a]	$Ratio_{(20/5)}$[b]	$Time_{0.9}$,[c] min.	Ref.
Carbon black + mineral oil + vegetable oil	Terg-O-Tometer	Built sodium alkylarylsulfonate	40	1.52	145	195
Lampblack + starch + mineral oil + shortening	Launder-Ometer	Built anionic	36	1.73	92	118
Carbon black	Launder-Ometer	None, 60°	22	2.5	66	181
		Sodium dodecyl sulfate[d]	12	3.25	98	181
		Sodium octadecyl sulfate[d]	22	2.7	54	181
			5-minute uptake, %[e]	$Ratio_{(20/5)}$[f]	$Time_{0.9u}$,[g] min.	
Carbon black	Launder-Ometer	None, 60°	12	4.1	81	181
		Sodium dodecyl sulfate[d]	6	4.0	107	181
		Sodium octadecyl sulfate[d]	11	3.0	120	181
Carbon black	Chopped fiber	Sodium alkylnaphthalene-sulfonate	62	1.28	28	148, 150

[a] Amount of soil removal during first 5-minute wash period, as percent of removable soil.
[b] Lowest ratio of soil removal after initial 20 minutes to soil removal after initial 5 minutes.
[c] Shortest time required to remove 90% of removable soil.
[d] At 60° and 0.5 g/liter surfactant.
[e] Amount of soil adsorbed during first 5 minutes, as percent of maximum uptake.
[f] Ratio of soil adsorption after initial 20 minutes to soil adsorption after initial 5 minutes.
[g] Shortest time required to adsorb an amount of soil equal to 90% of maximum uptake.

The exponent n was always positive and smaller than unity. The values of the two constants A and n depended on the nature of surfactant, soil, and fabric, and on the equipment used. Bacon and Smith thus showed that the effects on soil removal of force, time, and surfactant concentration are interchangeable, and that the product of these three variables is constant as long as all other factors remain unchanged (118).

Loeb and Shuck estimated the mechanical energy input during the washing process calorimetrically, by combining the stirring mechanism of a Terg-O-tometer with a two-liter Dewar flask. Since the mechanical energy is transformed into heat, it can be determined from the temperature increase of the thermally insulated bath. The fabric, cotton heavily soiled with colloidal carbon black plus mineral and cottonseed oils, was washed with a built nonionic detergent. They expressed the efficiency of the utilization of the mechanical energy as the ratio of soil removal to work input. In the absence of detergent, the efficiency was very low; alternatively, there was much redeposition. The efficiency increased considerably with increasing detergent concentration up to 0.15%, then levelled off. The detergent improved the utilization of the mechanical energy (195a). Loeb and Shuck then insulated a full-size washing machine to transform it into an adiabatic calorimeter (195b).

Wagner chose to represent the kinetics of soil adsorption and desorption by a two-parameter equation involving a hyperbolic sine function (196), after he observed that first-order kinetics prevailed only during the initial stages of adsorption (168) and desorption (197). As the data of Table 6.5 indicate, desorption of carbon black was faster in water than in a 0.05% sodium dodecyl sulfate solution. The percentage of soil removal as a function of time was found to increase in the following order of sodium dodecyl sulfate concentrations: 0.03, 0.06, 0, 0.25, 1.0, 0.5, and 4.0 g/liter. The percentage of soil adsorption by the fabric, based on the amount adsorbed at infinite time, as a function of time was found to increase in the following order of sodium dodecyl sulfate concentrations: 2.0, 1.0, 0.125, 0.06, 0.25 = 0.5, 4.0, and 0 g/liter, although several of the curves representing fractional soil concentration on the fabric versus time crossed over one another (181). This surprising behavior was observed not only for the adsorption and desorption of carbon black but also with the more hydrophilic particulate soils rutile (168) and ferric oxide (181), and with another surfactant, namely, sodium dodecylbenzenesulfonate (168). The specific rate constants of adsorption and desorption for these three soils as a function of surfactant concentration went through maxima in the vicinity of the critical micelle concentrations of the two surfactants. In the case of carbon black, these maxima of the rate constants for adsorption and desorption were lower than their values in water (168, 181, 197). In other words, under the conditions of Wagner's washing experiments, water apparently removed carbon black from cotton faster than sodium dodecyl sulfate or dodecylbenzenesulfonate solutions! This may have

TABLE 6.6

Experiments with Artificially Soiled Fabrics on Rate of Washing with Unspecified Kinetics

Fabric	Soil	Equipment	Detergent	5-minute removal, %[a]	Ratio (20/5)[b]	$Time_{0.9}$,[c] min.	Ref.
Cotton	Carbon black + mineral oil + vegetable oil	Terg-O-Ometer	Built anionic	72	1.39	8	198
Cotton	Magnetic iron oxide + shortening + mineral oil	Terg-O-Tometer	Soap	83	1.16	13	126
			Sodium dodecyl sulfate	65	1.30	29	126
Cotton	Carbon black + vegetable oil	Terg-O-Tometer	Soap	83	1.16	32	199
			Built nonionic	74	1.25	36	199
			Built sodium alkylarylsulfonate	69	1.30	38	199
			Sodium dodecyl sulfate	58	1.34	>60	199
Cotton	Carbon black	Launder-Ometer	Dodecanol$(EO)_{13}$, 60°	44	1.58	52	167
			Potassium laurate, 60°	9	3.4	∞	167
			Sodium dodecyl sulfate, 60°	11	2.4	220	167
Wool	Ferric oxide + kaolinite + carbon black	Domestic, impeller washing machine	Soap	82	1.21	7	200
Wool	Carbon black + stearic & oleic acids + lard + mineral oil	Domestic, impeller washing machine	Soap	92	1.09	3	200
Cotton	Graphite + complex fatty mixture	Chopped fiber	Soap + $Na_2CO_3/NaHCO_3$	50	1.61	20	190
			Nonionic + $Na_2CO_3/NaHCO_3$	56	1.10	12	190
			Sodium alkylarylsulfonate + $Na_2CO_3/NaHCO_3$	20	2.25	52	190
			Sodium sec. alkyl sulfate + $Na_2CO_3/NaHCO_3$	25	2.9	55	190
Nylon	Graphite + complex fatty mixture	Chopped fiber	Soap + $Na_2CO_3/NaHCO_3$	88	1.08	8	190
			Nonionic + $Na_2CO_3/NaHCO_3$	85	1.13	13	190
			Sodium alkylarylsulfonate + $Na_2CO_3/NaHCO_3$	84	1.07	22	190
			Sodium sec. alkyl sulfate + $Na_2CO_3/NaHCO_3$	76	1.18	21	190
Polyester	Graphite + complex fatty mixture	Chopped fiber	Soap + $Na_2CO_3/NaHCO_3$	74	1.18	13	190
			Nonionic + $Na_2CO_3/NaHCO_3$	68	1.37	17	190
			Sodium alkylarylsulfonate + $Na_2CO_3/NaHCO_3$	56	1.57	27	190
			Sodium sec. alkyl sulfate + $Na_2CO_3/NaHCO_3$	12	2.8	60	190

[a] Amount of soil removed during first 5-minute wash period, as percent of removable soil.
[b] Lowest ratio of soil removal after initial 20 minutes to soil removal after initial 5 minutes.
[c] Shortest time required to remove 90% of removable soil.

been an artifact, resulting from the fact that the assessment of soil concentration on the fabric was based on reflectance measurements, and that deflocculation of the adsorbed carbon black on the fabric by the surfactants increased its hiding power or tinting strength. Alternatively, deflocculation of the suspended secondary particles or aggregates by the surfactants reduced their particle size (113) to such an extent that they became much more strongly attached to the fabric (see Sec. VI,B,3).

Washing experiments for which no kinetic expression was derived are listed in Table 6.6. As was mentioned in Sec. VII,A,4 and 5 nonionic surfactants and soaps were superior to alkylarylsulfonates and alkyl sulfates in removing and preventing the redeposition of particulate soil. The kinetic experiments listed in Table 6.6 show that these surfactants are ranked in the same order according to rates of soil removal (190) as they were ranked when equilibrium or near-equilibrium removal and prevention of redeposition were the criteria of effectiveness.

REFERENCES

1. H. L. Sanders and J. M. Lambert, J. Amer. Oil Chem. Soc., 27, 153 (1950).
2. N. F. Getchell, Text. Res. J., 25, 150 (1955).
3. R. Hochreiter, Fette, Seifen, Anstrichm., 68, 31 (1966).
4. W. C. Powe, Text. Res. J., 29, 879 (1959).
5. C. A. Bowers and G. Chantrey, ibid., 39, 1 (1969).
6. D. Rivin, Rubber Chem. Technol., 36, 729 (1963).
7. J. J. Kipling, Adsorption from Solutions of Non-Electrolytes, Academic Press, New York, 1965, pp. 169-173.
8. F. Z. Saleeb and J. S. Kitchener, J. Chem. Soc., 1965, 911.
9. H. Schott and I. J. Kazella, J. Amer. Oil Chem. Soc., 44, 416 (1967).
10. G. A. Parks, Chem. Rev., 65, 177 (1965).
11. O. Oldenroth, Waescherei-Tech. und -Chem.,15 (5), 1 (1962).
12. O. Oldenroth, Fette, Seifen, Anstrichm., 64, 468 (1962).
13. J. W. McBain, Colloid Science, Heath, Boston, 1950, p. 270.
14. H. Schott, J. Amer. Oil Chem. Soc., 45, 414 (1968).
15. T. G. Jones, in Surface Activity and Detergency (K. Durham, ed.), MacMillan, London, 1961, pp. 108-116.
16. B. D. Cuming and J. H. Schulman, Aust. J. Chem., 12, 413 (1959).
17. K. J. Mysels, Introduction to Colloid Chemistry, Wiley-Interscience, New York, 1959, Chap. 15.

18. J. Th. G. Overbeek, in Colloid Science (H. R. Kruyt, ed.), Elsevier, Amsterdam, 1952, Vol. 1, Chap. 8.

19. J. H. Schenkel and J. A. Kitchener, Trans. Faraday Soc., 56, 161 (1960).

20. G. A. Johnson and K. E. Lewis, J. Appl. Chem., 17, 283 (1967).

21. J. L. Moilliet, B. Collie, and W. Black, Surface Activity, 2nd Ed., Spon, London, 1961, Chap. 3.

22. Ref. 7, Chap. 11.

23. J. W. Jordan, J. Phys. Colloid Chem., 53, 294 (1949).

24. L. Ter Minassian-Saraga, J. Soc. Chim, Phys., 57, 10 (1960).

25. A. M. Sookne and M. Harris, Text. Res. J., 11, 238 (1941).

26. F. H. Sexsmith and H. J. White, J. Colloid Sci., 14, 598 (1959).

27. J. Koning and K. G. van Senden, Chem. Weekbl., 62 321 (1966).

28. K. G. van Senden and J. Koning, Fette, Seifen, Anstrichm., 70, 36 (1968).

29. E. K. Goette, J. Colloid Sci., 4, 459 (1949).

30. H. Schott, Text. Res. J., 35, 1120 (1965).

31. M. E. Ginn, T. A. Schenach, and E. Jungermann, J. Amer. Oil Chem. Soc., 42, 1084 (1965).

32. H. Schott, Kolloid-Z. und. Z. Polym., 219, 42 (1967).

33. A. L. McClellan and H. F. Harnsberger, J. Colloid Interface Sci., 23, 577 (1967).

34. J. C. Abram and M. C. Bennett, ibid., 27, 1 (1968).

35. H. Schott, ibid., 23, 46 (1967).

36. W. R. Foster and J. M. Waite, in Symposium on Chemistry in the Exploration and Production of Petroleum, Amer. Chem. Soc. Division of Petroleum Chemistry, Dallas, Texas, April 1956, pp. 39-44.

37. H. Schott, Kolloid-Z. und. Z. Polym., 199, 158 (1964).

38. H. Schott, J. Colloid Interface Sci., 24, 193 (1967).

39. J. M. Corkill, J. F. Goodman, and J. R. Tate, Trans. Faraday Soc., 62, 979 (1966).

40. K. G. Mathai and R. H. Ottewill, ibid., 62, 750 (1966).

41. A. E. Alexander and P. Johnson, Colloid Science, Oxford Univ. Press-Clarendon, London, 1949, Vol. 1, pp. 549-550.

42. J. D. Skewis and A. C. Zettlemoyer, Proc. Intern. Congr. Surface Activity, 3rd, Cologne, 1960, Vol. 2, p. 401.

43. G. R. F. Rose, A. S. Weatherburn, and C. H. Bayley, Text. Res. J., 21, 427 (1951).

44. Ref. 17, Chaps. 2 and 4.

45. W. M. Urbain and L. B. Jensen, J. Phys. Chem., 40, 821 (1936).

46. L. N. Ray and A. W. Hutchison, J. Phys. Colloid Chem., 55, 1334 (1951).

47. W. Kling and H. Lange, Kolloid-Z., 127, 19 (1952).

48. M. E. Ginn, R. M. Anderson, and J. C. Harris, J. Amer. Oil Chem. Soc., 41, 112 (1964).

49. F. Z. Saleeb and J. A. Kitchener, Proc. Int. Congr. Surface-Active Substances, 4th, Brussels, 1964, Vol. 2, p. 129.

50. G. A. Johnson and R. A. C. Bretland, J. Appl. Chem., 17, 288 (1967).

51. C. P. Kurzendoerfer and H. Lange, Fette, Seifen, Anstrichm., 71, 561 (1969).

52. B. J. Rutkowski, J. Amer. Oil Chem. Soc., 45, 266 (1968).

53. H. Lange, Kolloid-Z., 169, 124 (1960); or J. Phys. Chem., 64, 538 (1960).

54. W. L. Courchene, J. Phys. Chem., 68, 1870 (1964).

55. M. J. Vold, J. Colloid Sci., 16, 1 (1961).

56. R. H. Ottewill, in Nonionic Surfactants (M. J. Schick, ed.), Dekker, New York, 1967, Chap. 19.

57. Y. M. Glazman, Discuss. Faraday Soc., 42, 255 (1966).

58. T. M. Doscher, J. Colloid Sci., 5, 100 (1950).

59. R. Abe and H. Kuno, Kolloid-Z. und. Z. Polym., 181, 70 (1962).

60. I. Reich and R. D. Vold, J. Phys. Chem., 63, 1497 (1959).

61. S. V. Vaeck and E. Maes, Tenside, 5, 4 (1968).

62. R. D. Vold and A. K. Phansalkar, Rec. Trav. Chim., 74, 41 (1955).

63. E. K. Fischer, Colloidal Dispersions, Wiley, New York, 1950.

64. Nineteenth Annual Coating Conference, Technical Association of the Pulp and Paper Industry, Miami Beach, Florida, May 1968.

65. H. E. Rose and H. B. Lloyd, J. Soc. Chem. Ind., 65, 65 (1946).

66. P. C. Hiemenz and R. D. Vold, J. Colloid Interface Sci., 21, 479 (1966).

67. L. Greiner and R. D. Vold, J. Phys. Colloid Chem., 53, 67 (1949).

68. R. D. Vold and C. C. Konecny, ibid., 53, 1262 (1949).

69. A. Moore and A. P. Lemberger, J. Pharm. Sci., 52, 223 (1963).

70. L. Hsiao and H. N. Dunning, J. Phys. Chem., 59, 362 (1955).

71. H. Schott, J. Colloid Interface Sci., 26, 133 (1968).

72. O. E. Ford, Text. Res. J., 38, 339 (1968).

73. I. Reich, Ph. D. Dissertation, Univ. Southern California, Los Angeles, 1955.

74. F. V. Nevoline, T. G. Tipissova, V. A. Poliakova, and A. M. Semionova, Proc. Int. Congr. Surface-Active Substances, 4th, Brussels, 1964, Vol. 3, p. 245.

75. M. E. L. McBain and E. Hutchinson, Solubilization and Related Phenomena, Academic Press, New York, 1955, pp. 204-209.

76. A. S. Weatherburn and C. H. Bayley, Text. Res. J., 22, 797 (1952).

77. H. Koelbel and K. Hoerig, Angew. Chem., 71, 691 (1959).

78. J. C. Harris, Text. Res. J., 18, 669 (1948).

79. A. S. Weatherburn, G. R. F. Rose, and C. H. Bayley, Can. J. Res., 28 (F), 51 (1950).

80. A. L. Meader and B. A. Fries, Ind. Eng. Chem., 44, 1636 (1952).

81. T. F. Boyd and R. Bernstein, J. Amer. Oil Chem. Soc., 33, 614 (1956).

82. A. Fava and H. Eyring, J. Phys. Chem., 60, 890 (1956).

83. G. S. Perry, A. S. Weatherburn, and C. H. Bayley, J. Amer. Oil Chem. Soc., 34, 493 (1957).

84. H. C. Evans, J. Colloid Sci., 13, 537 (1958).

85. G. G. Jayson, J. Appl. Chem., 9, 422 (1959).

86. S. R. Epton and J. M. Preston, Int. Congr. Detergency, 1st, Paris, 1954, Vol. 2, p. 471.

87. M. E. Ginn, F. B. Kinney, and J. C. Harris, J. Amer. Oil Chem. Soc., 38, 138 (1961).

88. W. J. Schwarz, A. R. Martin, B. J. Rutkowski, and R. C. Davis, Proc. Int. Congr. Surface Activity, 3rd, Cologne, 1960, Vol. 4, p. 37.

89. Idem, Fette, Seifen, Anstrichm., 64, 57 (1962).

90. W. J. Schwarz, A. R. Martin, and R. C. Davis, Text. Res. J., 32, 1 (1962).

91. H. Schott, ibid., 37, 336 (1967).

92. B. E. Gordon and W. T. Shebs, Proc. Int. Congr. Surface Activity, 5th Barcelona, 1968, Vol. 3, p. 155.

93. R. C. Mast and I. Benjamin, J. Colloid Interface Sci., 31, 31 (1969).

94. B. E. Gordon, G. A. Gillies, W. T. Shebs, G. M. Hartwig, and G. R. Edwards, J. Amer. Oil Chem. Soc., 43, 232 (1966).

95. H. N. Dunning, J. Chem. Eng. Data, 2, 88 (1957).

96. A. Waag, Proc. Int. Congr. Surface Activity, 5th, Barcelona, 1968, Vol. 3, p. 143.

97. Ref. 17, Chap. 16.

98. G. J. Biefer and S. G. Mason, J. Colloid Sci., 9, 20 (1954).

99. S. M. Neale and R. H. Peters, Trans. Faraday Soc., 42, 478 (1946).

100. D. A. I. Goring and S. G. Mason, Can. J. Res., 28 (B), 307, 323 (1950).

101. D. Stigter, J. Colloid Sci., 19, 268 (1964).

102. J. S. Stanley, J. Phys. Chem., 58, 533 (1954).

103. A. M. Sookne and M. Harris, Text. Res. J., 9, 437 (1939).

104. P. Karrer and P. Schubert, Helv. Chim. Acta, 11, 221 (1928).

105. E. Hageboeke, Dissertation, Bonn, 1956; through W. Kling, in Ullmanns Encyklopaedie der Technischen Chemie, 3rd. Ed., Vol. 18, Urban & Schwarzenberg, Munich, 1967, p. 312.

106. M. v. Stackelberg, W. Kling, W. Benzel, and F. Wilke, Kolloid-Z., 135, 67 (1954).

107. A. M. Sookne and M. Harris, Text. Res. J., 9, 374 (1939).

108. K. Kanamaru, Kolloid-Z., 168, 115 (1960).

109. M. Harris and A. M. Sookne, J. Res. Nat. Bur. Stand., 26, 289 (1941).

109a. Y. Iwadare and T. Suzawa, Bull. Chem. Soc. Japan, 43, 2326 (1970).

110. A. M. Sookne and M. Harris, J. Res. Nat. Bur. Stand., 26, 65 (1941).

111. J. A. Ciriacks and D. G. Williams, J. Colloid Interface Sci., 26, 446 (1968).

111a. Y. Iwadare, Bull. Chem. Soc. Japan, 43, 3364 (1970).

112. W. P. Evans and M. Camp, Proc. Int. Congr. Surface-Active Substances 4th, Brussels, 1964, Vol. 3, p. 259.

113. A. K. Phansalkar and R. D. Vold, J. Phys. Chem., 59, 885 (1955).

114. J. W. Hensley, M. G. Kramer, R. D. Ring, and H. R. Suter, J. Amer. Oil Chem. Soc., 32, 138 (1955).

115. J. W. Hensley and C. G. Inks, ASTM Special Technical Publication No. 268, 27 (1959).

116. B. E. Gordon and E. L. Bastin, J. Amer. Oil Chem. Soc., 45, 754 (1968).

117. J. M. Lambert et al., Nucleonics, 12 (2), 40 (1954).

118. O. C. Bacon and J. E. Smith, Ind. Eng. Chem., 40, 2361 (1948).

119. J. C. Harris, M. R. Sullivan, and L. E. Weeks, ibid., 46, 1942 (1954).

120. W. H. Rees, J. Text. Inst. (Trans.), 53, T230 (1962).

121. J. M. Lambert and H. L. Sanders, Ind. Eng. Chem., 42, 1388 (1950).

122. A. S. Weatherburn and C. H. Bayley, Text. Res. J., 25, 549 (1955).

123. W. J. Hart and J. Compton, ibid., 23, 418 (1953).

124. I. Reich, F. D. Snell, and L. Osipow, Ind. Eng. Chem., 45, 137 (1953).

125. W. Strauss, Kolloid-Z., 150, 134 (1957).

126. W. P. Utermohlen and M. E. Ryan, Ind. Eng. Chem., 41, 2881 (1949).

127. J. C. Harris, Detergency Evaluation and Testing, Wiley-Interscience, New York, 1954, Chap. 3.

128. E. Goette, W. Kling, and H. Mahl, Melliand Textilber., 35, 1252 (1954).

129. W. Kling and H. Mahl, ibid., 35, 640 (1954).

130. Ref. 127, Chap. 4.

131. J. Ilg, Forschungsber. Nordrhein-Westfalen, No. 1437, p. 21 (1965).

132. B. R. Porter, C. L. Peacock, V. W. Tripp, and M. L. Rollins, Text. Res. J., 27, 833 (1957).

133. W. H. Rees, J. Text. Inst. (Proc.), 45, P612 (1954).

134. J. W. Schappel, Text. Res. J., 26, 211 (1956).

135. J. Berch, H. Peper, J. Ross, and G. L. Drake, Amer. Dyest. Rep., 56 (6), 27 (1967).

136. A. S. Weatherburn and C. H. Bayley, Text. Res. J., 27, 199 (1957).

137. K. Durham, Proc. Int. Congr. Surface Activity, 2nd, London, 1957, Vol. 4, p. 60.

138. B. V. Derjaguin, Discuss. Faraday Soc., 18, 85 (1954).

139. A. Bierman, J. Colloid Sci., 10, 231 (1955).

140. R. Hogg, T. W. Healey, and D. W. Fuerstenau, Trans. Faraday Soc., 62, 1638 (1966).

141. S. Kratohvil, G. E. Janauer, and E. Matijevic, J. Colloid Interface Sci., 29, 187 (1969).

142. H. Krupp, Advan. Colloid Interface Sci., 1, 111 (1967).

143. H. Schott, Text. Res. J. 35, 612 (1965).

144. H. Schott, ibid., 40, 924 (1970).

145. D. G. Williams and J. W. Swanson, Tappi, 49, 147 (1966).

146. J. Compton and W. J. Hart, Ind. Eng. Chem., 43, 1564 (1951).

147. A. S. Weatherburn and C. H. Bayley, Text. Res. J., 27, 358 (1957).

148. W. J. Hart and J. Compton, ibid., 23, 164 (1953).

149. K. Nettelnstroth, Melliand Textilber., 48, 671 (1967).

150. W. J. Hart and J. Compton, Ind. Eng. Chem., 44, 1135 (1952).

151. E. F. Wagner, Text.-Rundsch., 18, 17 (1963).

152. V. W. Tripp, A. T. Moore, B. R. Porter, and M. L. Rollins, Text. Res. J., 28, 447 (1958).

153. L. W. Mazzeno, R. M. H. Kullman, R. M. Reindhardt, H. B. Moore, and J. D. Reid, Amer. Dyest. Rep., 47 (9), 299 (1958).

154. T. H. Shuttleworth and T. G. Jones, Proc. Int. Congr. Surface Activity, 2nd, London, 1957, Vol. 4, p. 52.

155. A. M. Schwartz, J. W. Perry, and J. Berch, Surface Active Agents and Detergents, Wiley-Interscience, New York, 1958, Chap. 21.

156. J. Tuzson and B. A. Short, Text. Res. J., 32, 111 (1962).

157. W. P. Utermohlen, M. E. Ryan, and D. O. Young, ibid., 21, 510 (1951).

158. J. Berch and H. Peper, ibid., 33, 137 (1963).

159. Idem, ibid., 35, 252 (1965).

160. Idem, ibid., 34, 844 (1964).

161. F. Fortess and C. E. Kip, Amer. Dyest. Rep. 42, 349 (1953).

162. T. Fort, H. R. Billica, and C. K. Sloan, Text. Res. J., 36, 7 (1966).

163. E. Howarth and L. P. S. Piper, ibid., 36, 857 (1966).

164. H. R. Billica, ibid., 36, 857 (1966).

165. J. Powney and R. W. Noad, J. Text. Inst., 53, T205 (1962).

166. T. H. Grindstaff, H. T. Patterson, and H. R. Billica, Text. Res. J., 37, 564 (1967).

167. W. Strauss, Kolloid-Z., 158, 30 (1958).

168. E. F. Wagner, Text.-Rundsch., 18, 295 (1963).

169. E. F. Wagner, Tenside, 3, 423 (1966).

170. J. Compton and W. J. Hart, Ind. Eng. Chem., 45, 597 (1953).

171. H. Stuepel, Fette und Seifen, 55, 583 (1953).

172. J. Ross, P. T. Vitale, and A. M. Schwartz, J. Amer. Oil Chem. Soc., 32, 200(1955).

173. P. T. Vitale, ibid., 31, 341 (1954).

174. R. C. Palmer and E. K. Rideal, J. Chem. Soc., 1939, 573.

175. H. Alter and W. Soller, Ind. Eng. Chem., 50, 922 (1958).

176. W. Kling, H. Lange, G. Boehme, H. Krupp, G. Sandstede, and G. Walter, Proc. Int. Congr. Surface-Active Substances, 4th, Brussels, 1964, Vol. 2, p. 439.

177. J. Visser, J. Colloid Interface Sci., 34, 26 (1970).

178. H. Schott, Text. Res. J., 36, 226 (1966).

179. J. Berch, H. Peper, and G. L. Drake, ibid., 34, 29 (1964).

180. O. Oldenroth, Z. Gesamte Textilind., 70 (12), 2 (1968).

181. E. F. Wagner, Fette, Seifen, Anstrichm., 67, 355 (1965).

182. H. G. Wagner, Proc. Int. Congr. Surface Activity, 2nd, London, 1957, Vol. 4, p. 113.

183. M. Kashiwagi and T. Tamura, Bull. Chem. Soc. Japan, 43, 558 (1970).

184. R. E. Wagg, J. Text. Inst., 43, T325 (1952).

185. H. Schott, J. Pharm. Sci., 58, 1443 (1969).

186. K. Goebell and W. Kling, Melliand Textilber., 42, 1026 (1961).

187. W. P. Utermohlen, E. K. Fischer, M. E. Ryan, and G. H. Campbell, Text. Res. J., 19, 489 (1949).

188. W. J. Hart and J. Compton, ibid., 23, 158 (1953).

189. J. J. Bikerman, The Science of Adhesive Joints, 2nd Ed., Academic Press, New York, 1968, Chap. 7.

190. R. E. Wagg and C. J. Britt, J. Text. Inst. , 53, T205 (1962).

191. T. H. Vaughn, A. Vittone, and L. R. Bacon, Ind. Eng. Chem., 33, 1011 (1941).

192. O. C. Bacon and J. E. Smith, Amer. Dyest. Rep., 43, P619 (1954).

193. W. P. Utermohlen and E. L. Wallace, Text. Res. J., 17, 676 (1947).

194. G. Kretzschmar, Fette, Seifen, Anstrichm., 66, 222 (1964).

195. L. Loeb, P. B. Sanford, and S. D. Cochran, J. Amer. Oil Chem. Soc., 41, 120 (1964).

195a. L. Loeb and R. O. Shuck, ibid., 46, 299 (1969).

195b. idem, ibid., 48, 25 (1971).

196. E. F. Wagner, Fette, Seifen, Anstrichm. 66, 696 (1964).

197. E. F. Wagner, Text.-Rundsch. 19, 317 (1964).

198. W. M. Linfield, E. Jungermann, and J. C. Sherrill, J. Am. Oil Chem. Soc., 39, 47 (1962).

199. J. C. Harris, ibid., 29, 110 (1952).

200. J. H. Brooks and J. R. McPhee, Text. Res. J., 37, 371 (1967).

Chapter 7

ROLE OF MECHANICAL ACTION IN THE SOIL REMOVAL PROCESS

B. A. Short
Whirlpool Corporation
Benton Harbor, Michigan

I. GENERAL REMARKS ON MECHANICAL ACTION AND SOIL REMOVAL

A. Introduction

It is the aim of this chapter to construct a general overall picture of the role of mechanical action in the soil removal process. Hopefully, it will provide the necessary framework for rational classification of information in this area in the future. In some instances only hypothetical statements can be made. It is hoped that future investigations will involve attempts to verify the most important of these hypotheses.

Although our primary concern is soil removal during the laundering process, it is evident that other cleaning processes, i.e., dry cleaning, dish washing, vaccum cleaning, etc., will be intimately involved in any overall investigation. Due to the great overlapping of all the cleaning processes, the scope of the discussion will at times become broader than the primary topic. It is felt, however, that the broader scope is justified by this overlapping of information and interests.

B. Classification of Cleaning Phenomena

The phenomena involved may be considered under three arbitrary size classifications. Large-scale effects are those which involve the directed motion of the surface to be cleaned and of the main fluid body. In the laundering process, the object of this directed motion is to mix the fabric and the free liquid. As a consequence of this mixing, the free liquid should retain a homogeneous detergent and soil concentration at any given instant. If this homogeniety is not maintained, portions of the main liquid body will lose their effectiveness in the soil removal process. The effectiveness of large-scale motion is essentially a function of cleaning device geometry and the velocity distribution within the liquid body.

Small-scale effects are due primarily to a mass transfer process (detergent to cloth-soil from cloth), the purely mechanical removal of microscopic dirt trapped between fibers, and the dissolution of solid materials near the cloth surface. The study of these effects is confined to the thin liquid film or boundary layer near the cloth surface. Whereas the concentrations and the large-scale velocities are assumed to be uniform in the main fluid body, they will show definite gradients across the boundary layer (see Fig. 7.1). Water retention and wetting properties of the fiber structure or the surface to be cleaned are other topics of interest in any study of small-scale effects.

Molecular-scale effects have to do with the forces, mainly of an electrical nature, which attach soil material to a surface. The distance at which these forces are active must be measured in angstroms (1 Å = 0.0001 μ). At such distances from the fiber surface, the liquid velocities become negligible, and the intensity of mechanical agitation has little influence on the breaking of soil retention forces, particularly when compared to the influence of chemical action. The contribution, however, of such special mechanical effects as fiber bending or liquid boiling to soil removal at this scale, while perhaps appreciable, is a matter of intense speculation.

1. Large-Scale Effects

As previously mentioned, the role of large-scale mixing is to maintain a uniform soil and detergent concentration throughout the free liquid body. The liquid body may be considered as a reservoir from which fresh detergent solution is removed and into which soiled solution is discharged. Ideal cleaning conditions are obtained if this reservoir is of infinite size, i.e., open circuit operation. In practical application, however, the closed circuit operation must be employed. Both open and closed circuits are illustrated in Fig. 7.2. Since the reservoir in the closed circuit must be of finite size, the ratio of the available liquid to the amount of soil to be

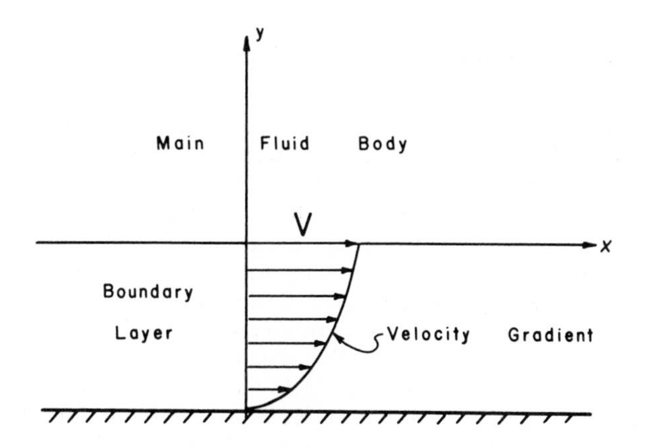

FIG. 7.1. Velocity gradient in the boundary layer; V is the velocity of liquid in the main fluid body.

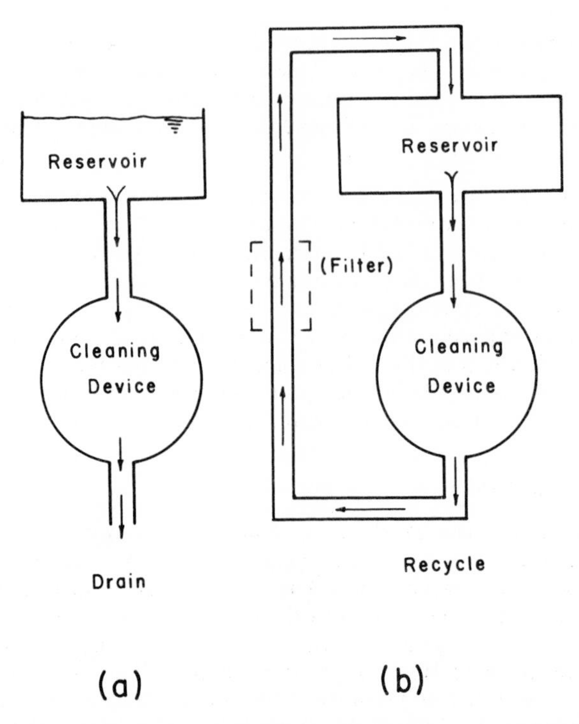

FIG. 7.2. Open circuit operation (a) and closed circuit operation (b) cleaning processes. A filter may be a part of the system in closed circuit operation.

removed will have a decided effect on the effectiveness of the cleaning process.

In large-scale fluid motion, the moving fluid has a velocity at every point in the liquid body, the time average of which is constant for steady motion. The flow will follow certain streamlines and as a first approximation energy losses can be neglected. In such an ideal streamline flow, Fig. 7.3(a), the fluid maintains its identity along a streamline and neighboring stream filaments do not diffuse in each other. In order to achieve a mixing effect, in streamline flow, fluid filaments of heterogeneous nature must be brought together. At the end of the mixing process one would obtain a superposition of thin sheets of different materials. Such mixing occurs when highly viscous, pasty materials are mixed. In such circumstances, the effectiveness of the mixing will depend entirely on the geometry of the flow. If the initial form of the interfacial surface and the type of deformation are known, then the deformed interfacial surface can be computed. Random fluctuations of the flow velocity result in the flow condition called

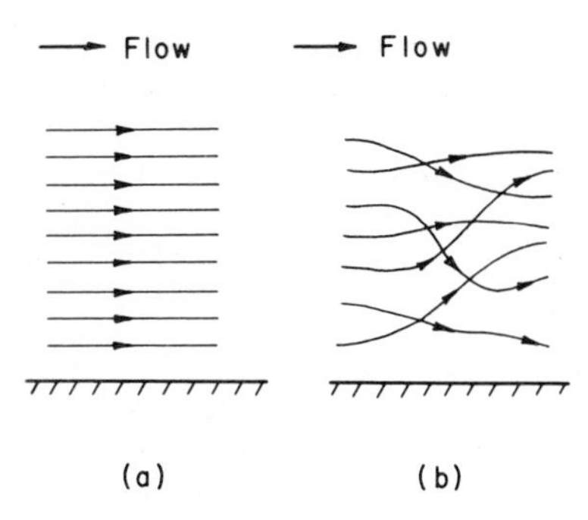

FIG. 7.3. Illustration of ideal (laminar) streamline flow (a) and turbulent flow (b) regimes.

turbulence, Fig. 7.3(b). These random fluctuations can be defined only by indicating their frequency, and associated amplitude spectrum. Turbulence is considered energy loss because the energy originally imparted to the liquid is intended to create directed motion. Motion in other than the intended direction is a loss when considered in this sense. The ultimate decay of motion is turbulence on the molecular level and is commonly identified as heat.

Turbulence, however, favors mass exchange perpendicular to the flow direction and is responsible for turbulent diffusion. Turbulence also brings about an apparent increase in the fluid viscosity. Viscosity arises from a momentum transfer process that is analogous to the other transport processes of mass transfer and heat transfer. In order to evaluate mass transfer in a turbulent fluid it is sufficient to know the overall turbulent transfer coefficient, without any reference to the detailed frequency spectrum of turbulence. Thus, controlled turbulence is a desirable condition from the point of view of mixing. The effectiveness of mixing is directly proportional to the turbulence level and therefore to the energy consumption.

The analogy between momentum transfer (viscosity), mass transfer (diffusion), and heat transfer is treated in several books (1, 2, 3) and provides information on overall transfer coefficients. For example, turbulent mixing can be evaluated by measuring the changes in temperature, i.e., the heat transfer effect, at different locations inside a mixing device. The subject of streamline mixing has been studied using granular solids (4, 5) and highly viscous liquids (6).

A purely theoretical treatment of a mixing process quickly becomes hopelessly involved. Although an ideal streamline motion can be studied analytically, the presence of turbulence modifies the velocity distribution. Extensive data can be found on turbulence in general (7); however, the turbulence level can be predicted only for a very simple flow pattern. On the other hand, dimensional analysis procedures lead to expressions which predict the effect of changes in viscosity, density, speed of agitation, and size of the mixing device on the overall mass transfer (mixing) performance of the device (8). It must be pointed out that whereas a heat transfer process involves well-defined quantities, no clear measure of the effectiveness of mixing is available.

Most published work reports on the performance of cylindrical mixers with different types of agitation. The power requirements of several such mixers have been tested and corresponding expressions developed (8). The results obtained, however, cannot be applied directly to the conventional washing process. A broad review of the mixing process, as covered briefly above, is given by Rushton (8, 9).

2. Small-Scale Effects

Mass transport is the first of the small-scale effects. In particular, we are interested in the transport of detergent to the cloth surface and of the dispersed or suspended soil from the cloth surface to the main body of liquid. An early step in this analysis is the determination of the velocity distribution in the liquid film or boundary layer near the fabric surface. To do this, it is necessary to assume a specific type of flow in the "free stream fluid" and then investigate the corresponding boundary layer condition near the fabric surface. An example of such a "free stream motion" might be a steady fluid flow of uniform velocity parallel to the surface to be cleaned. Such could be the case when cleaning a dish with a liquid jet, as illustrated in Fig. 7.4. In the case of fabric washing, we have a flow with random velocity fluctuations, partly parallel and partly perpendicular to the fabric surface. Such a random flow can be resolved into periodic elements, each with a characteristic frequency. The mass transfer process for either of the cases cited may be deduced from the existing flow conditions, since they will involve a momentum transfer process across the boundary layer. This momentum transfer process will be directly analogous to the process of mass transfer.

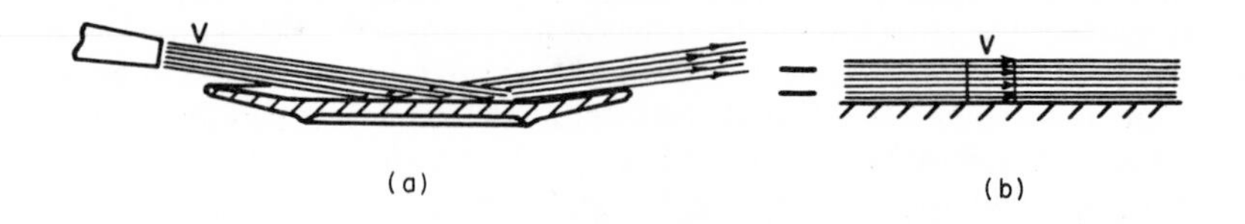

FIG. 7.4. Assuming a specific flow condition prior to analysis. Dish being cleaned by liquid jet (a) is assumed equal to uniform velocity parallel to the surface (b).

Separate consideration must be given to the parallel and perpendicular components of flow. The velocity distribution in the boundary layer for a uniform flow along a plate is well known and described in several references (3, 7, 10) and heat transfer rates have been measured. Some information is available on the case when the flow parallel to the surface is periodic (7, 10). In short, this is one of the best known cases. Consideration of flow perpendicular to the fabric surface, however, leads to two aspects. One is the wetting of the fabric due to capillary forces. The other is the flow through the fabric due to an external pressure differential or body force such as gravity or a centrifugal force field. The capillary water transport rate, water content, and fiber-water contact angle have been extensively studied (11, 12), mainly as related to fabric structure. The water surface tension, viscosity, contact angle, and the radius of the mean capillary size are related by a theoretical expression with the transport rate. Theory and experimental data agree except for the rather important influence of yarn roughness, for which no reliable measure could be derived. The total water-holding capacity of different fabrics has been measured (13).

A basic study relating to capillary liquid transport by Lange (14), deals with the kinetic changes of the surface tension. These changes are believed to be related to the migration speed of surface active agents from the main liquid body to the liquid-air interface or vice versa. The theoretical background of the problem was correlated with experimental data from the motion of a capillary meniscus. This work provides the background for the theoretical explanation of water extraction data obtained in a recent study on liquid extraction (15). The study reports information correlating water extraction and spin speeds in a centrifugal test device. Theoretical considerations predict that water retention (or extraction) will vary inversely with the square of the spin speed. This is in agreement with the results of the study except when centrifugal field strength exceeds 100 G. At field strengths exceeding 100 G, the water retention is consistently less than predicted by theory. The variation is felt to be due to the additional compression of the test load under its own centrifugal loading force.

Flow through porous media, a specific case of which is flow through fabrics, is treated extensively in an excellent book by Scheidegger (16). Several theoretical models are proposed to illustrate and explain the phenomenon of porous flow. The most suitable for our consideration seems to be the "drag theory" in which the flow relationship is derived from the individual drag of cylindrical bodies. These bodies in washing theory would represent the yarn. This consideration is valid when the distance between yarns is relatively large. An additional study of interest on flow through porous media is the work of Green and Duwez (17).

Depending on the value of the Reynolds number, two flow conditions can be distinguished. At low Reynolds numbers, i.e., $N_R < 2000$, laminar flow occurs in which viscous forces are predominant. At high Reynolds numbers, i.e., $\underline{N}_{\underline{R}} > 2200$, we will have turbulent flow and inertia forces will predominate. The flow through fabrics in particular has been studied experimentally (18). A semiempirical expression is given in this reference which combines the two flow conditions. More information is available, supporting the considerations above, on flow through wire grids (19) and fabrics (20).

Mechanical removal of macroscopic soil, another of the small-scale effects, will be dependent on the fluid drag forces acting in the vicinity of the fabric fibers. Macroscopic soil refers to particles of 2-μ mean diameter and above. Such particles may be cemented to the fiber, retained by the surface tension of a liquid film, or attached by the action of the van der Waals forces. Larsen (21), studied surface tension and van der Waals forces retaining glass beads (12-120 μ in diameter) to glass fibers (10-860 μ in diameter). He measured the retaining forces holding the beads to the fibers and the airstream velocity that was necessary to detach them from the fibers. Theoretical expressions were devised and later validated by experimental data.

Larsen (21) also investigated the effect of fiber vibration on particle removal. The results of this work on dust removal from individual fibers might be applied immediately to vacuum cleaning by analyzing the airflow conditions and the mutual influence of several fibers on each other. However, the validity of the expressions would have to be extended to fibers other than glass. A case of macroscopic soil removal has been studied in connection with dish washing (22). A water jet was used to remove soil from dishes. The data from this study indicate that soil removal depends on the quantity of work applied to the dish surface and not on the form, i.e., pressure, flow rate, time duration, etc., in which it is applied.

The removal of an oil film from a solid surface has been studied (23). By virtue of the surface activity of the detergent, the oil film contracts into droplet form and may even be removed completely. The mechanical work of oil contraction and the total work of removal are evaluated theoretically. Although this paper deals with the removal of a macroscopic soil, the oil film, the work involved does not result from the drag force of the liquid medium.

Mechanical work of emulsification, a third small-scale effect, can be estimated from the energy of the newly created surface. The actual work, however, includes the viscous resistance opposing any fluid motion and is largely dependent on the type of device and type of flow. Some information on emulsification power requirements can be found in the previously listed publications (8, 9). The evaluation of surface tension and the stabilization of an emulsion are considered in Sec. 3 under microscopic-scale effects.

3. Molecular-Scale Effects

The order of magnitude of the distance at which microscopic effects become predominant is well below 1 μ. At such small distances from the fiber surface the flow velocity is for all practical purposes nonexistent. The main features involved here are the electrical forces retaining soil particles, modification of these forces due to detergent activity, and molecular diffusion as a transport process.

In the absence of a fluid medium, other than air, the van der Waals attraction is responsible for soil retention. These forces become noticeable at distances of the order of 1/100 of the diameter of the fiber. Cotton fiber being of the size 10-15 μ, the largest active distance would be around 0.1 μ. At very small distances, repulsive forces (Born's forces) appear which counteract the van der Waals forces and maintain the soil particle in a stable position.

When the soiled fibers are wetted with an electrolyte, however, an electrical double layer is formed on the fiber and soil surface. Positive ions are rigidly attached to the solid surfaces and negative ions form a mobile layer further into the main body of liquid (see Fig. 7.5). A potential difference, called the zeta potential, is established across the double layer. As a consequence, a repulsive force appears between soil and fiber. The resultant of these three forces, i.e., the van der Waals attractive force, the Born repulsive force, and the repulsive force due to the zeta potential,

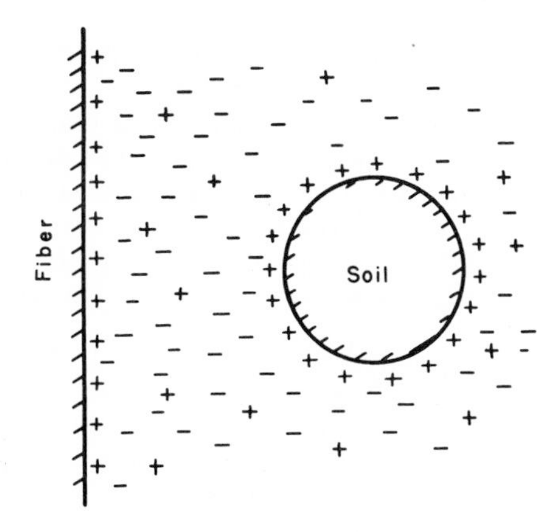

FIG. 7.5. The electrical double layer between a fiber surface and a soil particle.

will determine the potential field which a soil particle must overcome in order to be deposited on or removed from a fabric surface. The relationship of these forces is easily stated in the form of energy potentials. Combining the potential energy of the repulsive force due to the electrical double layer (zeta potential) with the potential energy due to the Born force we let their potential energy sum equal V_r. The potential energy of the van der Waals attractive force is represented by V_a. The resultant potential energy is then given by $V = V_a + V_r$, where V, V_a, and V_r are all functions of the distance from the particle to the fabric surface. These forces and energy potentials have been evaluated quantitatively by Lange (24, 25), and for an in-depth study, the reader is referred to Chap. 4.

The potential barrier resulting from the attractive-repulsive energy relationship may be overcome by the molecular motion (Brownian movement or diffusion process) in either direction, i.e., either soil particle removal or redeposition. The changes of the zeta potential and consequently of the potential barriers that occur during the washing process help to explain the soil redeposition that occurs in a heavily soiled fabric. According to Kling (26), the addition of a detergent will raise the wall potential by 50-100 mV. This change in potential will cut in half the energy required to overcome the potential threshold necessary to remove the soil particle.

Methods to measure the electrical double layer or zeta potential are described by Harris (27). Data on the nature of soil, fiber, electrolyte, temperature, and surfactant are included. One of the most interesting phenomena, involving the electrical double layer, is electroosmosis, i.e., the flow or migration of ions under an electrical potential.

Considerations similar to those above govern the stabilization of suspensions to prevent flocculation of suspended soil. These effects as well as solubilization of oily substances are treated in a series of papers by Harris (28-31). These take into account the particular role of surfactants. Harris also points out that there is a critical surfactant concentration which corresponds to the saturation of the liquid boundaries by a molecular layer of surfactant. At this critical concentration, flocculation of soil, emulsification, and even separation of the oil phase may occur. Above this critical concentration, micelles are formed, the soil is dispersed, oil solubilization occurs, and sudsing is enhanced. The zeta potential and many liquid properties show peculiar changes near the critical surfactant concentration. An excellent summary of the forces involved in detergency is given by Harris (32). The extensive literature sources existing on solubilization and emulsification of oils and on the suspension and flocculation of solid materials are not listed here. Two of note, however, would be the work of Becher (33) and Niven (34).

II. DIFFUSION PROCESS OF MACROSCOPIC PARTICLES

As established by the work of Chandrasekhar (35), the diffusion (mass transport) process of macroscopic particles is essentially the same as that of soluble matter. Starting with the random displacements of a particle resulting from the collisions with the molecules of the medium, he arrives at expressions predicting the probability of particle distribution in that medium. Subsequently, Chandrasekhar demostrates that the same expression is a solution of the differential equation for diffusion provided the diffusion constant is defined as:

$$D = kT/6a\eta \quad (7.1)$$

where k is a universal constant for the material being diffused, T is the absolute temperature, a is the mean diameter of the particle, and η is the viscosity of the medium. This expression, however, is the general expression for the diffusion constant, valid for soluble matter as well as for macroscopic particles (36). Thus, there is no essential difference between the diffusion of soluble matter and that of macroscopic particles.

Chandrasekhar goes further and derives expressions for the case where external forces are acting on the particles during the diffusion process. Of particular interest is the case of particles escaping from an equilibrium position over a potential barrier as a consequence of Brownian motion. This study allows us to evaluate not only a simple diffusion process of macroscopic particles but also transport processes which take place across a potential force barrier.

To obtain a quantitative estimate of the rate of removal, the potential field must be known in detail. According to the article by Lange (24), pigment particles are retained on the fiber surface by a potential field resulting from the Born repulsive and van der Waals attractive forces. This force is modified by the presence of an electrolyte because of the electrical double layer. As a consequence the particles must overcome a potential barrier in order to be removed. Lange estimated the numerical value of this potential barrier and the effect of the electrolyte concentration. This same problem is treated in a more general form by Frenkel (37). He develops the connections between the surface tension and intermolecular forces and treats the removal process as a stability problem of the thin liquid film between two interfaces. No restrictions are made as to the state (solid, liquid, or gas) of the two media which are separated by the liquid film. Therefore, his conclusions apply to forces retaining solid particles as well as liquid droplets or gas bubbles.

Actual soil removal curves have been correlated in a form similar to mass transfer data (38, 39). These data indicate that in some cases the mass transfer process is controlling the washing process and that the transport process of particulate matter can be described in the same manner as that of soluble materials.

III. STUDIES ON SOIL REMOVAL

A. Concept of Soil Removal

The majority of published data on washing are concerned with the chemical aspects of the process. Only a few papers deal with the effect of work input and more generally with the kinetics of the process (40). One of the reasons for this lack of data is that in most experiments cleaning action was measured by the reflectance of the cloth, a method which does not permit continuous monitoring of the process. In some instances soil removal has been correlated with turbidity measurements of the wash solution and continuous records were obtained. When a continuous measurement of soil versus time is made, the resulting curves show an asymptotic approach to an ultimate level of removal. Such a curve is shown in Fig. 7.6.

This type of curve suggests that the soil removal could be the result of a mass transfer process. As stated in the previous section, the correlations of Vaughn et al. (38) and Utermohlen and Wallace (39) would seem to justify this view.

Mass transfer in the washing process is understood to be the process by which fresh detergent is conveyed from the liquid bath to the fabric surface and by which the soil materials are transported away into the detergent bath. When molecularly divided matter is transported, mass transfer

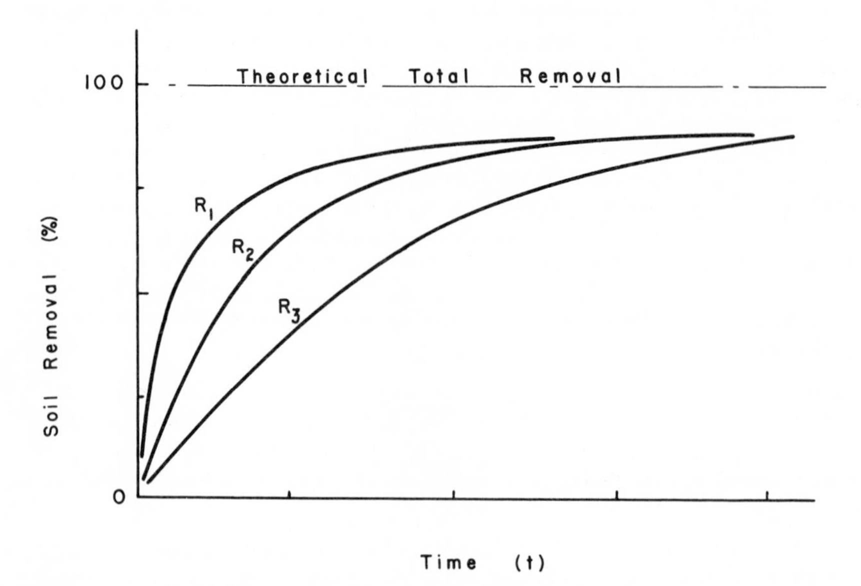

FIG. 7.6. Typical plot of percent soil removal vs. time, where $\underline{R}_1$, $\underline{R}_2$, and $\underline{R}_3$ are decreasing mechanical action levels.

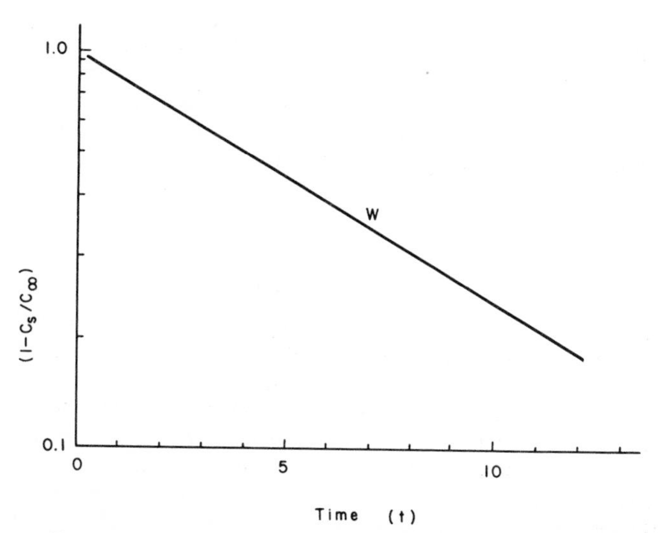

FIG. 7.7. Typical plot of soil concentration, C_S, in the wash bath vs. time. Value of curve slope, w, is the transfer coefficient for the process. C_∞ is the equilibrium bath concentration after infinite time.

becomes synonymous with diffusion. As defined in this discussion, mass transfer also includes the transport of colloidal, liquid, or solid macroscopic materials. In all cases the transfer rate, independent of the materials. In all cases the transfer rate, independent of the materials involved, is proportional to the concentration gradient present in the dispersing medium. When the material to be diffused is initially concentrated on a solid surface and is subsequently diffused into a liquid medium, it is usually assumed that the concentration gradient is restricted to a film in the close vicinity of the solid surface. Under this assumption the concentration of the remaining portion of the liquid medium may be considered essentially uniform. This assumption is particularly applicable when the process occurs under conditions of strong agitation. A theoretical analysis of such a mass transfer process predicts that the relative soil concentration remaining on the cloth should follow an exponential decay curve with elapsing time (41):

$$(C_\infty - C_S)/C_\infty = \exp -wt \qquad (7.2)$$

where C_S is the instantaneous soil concentration in the bath; C_∞ is the asymptotic value of soil concentration in the bath as the value of time, t, approaches infinity; and w is the transfer coefficient. A typical curve is shown in Fig. 7.7. In Eq. (7.2), the transfer coefficient is the inverse of the time constant of the process and is dependent on the configuration, the power input, and the loading of the machine. Although the transfer coefficient is dependent on the nature of the material to be transported, other variable effects should not be affected by a change of this material unless other processes, i.e., breaking of soil-retaining bonds, dispersion of agglomerates, emulsification, or other chemical reactions, take place and become controlling. Whereas it is unknown to what extent mass transfer influences soil removal, there is little doubt that such an influence is present. The rinsing process, however, is clearly dominated by mass transfer.

B. Soluble Soil Study

On the basis of the soil removal concept developed above, the mass transfer process for a conventional agitator washer has been investigated (41), under a variety of conditions. In this study, a 3% by weight solution of cupric sulfate in deionized water was selected as the "soil" material. In actual testing, swatches of AHLMA test material were soaked in the "soil" solution. The swatches were then inserted into a wash bath of deionized water and the rate of diffusion measured by an electrical conductivity device. Whereas strict identity between mass transfer and actual soil removal is not claimed, important implications and conclusions were drawn from this study. The following points reprinted from the work of Tuzson and Short (41) summarize this work [reprinted by permission from (41, p. 988), by courtesy of the Textile Research Journal]:

1. The time dependence of the bath concentration in the mass transfer process considered here can be well approximated, at least for pure substances, by a straight line, provided $(1 - C_S/C_\infty)$ is plotted against time on a semilog plot. The slope of the line, w, is then the transfer coefficient.

2. The relationship between the transfer coefficient and the essential variables can be expressed in dimensionless form, i.e.,

$$(wD^2/B)(\nu/B)^{-s} = M(D^2\omega/\nu)^x \tag{7.3}$$

where w = transfer coefficient, D = diameter of test device, B = diffusion coefficient, ν = kinematic viscosity of the liquid, M = dimensionless numerical constant, ω = angular velocity, and s and x are experimental exponents.

3. The relation between transfer coefficient, w, and Reynolds number, N_r, is extremely dependent on the condition of the load as it is introduced into the test apparatus. When the load is introduced in a distributed condition the relation between w and N_r is a direct proportionality. However, wide deviation from this proportionality is evident when the load is introduced in a nondistributed manner. (See Fig. 7.8.)

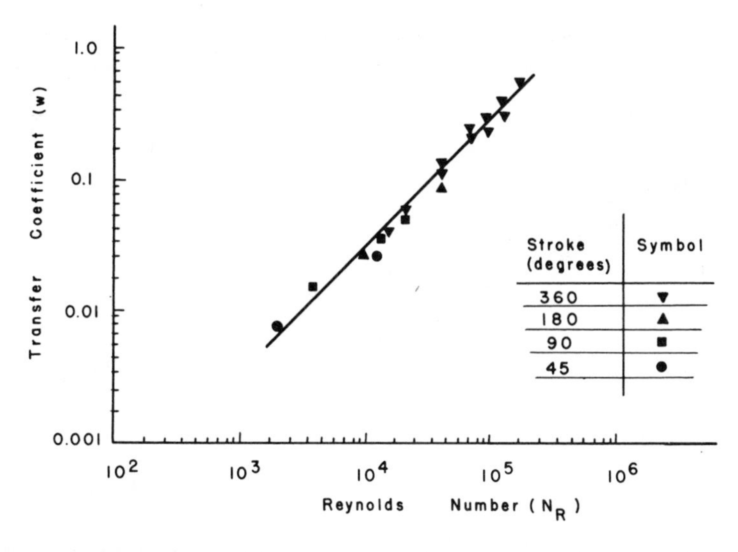

FIG. 7.8. Transfer coefficient, w, vs. the Reynolds Number, N_R, with well-distributed test load. Stroke angle constant at 180°. [Reprinted from (41, p. 986), by courtesy of Textile Research Journal.]

4. The transfer coefficient, w, is the same for similar washers of different size operated under identical conditions. (See Fig. 7.9.)

5. No clear indication of water level effect on the mass transfer process can be determined, since the available data show little or no effect whatsoever.

6. An inverse relationship exists between the transfer coefficient, w, and the amount of cloth loading.

C. Phenomena Related to Soil Removal

After having investigated mass transfer in an agitator washer in pure form, efforts have been made to investigate the influence of other phenomena which could control the soil removal rate. Particular attention has been given to forces retaining particulate matter to a surface. An estimate has been made of the order of magnitude of these forces (42). The results of this work indicate that the forces retaining small particles, i.e., 1 μ square or 1 μ mean diameter, are too strong to be counteracted by hydrodynamic forces alone. Chemical conditioning of the bath is therefore a necessity before these soils can be removed. The only possible exception might be the pressure waves resulting from the implosion of cavitation bubbles or boiling.

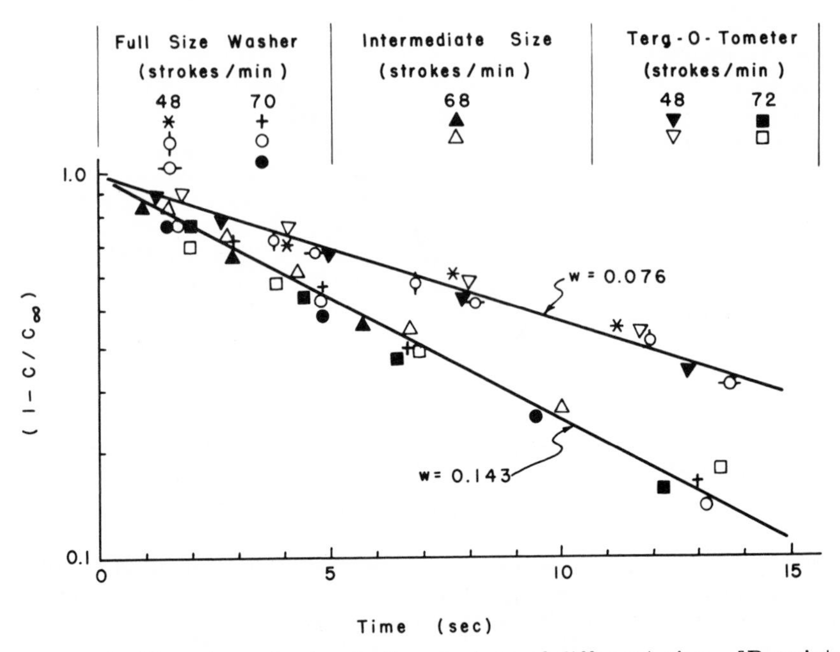

FIG. 7.9. Mass transfer in similar devices of different size. [Reprinted from (41, p. 988), by courtesy of Textile Research Journal.]

However, the most important insight gained from a study of soil-retaining forces must be that the soil deposition-removal process and the agglomeration-dispersion process of these particles are due to the same forces. Data on the composition of particulate soil have been developed (43). Although the material identification provided in this study is very valuable, the original composite nature of the elements is not disclosed. Work by Powe (44), however, has identified clay particles on fabric after washing. Data on the agglomeration of clays (45) point out the particular nature of these materials. In kaolin, because of its platelike crystal structure, the surface charges on different faces may have very different positive or negative values. Because of this variation in surface charge, particle agglomeration is favored. In addition, the large ion-exchange capacity of clays will contribute to the unusual behavior of clay in suspension.

A distinction should be made between colloids consisting of very fine particles and coarser particle suspensions. Whereas thermal agitation above will promote the agglomeration of very fine particles, large particles, because of their relative insensitivity to thermal agitation, require some form of mechanical agitation. The agglomeration and dispersion of carbon and ferric oxide suspensions have been studied experimentally (46). In this work, the qualitative influence of the agitation history on the size distribution of the suspension has been demonstrated. For small-size particles, thermal agitation (Brownian motion) is not sufficient to disperse flocculates and irreversible agglomeration occurs. For large particles, simultaneous agglomeration and dispersion takes place. Thus an equilibrium size distribution is established for any given agitation intensity level. An increase in the agitation intensity produces a relatively more dispersed condition. In this sense, the agglomeration of large particles is reversible. Theoretical work, including a good bibliography, on the agglomeration process has been done by Rajagopal (47).

D. The Coulter Counter

Although in any study of particulate soil removal, the selection of a specific soil material and a particular test fabric is the cause of initial concern, it will soon be apparent that the real cause for concern is the development of a reliable experimental technique. Some of the better known methods of measuring soil removal are reflectivity measurement (39, 48-50), chemical quantitative analysis (51, 52), turbidity measurement (38), and radiotracer techniques (53). The first method (reflectivity) is restricted to soil materials of dark color. The second has been used exclusively with iron oxide pigment soils. None of these methods will permit the study of particle size, since they measure only the overall soiling effect. None of them are particularly simple and, in some instances, doubt has been expressed as to their significance. In particular, none of the methods above can be used to measure soiling with clay (kaolin) particles. For these reasons, other methods have been developed. One of these, the application of the Coulter Counter (54), allows a continuous record of soil removal to be obtained. By this technique, changes in both particulate soil concentration and size distribution can be monitored. The same procedure can be used to study soil deposition or to investigate possible agglomeration taking place in an agitated suspension of fine particles. No restriction is placed on the types of soil which can be studied and this method may be used to study the removal of oil droplets. Anyone interested in utilizing the Coulter Counter should become familiar with the principle of its operation (55, 56).

E. Particulate Soil Study

In previous work, the principle of mass transfer has been applied to general washing theory (41). It was indicated that mass transfer dominates the soil removal process when the soil-retaining forces are very weak. The application of the mass transfer theory was experimentally substantiated using soluble soil material. This work has been extended, however, into the field of particulate soil materials where soil-retaining forces become important (54).

Many materials have been used in previous investigations as substitutes for natural soil (38, 39, 49, 57). Through the use of the electron microscope, Powe (44) has obtained visual evidence of the presence of clay materials in the soiling process. It was shown to be particularly true in the case of known soil redeposition. On this basis, a clay material, kaolin (ASP-400) (58), was selected for use in the particulate soil removal studies previously cited (54).

Results of this study imply that the molecular surface forces responsible for agglomeration and soil-retaining forces are of the same nature. Data obtained on agglomeration in this study aid in illustrating the role which mechanical agitation plays in the dispersion (or removal) and agglomeration (or deposition) process of clay particles. Further studies are needed in the broad area of soil retention forces and the modification of these forces by chemical means. It is important to note, however, the need for closely controlled soiling conditions in such a study. Low agitation speeds during the deposition period results in agglomerates of particles being deposited on the cloth surface. Soiling by this process appears to be of the "entrapment" type and very easy to remove. At high agitation speeds during deposition, primary particles with strong surface force attachment are deposited. A large portion of material attached in this fashion becomes unremovable and thus would correspond to a "redeposition" type of soil retention.

In general, a first-order mass transfer process, should it exist in the very early washing stage, is quickly clouded over by agglomeration and redeposition as the wash process progresses. Indications are that the simple first-order concept of mass transfer does not apply and that a higher-order mass transfer model would be necessary for quantitative analysis of the particulate soil removal data presently available (59). Although the removal speed of particulate soils clearly is increased by increases in agitation speed during the first few minutes, the ensuing multiple processes, i.e., agglomeration, removal, and redeposition, make it difficult to distinguish the role of mass transfer in the overall process for particulate soil materials. These effects are evident in Fig. 7.10.

The close similarity between the features of the agglomeration process and the deposition and removal process seems to confirm the hypothesis that soil-retaining forces and those responsible for agglomeration are of the same type. However, the effects of the nature of the soil and the chemical variables of the suspension would have to be investigated more thoroughly in order to shed light on the identity of these phenomena.

F. Multiple Bath Systems

Later work (60) in the area of particulate soil removal in multiple bath systems has led to results which may be summarized under the following points [reprinted by permission from (60, p. 475), by courtesy of Textile Research Journal]:

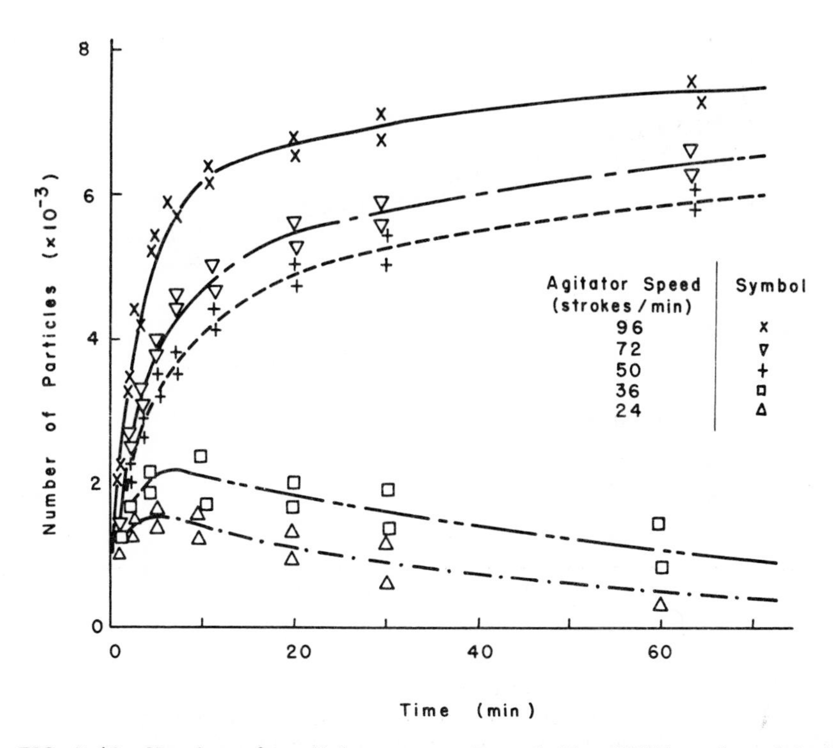

FIG. 7.10. Number of particles removed per half-milliliter of wash bath volume vs. washing time. Mean particle diameter greater than 4 μ. [Reprinted from (54, p. 114), by courtesy of Textile Research Journal.]

1. The removal process of particulate soils may be considered as a series of removal stages. In each of these stages, a different attachment mechanism is involved and thus, a different removal mechanism is required.

2. One portion of the removal process (the microocclusion phase) may, however, be considered a rate process in which particle diffusion is a dominant factor.

3. The complete process of particulate soil removal is not a rate process in the conventional sense, but a combination of processes in which one may tend to dominate at some particular time in the overall removal process. (See Fig. 7.11.)

4. Multiple bath washing may be considered a series of single-bath washes, the results of which may be superimposed to provide a cumulative removal curve. Such a curve, however, is not continuous. (See Fig. 7.12.)

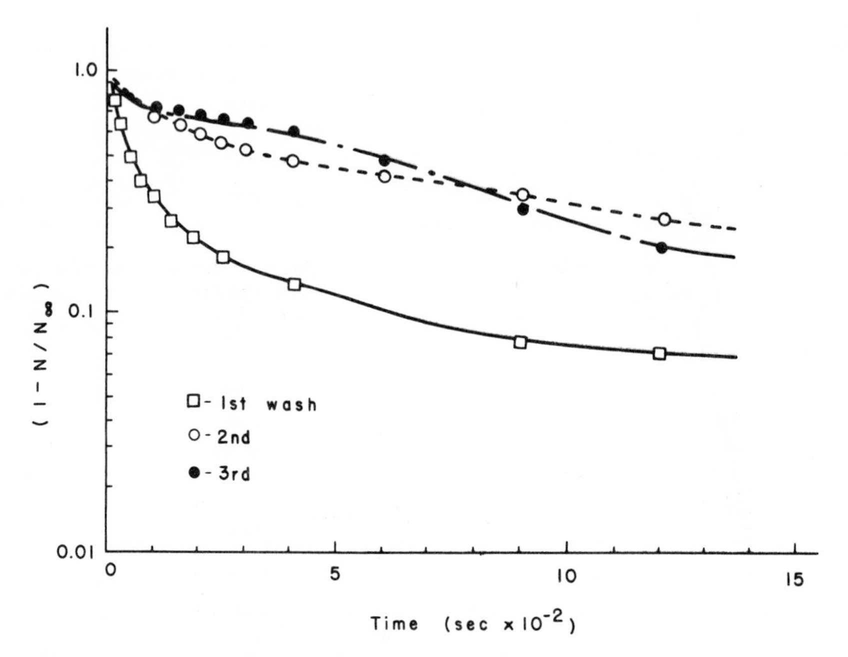

FIG. 7.11. Mass transfer of particulate soil vs. time with the number of washes as a parameter. Duration of each wash was 20 min with stroke speed (96 strokes/min) and stroke angle (360°) constant. All tests conducted in a Terg-O-Tometer with a wash bath volume of 500 ml deionized water. [Reprinted from (60, p. 475), by courtesy of Textile Research Journal.]

G. Attachment Mechanisms for Particulate Soil

In dealing with particulate soils, it is of interest to speculate as to the possible attachment mechanisms involved. Particulate soiling phenomena have been studied in great detail by Compton and Hart (61-63). In the absence of any oil or grease, they concluded that geometric bonding was the primary mechanism in particulate soiling. Additionally, they divided this mechanism into two parts which are related by size or scale factors. The first of these parts, macroocclusion, becomes active when soil particles are trapped between fibers in the yarn and between fibers in the weave. The second, microocclusion, is active when fine soil particles become entrapped in the smaller cracks and crevices of individual fibers. Of the two, microocclusion was found to be the important soiling mechanism for highly textured, natural fibers such as cotton and wool. Macroocclusion was the most important dry soiling mechanism for nylon fabric. Kling and Mahl (64) found that significant amounts of particulate soil adhered to the smooth portions of the fiber surfaces. This suggested that chemical-physical bonding as well as geometric bonding has a definite role in particulate soiling. The addition of perfluorodecanoic acid or its metal salt greatly reduced the particulate soiling of cotton. Since no evidence of visible change in the fiber could be found, the effect was attributed solely to a reduction in the surface energy of the cotton fibers (65).

Relating these studies to the particulate soil removal curves of Fig. 7.11, it is evident that at least three mechanisms of soil removal may occur:

1. Macroocclusion in which the particle is quickly carried away by the combination of mechanical action (flexing) and bulk fluid flow. This would correspond to the initial 30-60 sec of the removal process shown in Fig. 7.11.

2. Microocclusion in which particles must migrate (diffuse) from the fabric surface through the boundary layer of the fluid to the bulk solution. This would be the typical time-dependent or rate process portion of the removal process and would correspond to approximately the next 180-240 sec of Fig. 7.11.

3. Combined chemical-physical bonding in which the addition of energy in the chemical or thermal form is required. In this instance, mechanical

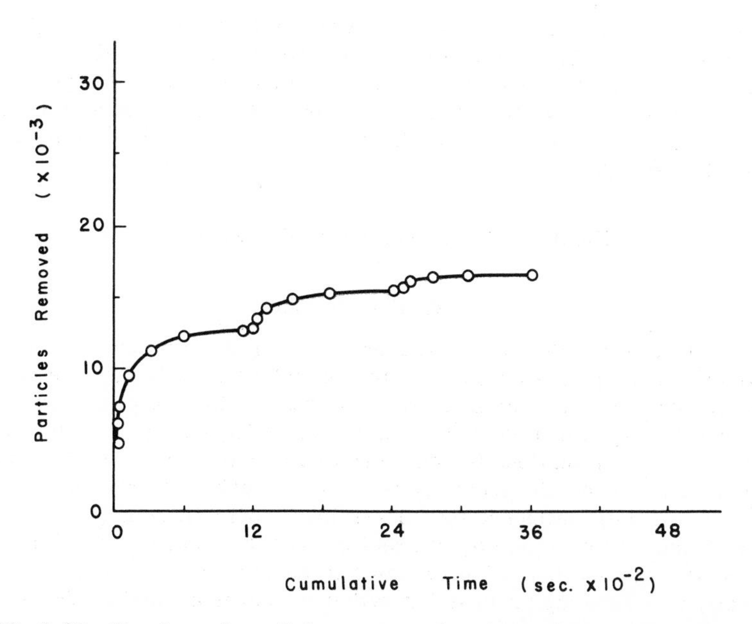

FIG. 7.12. Number of particles removed vs. cumulative time. Mean particle diameter $4\frac{1}{2}$ μ. [Reprinted from (60, p. 474), by courtesy of Textile Research Journal.]

energy serves only to provide a transport mechanism. Before the transport mechanism can be effective, however, the soil attachment force must be overcome and the soil particle moved into the boundary film layer where the diffusion process can occur. This would apply to the remainder of the removal process period.

In papers by Porter et al. (66) and Tripp et al. (67), the effect of finishes on particulate soil deposition was studied. In general, hard film-forming finishes will substantially reduce the retention of particulate soil by cotton. The surface crevices will be coated over by the finish and thus micro-occlusion will be prevented. Under certain conditions, however, hard surfaces may crack after being applied to the fiber. This will introduce new crevices to act as host sites for soil particles (68). Soft, rubbery finishes serve to enhance particulate soiling, presumably because the surface film is easily deformed and particles become embedded in the surface film.

Many natural soiling environments are made up of composites of both fluids and particles. In such cases, the fluid may act as the carrier for the particle and give rise to "sorptive" bonding of the particle to the textile (69). Under these conditions, the major problem will be the removal of the fluid (usually oil) carrier.

Others who have discussed the subject of soiling are Getchell (70), Masland (71), and Snell et al. (72). The influence of oily soil on the particulate soil removal process has been studied by Utermohlen et al. (51). The importance of electrostatic bonding between fibers and soil particles has also been studied; however, the results of these studies have been inconclusive (73).

IV. HIGH-SPEED AGITATION AND SOIL REMOVAL

A. Background

It has long been felt that high-speed agitation or vibratory energy should have excellent application in the washing and cleaning fields. Remarkable cleaning has been accomplished in industrial applications, particularly through the use of ultrasonics. Generally, perhaps because so much was expected, the applications to commercial and home cleaning problems to date have produced disappointing results. All efforts to revolutionize or even significantly influence the conventional home cleaning processes have yielded little in the way of practical results. Indeed one of the most interesting applications of ultrasonics in laundry research does not relate directly to a cleaning process but rather concerns a rapid method of evaluating detergency (74).

A natural question in this section concerns the meaning of low frequency, high frequency, etc. Using the present normal washer agitation speed as a reference, i.e., allowing 70 to approximately equal 10^2 cycles per minute (cpm), we can make the following distinctions:

Type of agitation	Order of magnitude, cpm
Normal	$10 - 10^2$
Low frequency	$10^2 - 10^3$
Sonic frequency	$10^3 - 10^6$
High frequency (ultrasonics)	$10^6 - 10^{10}$

Accepting this arbitary classification, we can proceed to discuss the various types of high-speed agitation, i.e., agitation rates greater than 10^2 cpm.

B. Low-Frequency System

Two quite different principles of washing utilizing low-frequency vibratory means can be visualized. One method would make use of the inertial forces set up in the main body of liquid by a vibrating mechanism. In this method, the mass of the soil particle becomes very important since it is the product of particle mass and applied acceleration that determines the value of the inertia force acting on the particle. It is the relative difference between the inertia force acting on the particle and the inertia force acting on the surface to be cleaned that is relied upon to produce the cleaning action. Since the mass of soil particles whose mean diameter is 100 μ or less is extremely small, the ability of inertial forces to successfully remove such particles from a supporting surface is highly speculative at low-frequency energy inputs.

A second method would rely on the shear stress developed in the body of liquid by the vibrating mechanism. Such shear stress conditions will be the result of flow velocity differences in the liquid. Since the flow velocity at the surface to be cleaned has previously been assumed to be zero, it is logical to further assume that a velocity gradient and, therefore, a shear stress field will extend from the surface into the liquid body. The ability of this shear field to remove soil particles from a surface will depend directly on the ratio of soil particle diameter and boundary layer thickness. For values of this ratio significantly greater than 1, the chances of removal by shear forces are enhanced. For values significantly less than 1, the chances of such removal are appreciably diminished.

Using 60-cycle household voltage and a magnetostrictive transducer, a source of 120 cps vibratory energy can be achieved. Applying the same soluble soil removal test technique outlined in Sec. III, B of this chapter, interesting test results were achieved in a Terg-O-Tometer-type test situation. The findings indicate that the higher the agitation frequency (assuming cavitation is not achieved), the more localized is the agitation effect. Consequently, as the frequency increases the bulk mixing effect becomes less efficient and the overall mass transfer rate falls off (75). While these results were obtained utilizing a soluble soil, there is no reason to believe that the findings cannot be applied to the particulate soil condition.

C. Sonic Frequency Systems

No applicable literature that relates to vibratory energy sources operating in the range of 10^3-10^6 cpm has been noted in the vast amount of work on the cleaning processes. This can be explained by the fact that vibrations in this range, while they can be highly pleasing (such as those produced in a high-fidelity music system), are clearly audible to the human ear. Since it is extremely doubtful that such energy applied in this range to a cleaning process would be anywhere near as pleasing, the lack of such application is quite understandable. In addition, sonic frequency vibrations have relatively long wavelengths and are difficult and expensive to generate. Moreover in this era of concern over environmental pollution (and unwanted noise is a pollution problem) it is safe to conclude that such an application will not be made until the marriage of efficient vibratory cleaning and pleasing audio output is achieved.

D. Ultrasonic Systems

Ultrasonic cleaning systems employ high-frequency, mechanical vibrations transmitted into and through a suitable cleaning solution. The material to be cleaned is immersed in the solution where the cleaning action occurs. Although similar to audible sound (sonic frequency), ultrasonic vibrations are at a frequency above a level that can be heard, i.e., greater than 20,000 cps. Since sound is a compressional wave, the action of this high-frequency energy in a liquid may be considered as the rapid generation and violent collapse of minute bubbles. Countless small but intense impacts erode surface soil from immersed parts. This action is called cavitation. A greater discussion of cavitation in liquids will be found in Sec. IV,E.

An ultrasonic cleaning system consists of the following components:

1. an ultrasonic generator capable of producing high-frequency electrical energy;
2. transducers to convert the electrical energy from the generator into mechanical energy;
3. a properly selected cleaning solution;
4. a container to hold the cleaning solution and to which are attached the transducers.

The transducers may be attached to the bottom or the sides of the container. Hermetically sealed transducers are simply immersed in the cleaning solution.

Conventional ultrasonic generators utilize a simple power oscillator to convert line current to rf power at the desired frequency. Such generators normally require periodic tuning to compensate for such cleaning variables as solution temperature, solution level, and type of cleaning load. While manual tuning is suitable for applications where the generator is readily accessible, it does require the attention of an operator. Automatic tuning systems for ultrasonic generators have been developed (76).

Transducers fall into two main types, electrostrictive and magnetostrictive. Electrostrictive transducers are those which change dimension in the presence of an electrical field. In magnetostrictive transducers the material is influenced by a magnetic field. The ability of a transducer to convert electrical energy to mechanical energy in an efficient manner is very important, since the higher the conversion efficiency, the lower will be the generator size requirement for a given output of mechanical energy. And it should be remembered that the generator is the major cost item of an ultrasonic system. Electrostrictive transducers operate with conversion efficiencies in the range of 70-90% and magnetrostrictive transducers in the range of 20-50%. It is for this reason and the fact that newer materials (lead zirconate titanate) have raised the operating temperature limit (500°F), that most ultrasonic systems utilize electrostrictive transducers.

The selection of cleaning liquids is determined by the nature of the soil to be removed and the material to be cleaned. Obviously the cleaning solution should act only on the soil and should not attack the material to be cleaned. Physical and chemical properties of the cleaning liquid become important because the liquid not only must transmit the energy from the transducer but also must take an active part in the cleaning process.. The liquid acts as a solvent chemically to attack the soil and physically to supply mechanical action through cavitation or relative acceleration forces. Considerations in the selection of cleaning liquid include the nature of the soil to be removed, the type of material to be cleaned, cost, availability, and any possible fire or health hazards.

A method and apparatus utilizing ultrasonic energy for washing textiles were proposed in 1953 by Goldwasser (77). While generating a great deal of interest in the ultrasonic approach to conventional cleaning, the method was not carried to commercialization.

E. Cavitation in Liquids

When a sound wave applied to a liquid has sufficient pressure amplitude in its negative phase to bring the total pressure below the vapor pressure of the liquid at any given temperature, the liquid will start local boiling (see Fig. 7.13). At various nucleation sites, bubbles will begin to grow in the liquid. As long as the total pressure remains below the vapor pressure of the liquid, the bubbles will continue to grow. However, when the total pressure rises above the vapor pressure, the bubbles will begin to shrink and finally will collapse. If there is an appreciable amount of some gas, such as air, dissolved in the liquid, the collapse will be buffered. The dissolved gas will diffuse into the otherwise vapor-filled bubbles. Because gas diffuses into the growing bubbles faster than it goes into solution when the bubbles collapse, the gas content of such periodically expanding and contracting bubbles will increase. As the gas content increases, the bubble becomes large enough to rise to the surface, i.e., the liquid will be degassed.

When the gas content of the liquid is sufficiently reduced, the type of cavitation changes from gaseous cavitation to vaporous cavitation (see Fig. 7.13). In the low-pressure phase of the sound wave, the bubbles grow as before. The subsequent collapse, however, is not buffered since the bubble contains only vapor which condenses at once as the bubble shrinks. Therefore, the final effect of the collapse might be likened to a slowly closing door that is abruptly slammed shut with a bang. The collapse

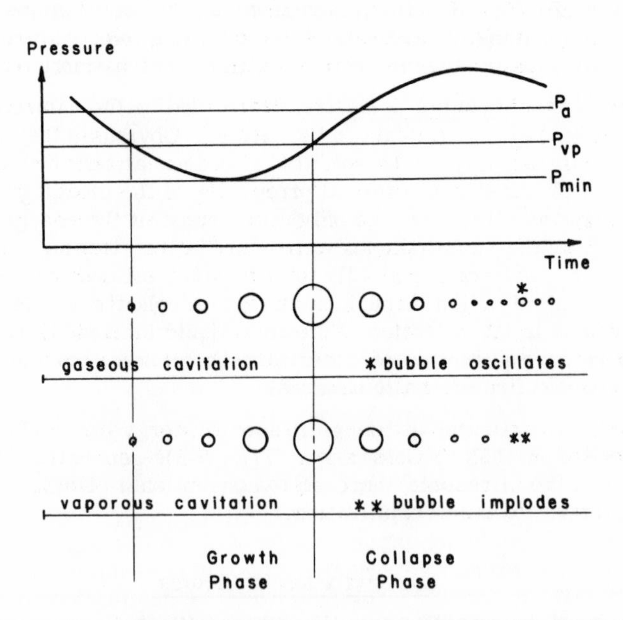

FIG. 7.13. Vaporous and gaseous cavitation in a liquid; P_a is the ambient pressure, P_{vp} is the vapor pressure of the liquid, and P_{min} is the minimum pressure achieved in the liquid.

ends in a sharp pressure rise which sets forth a small shock wave in the liquid. This shock wave has very strong mechanical action effects on the surface of objects immersed in the liquid.

The mechanical action of gaseous cavitation is essentially independent of frequency. However, the activity of vaporous cavitation increases with diminishing frequency. Thus, where intense surface activity is needed, it will be more advantageous to use frequencies lower than those normally employed for gaseous cavitation. Ultrasonic cleaning equipment working in the frequency range of 20-40 kilocycles will be more suitable for such an application than equipment working at higher frequencies.

In-depth studies on the mechanics of cavitation and the physics of ultrasonic cleaning may be found in the work of Ziegler (78), Briggs *et al*. (79), Richardson (80), and Gollmick (81). For those interested in applications, particularly engineering applications, the book on ultrasonics by Carlin (82) is highly recommended.

F. Conclusions

Generally, it is felt that cavitation plays an essential role in the mechanism by which soil particles are removed during ultrasonic washing (83). The accepted way to clean *metals* with ultrasonics is to cause cavitation in the cleaning solvent near the surface to be cleaned. The metal surface reflects the ultrasonic pressure "rarefactions", and in the process the water pressure near the metal surface is lowered. It is in this low-pressure region that vaporous cavitation will occur. Fabric surfaces, however, reflect these "rarefactions" as "compressions" and in the process, inhibit the formation of cavitation bubbles. The reason for this difference in reflection characteristics can be explained by a physical property called acoustical impedance, i.e., the product of material density and the speed of sound in that material. Sound waves are reflected when there is an acoustical impedance mismatch. A list of acoustical impedance values for some selected materials is shown in Table 7.1.

Sound waves tend to be transmitted, however, when there is an apparent matching (within approximately one order of magnitude) of acoustical impedance. Fabrics saturated with water appear to approach this condition of matching acoustical impedance. In line with this reasoning, cavitation would have to be ruled out as a mechanism for cleaning fabrics with ultrasonic energy.

Accepting the conclusion that cavitation will not occur near a fabric surface does not rule out ultrasonics as a fabric cleaner! It has been postulated that ultrasonic cleaning may well be due to a combination of cavitation *and* acceleration of the cleaning fluid (82). The acceleration referred to is the longitudinal vibration of the solvent in the direction of the sound wave at the resonant frequency of the generator. It is this

TABLE 7.1

Acoustical Impedance Values for Selected Materials

Material	Acoustical impedance (10^6)
Steel	4.76
Lead	2.73
Glass	1.80
Aluminum	1.70
Bakelite	0.36
Polystyrene	0.29
Water	0.14
Air	0.000041

accleration mode, rather than cavitation, which may be one of the keys to the successful cleaning of fabrics.

Rosenberg of the USSR Academy of Sciences in Moscow, using a high-speed movie camera, compared the effects of cavitation and acceleration modes in removing rosin and graphite from a glass slide (84). The results demonstrated that cavitation was much more effective in soil removal than acceleration. However, this result was obtained using a chemically inactive solvent (water) with no detergent, soap, or surfactant. Goldwasser has hypothesized that the selection and use of the proper active solvent may well be critical to effective cleaning action (85). Thus, the experimental results of Rosenberg do not rule out the conclusion that ultrasonic cleaning without cavitation requires both mechanical and chemical action.

REFERENCES

1. R. C. L. Bosworth, Transport Processes in Applied Chemistry, Macmillan, New York, 1957.
2. M. Jakob, Heat Transfer, Vols. I and II, Wiley, New York, 1949.
3. W. H. McAdams, Heat Transmission, 3rd ed., McGraw-Hill, New York, 1954.
4. N. K. Maitra and J. M. Coulson, J. Imp. Coll. Chem. Eng. Sci., 4, 135 (1948).

5. J. M. Coulson and N. K. Maitra, Ind. Chem., 26, 55 (1950).

6. R. S. Spencer and R. M. Wiley, Colloid Sci., 6, 133 (1951).

7. S. Goldstein, Modern Developments in Fluid Dynamics, Vols. I and II, Oxford Univ. Press (Clarendon), London, 1938.

8. J. H. Rushton, Chem. Eng. Progr., 47, 485 (1951).

9. J. H. Rushton, Ind. Eng. Chem., Annual Review of the Subject of Mixing (1958).

10. S. Pai, Viscous Flow Theory, Vols. I and II, Van Nostrand, Princeton, N.J., 1956.

11. N. R. S. Hollies, M. M. Kaessinger, and H. Bogaty, Text. Res. J., 26, 829 (1956).

12. N. R. S. Hollies, M. M. Kaessinger, B. S. Watson, and H. Bogaty, ibid, 27, 8 (1957).

13. L. Fourt, A. M. Sookne, D. Frishman, and M. Harris, ibid., 21, 26 (1951).

14. H. Lange, Kolloid-Z, 136, 103 (1954).

15. B. A. Short and W. J. Gertz, Text. Res. J., 34, 532 (1964).

16. A. E. Scheidegger, The Physics of Flow through Porous Media, Macmillan, New York, 1957.

17. L. Green and P. Duwez, J. Appl. Mech., 18, 39 (1951).

18. M. J. Goglia, H. W. S. LaVier, and C. D. Brown, Text. Res. J., 25, 296 (1955).

19. S. F. Hoerner, Air Force Tech. Rep. No. 6289, 1950.

20. S. F. Hoerner, Text. Res. J., 22, 286 (1952).

21. R. I. Larsen, Am. Ind. Hyg. J., 19, 265 (1958).

22. J. Ruspino, unpublished work, 1956.

23. W. Kling and H. Lange, Kolloid-Z, 142, 1 (1955).

24. H. Lange, ibid., 154, 103 (1957).

25. H. Lange, ibid., 156, 103 (1958).

26. W. Kling, Waescherei - Tech. Chem., 12, 542 (1959).

27. J. C. Harris, Text. Res. J., 28, 912 (1958).

28. J. C. Harris, J. Amer. Oil Chem. Soc., 35, 428 (1958).

29. J. C. Harris, Soap Chem. Spec., Pt. 1, June, p. 50, Pt. 2, July, p. 47 (1958).

30. J. C. Harris, Amer. Dyest. Rep., 47, 435 (1958).

31. J. C. Harris, J. Amer. Oil Chem. Soc., 35, 670 (1958).

32. J. C. Harris, Soap Chem. Spec., 68 (May 1961).

33. P. Becher, Emulsions: Theory and Practice, Reinhold, New York, 1957.

34. W. W. Niven, Jr., Fundamentals of Detergency, Reinhold, New York, 1950.

35. S. Chandrasekhar, Rev. Mod. Phys., 19, 20 (1943).

36. W. Jost, Diffusion, Academic Press, New York, 1952, p. 462.

37. J. Frenkel, Kinetic Theory of Liquids, Dover, New York, 1955, p. 332.

38. T. H. Vaughn, A. Vittone, Jr., and L. R. Bacon, Ind. Eng. Chem., 33, 1011 (1941).

39. W. P. Utermohlen, Jr., and E. L. Wallace, Text. Res. J., 17, 670 (1947).

40. G. Kretzschmer, Fette, Seifen, Anstrichm., 66, 222 (1964).

41. J. Tuzson and B. A. Short, Text. Res. J., 30, 983 (1960).

42. J. Tuzson and B. A. Short, unpublished work, 1959.

43. H. L. Sanders and J. M. Lambert, J. Amer. Oil Chem. Soc., 27, 153 (1950).

44. W. C. Powe, Text. Res. J., 29, 879 (1959).

45. C. E. Marshall, J. Soc. Chem. Ind., 56, 457 (1931).

46. I. Reich and R. D. Vold, J. Phys. Chem., 63, 1497 (1959).

47. E. S. Rajagopal, Kolloid-Z, 167, 17 (1959).

48. O. C. Bacon and J. E. Smith, Ind. Eng. Chem., 40, 2361 (1948).

49. I. Reich, F. D. Snell, and L. Osipow, ibid., 45, 137 (1953).

50. O. Oldenroth, Waescherei-Tech. Chem., 17, 771 (1964).

51. W. P. Utermohlen, Jr., E. K., Fischer, M. E. Ryan, and G. H. Campbell, Text. Res. J., 19, 489 (1949).

52. W. P. Utermohlen, Jr., and M. E. Ryan, Ind. Eng. Chem., 41, 2881 (1949).

53. W. J. Diamond and H. Levin, Text. Res. J., 27, 787 (1957).

54. J. Tuzson and B. A. Short, ibid., 32, 111 (1962).

55. R. H. Berg, Amer. Soc. Test. Mater., Tech. Pub. No. 234, 1958.

56. H. E. Kubitschek, Res., 13, 128 (1960).

57. F. H. Rhodes and S. W. Brainard, Ind. Eng. Chem., 21, 60 (1929).

58. Minerals and Chemicals Corp. of Amer., Tech. Bull. No. 1004, Menlo Park, New Jersey.

59. L. Loeb, P. B. Sanford, and S. D. Cochran, J. Amer. Oil Chem. Soc., 41, 120 (1964).

60. B. A. Short, Text. Res. J., 35, 474 (1965).

61. J. Compton and W. J. Hart, ibid., 23, 158 (1953).

62. J. Compton and W. J. Hart, ibid., 23, 418 (1953).

63. J. Compton and W. J. Hart, ibid., 24, 263 (1954).

64. W. Kling and H. Mahl, Melliand Textilber., 35, 640 (1954).

65. V. W. Tripp, R. L. Clayton, and B. R. Porter, Text. Res. J., 27, 340 (1957).

66. B. R. Porter, C. L. Peacock, V. W. Tripp, and M. L. Rollins, ibid., 27, 833 (1957).

67. V. W. Tripp, A. T. Moore, B. R. Porter, and M. L. Rollins, ibid., 28, 447 (1958).

68. R. Tsuzuki and N. Yabuuchi, Amer. Dyest. Rep., 57, 472 (1968).

69. T. Fort, Jr., H. R. Billica, and C. K. Sloan, Text. Res. J., 36, 7 (1966).

70. N. F. Getchell, ibid., 25, 150 (1955).

71. C. H. Masland, Rayon Text. Mon., 20, 573 (1939).

72. F. D. Snell, C. T. Snell, and I. Reich, J. Amer. Oil Chem. Soc., 27, 62 (1950).

73. New York Section of A.A.T.C.C., Amer. Dyest. Rep., 41, 322 (1952).

74. J. S. Sherrill and W. C. White, J. Amer. Oil Chem. Soc., 33, 23 (1956).

75. J. Tuzson and B. A. Short, unpublished work, 1960.

76. Branson Instruments, Inc., Bulletin S-655a, Stanford, Conn.

77. S. Goldwasser (to Lever Bros.), U.S. Pat. 2,650,872 (1953).

78. G. Ziegler, Maschinenban und Warmewirtschaft, 9, 343 (1954).

79. H. B. Briggs, J. B. Johnson, and W. P. Mason, J. Acoust. Soc. Amer., 19, 664 (1947).

80. E. G. Richardson, Wear, 2, 97 (1958).

81. H. J. Gollmick, Metal Finishing J., May 1962, p. 161.

82. B. Carlin, Ultrasonics, 2nd ed., McGraw-Hill, New York, 1960.

83. J. Olaf, Acust., 7, 253 (1957).

84. L. D. Rosenberg, Ultrasonic News, 4, Winter 1960, p. 16.

85. S. Goldwasser, C. W. Haefle, Jr., and C. Padden (to Lever Bros.), U.S. Pat. 3,402,075 (1968).

Chapter 8

SOIL REDEPOSITION

Richard C. Davis
Whirlpool Corporation
Benton Harbor, Michigan

I. PREFACE

A. Scope

Soil deposition and redeposition in detergent baths have been studied almost continuously since the early 1900s. Literature involving these phenomena is so voluminous that even bibliographies accumulate into sizable books. For this reason, and to keep this chapter to a reasonable length, certain limitations have been applied by the author.

First, the chapter will deal with the subject of redeposition only in relation to systems involving soil and textiles. Soil redeposition on metals, ceramics, and other articles normally washed will be left to other authors in other chapters of this book.

Second, in most cases, the soil-textile interface will be studied only in aqueous baths. Soil redeposition is as serious a problem in drycleaning as it is in laundering; but the author prefers to leave study of the former to another writer. Certain data in this chapter were obtained from studies in nonaqueous media. Where these data are used, they are plainly labeled as such and parallels with aqueous systems are drawn.

Third, as in all science and technology, soil redeposition studies have been accelerating rapidly in number during the past 20 years. This acceleration has been made possible by the sheer numbers of researchers working in the field, new methodologies, new instrumentation, etc. For this reason, greatest stress will be placed in this chapter on works and methods devised since the end of World War II.

Finally, in regard to materials and instruments used in obtaining the test data reported in this chapter, the author will, on occasion, indicate instrumentation used in such tests, but will offer no description. These instruments are reported in detail in another chapter. In regard to chemical species, the author will always use chemical nomenclature in his script; but in certain graphs, tables, illustrations, etc., trade names will appear without change. In these cases, the reader is referred to such compilations as that of J. W. McCutcheon for clarification.

B. Philosophy of Chapter

It is the intention of the author to give as objective a view as possible for each of the systems and methods discussed in this chapter. But it is near impossible to expect a soldier to stand unattached in the midst of conflict; and certainly it can be said that many of the test methods, test data, and conclusions drawn therefrom are controversial. The author will, from time to time, take a stand in support of or in opposition to a given method, or conclusion. Where this is the case, the author has attempted clearly to separate news from editorial comment. The author admittedly has a bias in opposition to certain soiling materials, certain laboratory instruments, etc., but he has attempted to prevent this bias from interfering with his presentation of the data of others. If he has not succeeded in total prevention of this, he begs the reader's indulgence in advance.

II. INTRODUCTION TO SOIL REDEPOSITION

A. Definition

In its simplest terms, soil redeposition can be defined as the precipitation of soil, which has once been removed from a substrate, back onto that substrate, or onto accompanying substrates. The reprecipitation results from some inherent instability in the suspending medium or from a sudden, dramatic change in the suspending medium. A typical bath instability might be the presence of too much soil in the bath for the detergent system to suspend completely. A typical dramatic change might be the sudden dilution of the suspending detergent solution such as is realized during the rinsing operation. In either case, the results are the same. Soil is dropped out of suspension and is redeposited back onto the substrate. The result of repeated cycles of laundering involving soil redeposition is a gradual discoloration of white clothing leading ultimately to an unacceptable dingy gray.

B. Synonyms

Terms used synonymously for soil redeposition seem to vary with the particular industry using the term. "Soil redeposition" or "redep" seems to prevail in those industries closely connected with research in the area of surfactants and built detergent systems. In these areas, differentiation between unremoved soils and redeposited soils is of critical importance. Another term one sees on occasion is "tenaciously bound" soil; and while it is true that redeposited soils are "tenaciously bound," the term itself fails to distinguish between unremoved and redeposited soils.

In the textile industry, "wet soiling" appears to be the term used interchangeably with soil redeposition. As is implied, this term defines soil which will adhere to a fabric when that fabric is subjected to washing. Tests of this "wet soiling" property of fabrics of various fibers have become increasingly important in recent years because of developments of highly resinated finishes.

"Whiteness retention" or "whiteness maintenance" are descriptors widely used in the laundry industry to define a phenomenon which is actually the inverse of soil redeposition. Specifically, a laundry measures the whiteness of a standard cloth after an extended number of washes in its production machinery. The higher this "whiteness retention," the more efficient is the process in the prevention of soil redeposition. Bleaching may be an integral part of this "whiteness maintenance" program, or it may be anathema.

Finally, Madison Avenue has entered the soil redeposition picture with two terms still heard on occassion; these are: "tattle-tale gray" and "detergent wash-back." Other advertising descriptions such as "goes beyond white—all the way to bright" have not as yet become so universally used, nor do they apply to soil redeposition. Instead they apply to the masking effect of fluorescent whitening agents, possibly minimizing the effects of redeposition. They are covered in great detail in another chapter.

III. TEST METHODS

Methods employed in evaluating the soil redeposition proclivities of a given soil-detergent-water-fabric system fall into three general categories:

1. soil suspendability tests,
2. deposition tests and redeposition tests,
3. practical wash tests.

A. Soil Suspendability

The simplest soil suspendability tests are designed to measure the ability of a given system to suspend the soil and prohibit its plating out onto a substrate surface. Some of the early work in this general area was done by McBain and co-workers in 1923 (1). They determined the suspending qualities of various true soaps by measuring "carbon numbers." They made a slurry of carbon in the proper soap solution and filtered the resultant mixture through No. 5 Whatman filter paper. The "carbon number" was a measure of the amount of carbon carried through the filter versus that retained by the filter. The higher the carbon number, the more efficient was the suspending capacity of the soap solution considered to be. They found in this study that the addition of small amounts of electrolytes, preferably alkalis, served to increase the carbon number.

Snell and Reich (2) investigated the suspending power of detergent solutions using burnt umber as soil. The umber was carefully coated with a mixture of cottonseed oil, mineral oil, and oleic acid, and was agitated by stirring into the detergent solution. The mixture was poured into a Nessler tube, agitated 25 times and then allowed to stand, free of vibrations, for 2 hours at 50°C. After sedimentation, 50 ml of suspension were decanted into another Nessler tube, which was mounted into a Parr turbidimeter. The turbidimeter was adjusted upward until the filament in the bulb below the suspension "just vanished." The height to which the tube of the turbidimeter had to be raised was said to be a measure of the soil-suspending power of the detergent solution.

Mankowich (3) extended these basic sedimentation studies. He used various nonionic surfactants, anionic surfactants, and soaps in measuring the stability of suspensions of carbon, manganese dioxide, and certain dyestuff as "soil" in some later suspension studies (4).

One specific charge leveled against all sedimentation studies is that they are not related in any way to detergency investigations. Powney and Noad (5) seem to have been the first to point this out; they asserted that the continuous mechanical action present in all laundry processes precluded the application of passive soil suspension data to the study of detergency and washing. Thus, sedimentation and soil-suspending power in passive systems appear to have foundered on the hard reefs of realism, as so many tests have in the detergency area.

It is to be noted, however, that Powney and Noad precluded use of data derived only from static soil suspension tests. Weatherburn and coworkers (6) are among many researchers who have shown definite correlation between suspending power of dynamic detergent solutions and the wash process. From use of Eq. (8.1), they have shown that the suspending power of a detergent solution is of considerable importance in measuring the detersive efficiency, and in certain cases may have greater influence on the final result than actual soil removal power.

$$\text{Suspending power } \% = \frac{B - Bw}{Bo - Bw} \times 100 \tag{8.1}$$

where B is the average reflectance of fabric at the end of test; Bw is "water value" average reflectanee of fabric after shaking in a suspension of carbon in distilled water; and Bo is the original reflectance of white cloth.

Hart and Compton (7, 8) used this same equation for measurement of suspending power in a broad series of experiments which are discussed later in this chapter.

B. Soil Deposition and Redeposition Tests

Combinations of these two types of tests constitute the vast bulk of the literature devoted to the study of soil redeposition. Soil deposition is defined as the precipitation of soil from a suspension onto a substrate. It is differentiated from redeposition in this chapter, by the fact that "deposited" soil is not soil removed from one substrate and plated back either onto the same or an accompanying substrate. It is simply soil added to a bath and deposited onto the substrate during the course of the investigation. In recent literature, some doubts concerning the efficacy of deposition tests in the measurement of redeposition have been expressed. These doubts are covered elsewhere in this chapter.

Detergency literature is so replete with deposition and redeposition tests that the field must be subdivided in order to clarify a chaotic situation. For purposes of this chapter, the following subdivisions have been used:

Tests Performed for:

1. Evaluations of Surfactants
2. Evaluations of Builders with and without Surfactants
3. Evaluations of Antiredeposition Agents
4. Evaluations of Substrates
5. Evaluations of Various Soils
6. Evaluation of Mechanical Action
7. Evaluations of Test Methods

1. Evaluations of Surfactants

In a balanced laundry system, the surfactants employed serve the dual functions of, first, removing soil and, second, holding it in suspension so that it can be carried away from the fabric, thus preventing redeposition. It is natural then that deposition and redeposition tests be used to evaluate surfactants.

Powney and Noad (5) evaluated soaps as suspending agents for ilmenite black soil in the presence of cotton fabric. Their reason for selection of ilmenite was the difficulty in obtaining reproducible results with various carbon blacks. In this series of experiments they showed that the presence of small quantities of inorganic salts in the soap-ilmenite-cotton milieu served to decrease the suspending power of the soap as measured by increased darkening of the cotton. They found silicates least offensive among these inorganic salts. They also found that suspending power of soaps against ilmenite was a direct function of the chain length of the fatty acid used. The long-chain synthetic surfactants evaluated were found to be inferior to the soaps.

Working with a wide selection of carbon blacks, Weatherburn et al. (6) also investigated the suspending power of various soaps in the presence of cotton cloth. Their results are shown in Fig. 8.1.

These results tend to be in agreement with those of Powney and Noad (5) in that the longer-chained soaps show better protection than the sodium laurate. The curious inflections of the suspension curves of the saturated soaps were thought to be real and to be related to the critical micelle concentrations of those soaps.

Suspension power of 0.1% soap solutions at various temperatures is shown in Table 8.1 (6).

These results are similar to those obtained by Caryl (9) who worked with sulfosuccinic acid esters. Caryl showed that the more soluble esters

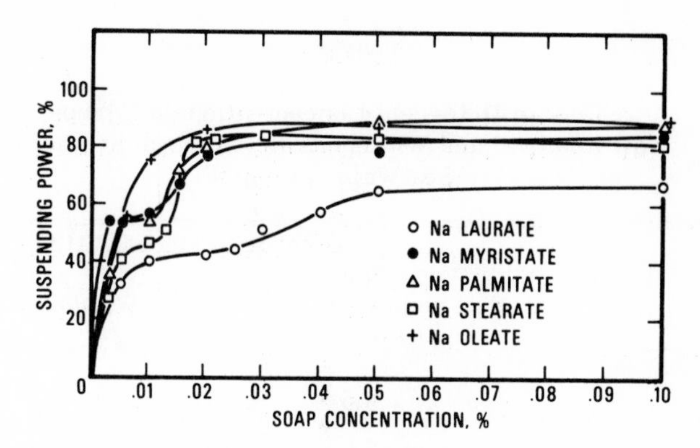

FIG. 8.1. Suspending power of pure soaps at 60°C.

TABLE 8.1

Effect of Temperature on the Suspending Power of 0.1% Soap Solutions

Soap	Suspending power, %			
	25°C	40°C	60°C	80°C
Sodium laurate	80.0	69.7	67.5	52.2
Sodium myristate	—	85.4	84.0	77.6
Sodium palmitate	—	—	88.2	94.5
Sodium stearate	—	—	82.0	95.6
Sodium oleate	88.7	89.9	87.9	88.5

(to be compared to sodium laurate soap) lost their wetting power at elevated temperatures while less soluble esters (to be compared to palmitates, stearates, etc.) seemed to gain in wetting power. Weatherburn assumes the negative protectivity of laurate soap to be analogous to this change in wetting power.

The universality of the deposition method is exhibited in the work of Vaughn and co-workers (10). In this program, the investigators were studying soil removal and whiteness retention in sea-water laundering of clothes. They used 0.25% surfactant solutions in synthetic sea water, and evaluated whiteness retention of cotton cloth against a suspension of carbon black. Typical results are shown in Table 8.2 (10).

TABLE 8.2

Effect of Detergent Concentration on Carbon Soil Removal and Whiteness Retention in Synthetic Sea Water at 120°F

	Detergent concentration			
	0.25	0.5	0.25	0.5
Product	Soil removal, mg/liter		Whiteness retention, %	
C[a]	3.5	3.5	13.1	18.2
E[b]	5.5	7.6	32.7	37.3
F[c]	5.4	5.7	45.8	36.3
G[d]	5.0	5.6	24.0	24.4

[a]Na salt of fatty acid sulfonated amide.

[b]Ethoxylated fatty acids mixture.

[c]Alkylaryl polyether alcohol.

[d]Sorbitol monooleate ethoxylate.

Vitale (11) used deposition techniques to investigate soil buildup of Aquadag, a suspension of graphite in water, on cotton cloth. He investigated various pure soaps, built soaps, and builder materials in this system. His results with pure, unbuilt soaps tend to verify those of Powney and Noad (5) who worked with ilmenite and Weatherburn et al. (6) who worked with other forms of carbon. Evaluating Aquadag deposition versus redeposition results obtained in washing naturally soiled shop towels, Vitale concluded that the lab test and natural soil tests gave correlatable results. He also showed fatty acid soaps to be superior to pure synthetic detergents in soil-suspending power.

Some of the doubt which today centers about the use of carbons as model soils was reported in 1960 by Weatherburn and Bayley (12). In these experiments, two carbons prominent in detergency research-Micronex, a channel black, and Aquadag, a graphite—were selected for investigation. Figures 8.2 and 8.3 show their results of deposition tests on cotton fabric.

The slopes of curves (A) and (B) in each figure show why the indiscriminate and nonselective use of carbon as soil can lead to conclusions of questionable validity. Weatherburn and Bayley blame the sodium sulfate in the Naccanol NR (sodium alkylbenzene sulfonate) for the discrepancies

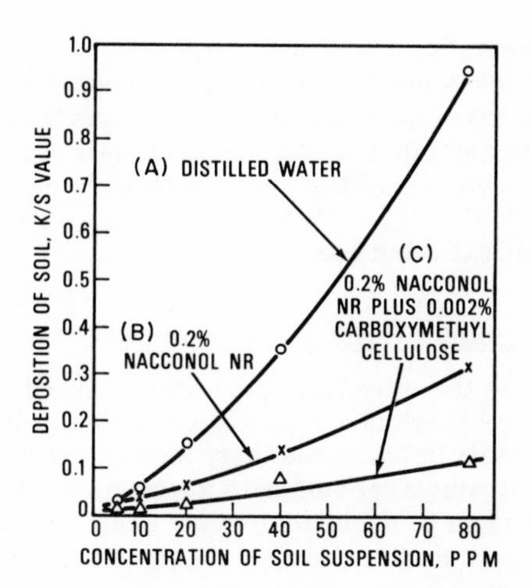

FIG. 8.2. Deposition of Micronex soil: (A) distilled water; (B) 0.2% Nacconol NR; (C) 0.2% Nacconol NR plus 0.002% carboxymethyl cellulose.

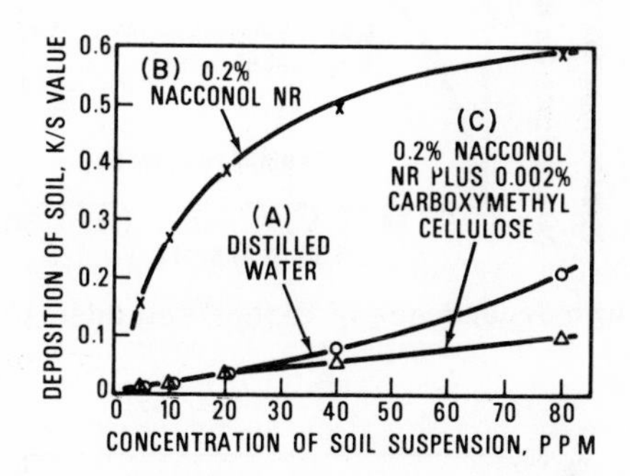

FIG. 8.3. Deposition of Aquadag soil: (A) distilled water; (B) 0.2% Nacconol NR; (C) 0.2% Nacconol NR plus 0.002% carboxymethyl cellulose.

between the two curves in each graph. The significance of curve (C) in each figure can be inferred from material presented later in this chapter.

Strauss (13) used deposition techniques to study the adsorption and desorption of carbon soil onto and from cotton cloth. Figure 8.4 is typical of his results. Here he shows that in an agitated system at 95°C in the presence of a high concentration of surfactant, equilibrium between soil in suspension and soil on fabric is attained in about 100 min.

Figure 8.5 shows the effect of surfactant concentration on the adsorption of carbon onto cotton, according to Strauss (13).

Thus, it can be seen that numerous investigators have used deposition and redeposition techniques to determine the role of the surfactant in soil suspension. They have investigated types of surfactants, concentrations, time, temperature, and other facets of the overall detergency process; but they have left unanswered the question brought out, but not asked, by Weatherburn and Bayley (12): "Why do different carbons show differing results under identical conditions?"

2. Evaluations of Builders with and without Surfactants

Sodium sulfate was the offending agent which caused the differences in deposition of channel black and graphite according to Weatherburn and Bayley (12). This salt is found routinely in built detergents, particularly those containing the alkylaryl sulfonate surfactants. In this case the salt occurs as a by-product of manufacture; but numerous inorganic salts are added to the detergent mixture to alter in some way the total performance

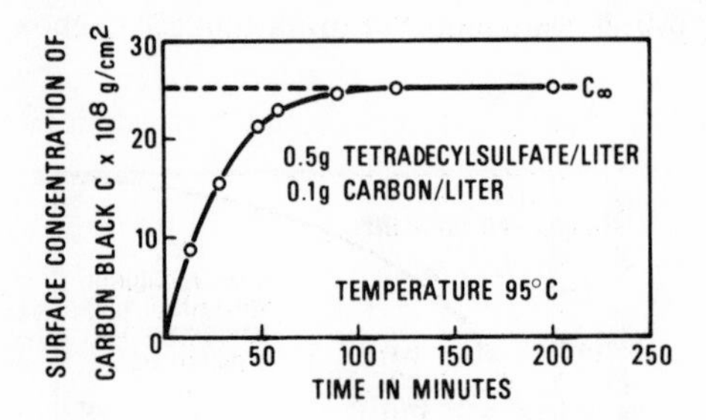

FIG. 8.4. Time dependence of carbon soil adsorption on cotton.

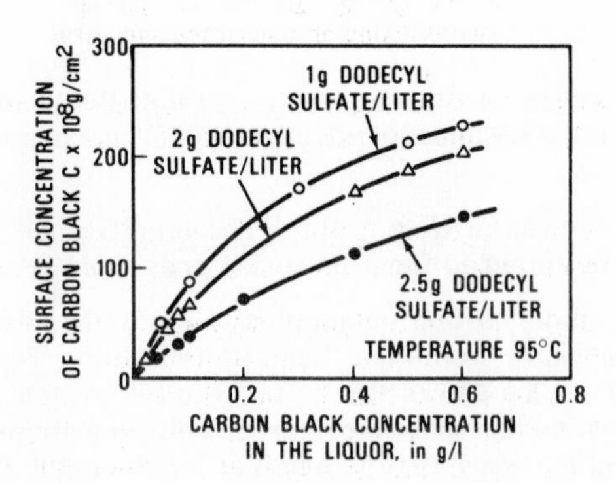

FIG. 8.5. Adsorption isotherms for carbon soil on cotton in increasing quantities of dodecyl sulfate.

of the blended product. Improvement over the performance of the surfactant alone is called "synergism," and products added to achieve "synergism" are called builders. Common builders in present-day detergent systems are the carbonate, silicate, and polyphosphate salts of sodium. In very recent times chelating agents such as tetrasodium ethylenediamine tetraacetate and trisodium nitrilo-triacetate have begun to appear as builders in detergent formulations also.

Deposition methodology has been used to evaluate builder-surfactant synergism for many years. Among the early researchers was Carter (14) who, using a variety of soils including ferric oxide, carbon black, yellow ocher, raw umber, burnt umber, and vermillion, investigated the value of silicates as detergents. He concluded that while soap was best for removing soils from cotton, a mixture of soap and silicate "seemed to combine the good properties of the two" in regard to deposition prevention.

Carter and Stericker (15) carried this study of silicates forward by studying the role of hard water in deposition onto cotton of the same general class of soils. They concluded that mixtures of sodium oleate and sodium silicate (1 Na_2O: 3.25 SiO_2) were superior to sodium oleate alone in hard water. Again, synergism has been demonstrated by deposition techniques.

The importance of the ratio of silicic acid to sodium oxide in silicate builders was shown by Powney and Noad (5). Their work indicated that those silicates with higher SiO_2 contents were more efficient than others in minimizing deposition of ilmenite. Weatherburn et al. (16) confirmed this finding.

As World War II erupted in 1939, builders became of prime importance because of shortages of fats for soap manufacture. Powney and Noad (5) investigated numerous such builders using ilmenite soil and cotton fabric. They concluded that simple alkalis such as sodium hydroxide and sodium carbonate caused increased deposition. They attributed this increased deposition to the sodium ion effect rather than pH. These results had been reported earlier by Snell (17) and Morgan (18). They showed that certain silicates and phosphates exerted considerable protective action due to "selective adsorption of the anions." Sodium hexametaphosphate and tetrasodium pyrophosphate were found to be particularly good in deposition retardation.

Vaughn and Vittone (19) verified the work of Powney and Noad. Figure 8.6 shows results which they obtained using 0.1% soap solutions and 0.05% builder. The letters "P.B." in Fig. 8.6 are an abbreviation for the term "proprietary builder." P.B. No. 1 was a mixture of silicates ($2.6Na_2O$: $1SiO_2$), carbonate, and tetrasodium pyrophosphate; P.B. No. 2 was a mixture of silicate ($2.0Na_2O$:$1SiO_2$), carbonate, and "minor compounds"; P.B. No. 3 was silicate ($3.4Na_2O$:$1SiO_2$), tetrasodium pyrophosphate, and

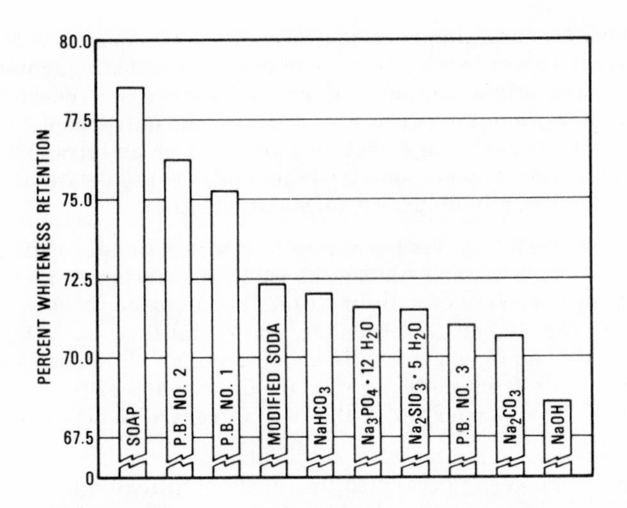

FIG. 8.6. Comparative whiteness retention of various built soap solutions.

"moderate amounts of organic matter." Again the role of silicates shows in whiteness maintenance, but the better results obtained by other researchers with polyphosphates are absent. The soil in these experiments consisted of a mixture of carbon black, cooking oil, and lubricating oil.

Vaughn and co-workers (10), working in the ultimate of built systems, sea water, were able to devise a detergent system to maintain whiteness on cotton uniforms. They used deposition techniques to develop the method in the laboratory.

Weatherburn and associates (16), working with soil suspension techniques in the absence of fabric, reported sodium hexametaphosphate as far superior to other builders in respect to static soil suspension. They showed little difference between other builders save sodium hydroxide and sodium orthosilicate which were rated as being very poor. They hastened to say that this type of test was "not necessarily indicative of the overall efficiency of builders" in the wash process.

It has been said (20) that soil redeposition was not a problem until synthetic detergents replaced soap after World War II. While this is probably an overstatement of fact, it has certainly been shown that soaps, in widest use before 1940, were superior to the unbuilt synthetic products which achieved prominence after the war. Because of availability at low, stable prices, these products replaced soaps quickly in commercial detergents made for the home. Their growth coincided with the rapid growth of the home washing machine also. Superior builder systems were required to make the new products operative in the home washer, and research was speeded to achieve the proper builder systems.

Using deposition techniques, Merrill and Getty (21) compounded a salt-free alkylaryl sulfonate with various builders to achieve a workable compromise. Silicates and tetrasodium pyrophosphate were found preferable to carbonates against a variety of soils including raw umber, ilmenite, and iron oxide.

Ross et al. (22) assert that the primary causes of soil redeposition are: (1) the presence of polyvalent ions, calcium, and magnesium in hard water; and (2) the introduction of high concentrations of sodium ion by the washing compound. In support of this, they present Fig. 8.7.

With these curves, they show that the deposition of graphite (Aquadag) onto cotton cloth follows the Schultz-Hardy rule. They continued their study to show that polyphosphates and tetrasodium ethylenediamine tetraacetate were particularly effective in suppression of deposition because they removed the polyvalent ions from solution and thus assisted in maintaining the colloidal stability of the detergent bath.

Rutkowski and Martin (23) extended the pioneering work of Ross, Vitale, and Schwartz by utilizing radiotracer techniques to study the deposition of a clay onto cotton. ^{45}Ca-tagged hard water and ^{32}P-tagged sodium tripolyphosphate were the radioactive ingredients used in their studies, as was ^{32}P-tagged tetrasodium pyrophosphate.

Their results showed that the adsorption of calcium and polyphosphate by cotton in a hard-water wash was dependent upon the calcium polyphosphate phase extant in the bath. A heterogeneous (precipitate) phase results in considerable deposition of the calcium polyphosphate salt, while little adsorption occurred in a homogeneous (solution) phase. Deposition of clay was not influenced by the presence of the insoluble calcium polyphosphate salt; however, there appears to have been a direct correlation between clay deposition and the number of unsequestered or unprecipitated

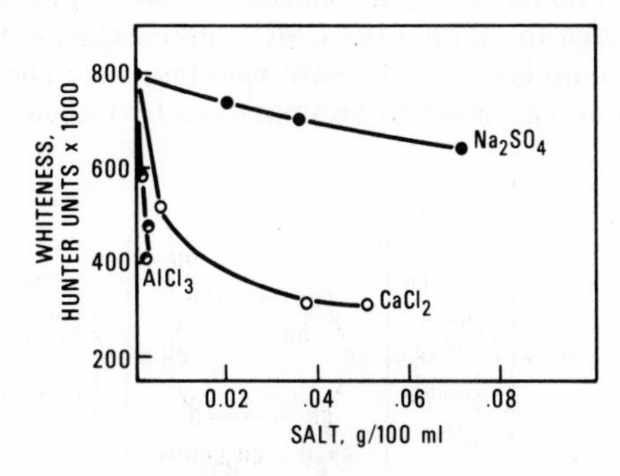

FIG. 8.7. Deposition at 120°F with mono-, di-, and trivalent cations, 0.035% sodium alkyl sulfate, 0.03% Aquadag.

calcium ions remaining in the bath. Tetrasodium pyrophosphate is shown to be somewhat more effective than sodium tripolyphosphate in minimizing clay deposition onto cotton.

3. Evaluations of Antiredeposition Agents

The shortage of fats during World War II, in both allied and axis nations, brought forth the wide usage of synthetic detergents usually derived from petrochemicals. Research showed almost at once that these products were inferior to soaps in regard to whiteness maintenance. Thus was born the necessity for the "antiredeposition" agent.

The first antiredeposition agent to be used commercially, apparently, was Tylose HBR (24). This product was composed of carboxymethyl cellulose, commonly abbreviated "CMC." Its chemical structure is shown in Fig. 8.8 (25). Nieuwenhuis (26) appears to have been among the earliest researchers to investigate the efficiency of CMC and to publish his results, although numerous publications had been made of German usage of this material by allied scientific investigating teams. The earliest known use of CMC as a detergent additive was covered in a French patent in 1936 (27). Nieuwenhuis showed that a soap-CMC mixture was more efficient than a soap-sodium carbonate mixture at higher use concentration. He also used CMC with a synthetic detergent, Teepol, and showed that white redeposition fabrics were whiter after washing than before. His soil was a complex mixture of oils, soot, and finely ground silica.

Some of the early American work with CMC was done by the Mid-West Section of the American Association of Textile Chemists and Colorists through a committee headed by Armstrong (25). This group investigated the reasons for the antiredeposition properties of CMC. They showed that the solubility of CMC was a function of the degree of substitution of the (OH) groups on cellulose by the glycollic acid group. They further showed that the viscosity was a function of the length of the cellulose molecule employed in making the CMC. Investigating the combination of CMC with ten surfactants of all ionic species, they showed deposition properties of all save cationics to be improved by the use of CMC (see Fig. 8.9).

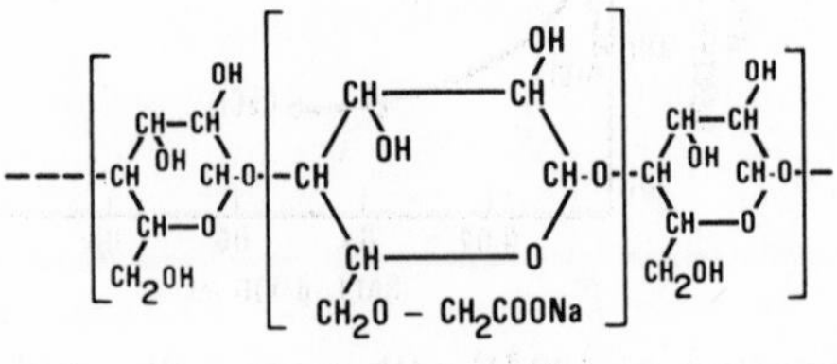

FIG. 8.8. Chemical structure of carboxymethyl cellulose (CMC) (25).

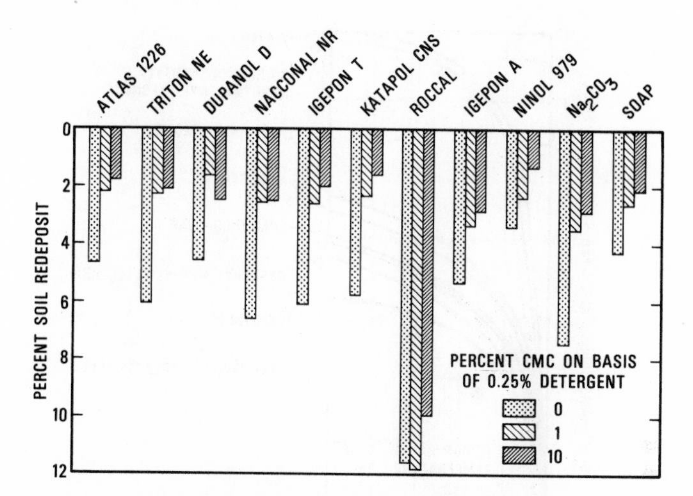

FIG. 8.9. Effect on soil redeposition of various concentrations of CMC in various surfactant solutions.

Bayley et al. (28), using Micronex carbon black, showed that use of CMC with soap was of no particular advantage; but with built soaps, wherein some suspending power of the soap was lost due to the builders, CMC was of value in recovering that lost suspending power. Differences between the effectiveness of CMC in soap and synthetic detergent solutions was shown by Law (29). With nonionic surfactants, improvements in detergency performance were detectable from the outset of experimentation; whereas with soap, improvements were noted only after four washes.

Figure 8.10 shows the efficiency of CMC and other antiredeposition agents against a carbon soil in the presence of cotton cloth according to Bartholomé and Buschmann (30).

These same authors show CMC to be more effective than other agents regardless of soil concentration. Of interest in this graph is the use of polyvinyl alcohol—a polymer currently used for its antiredeposition properties. Use of polyvinyl alcohol as an antiredeposition agent is covered in U.S. Patent 3,144,412 (31). Figure 8.11 (30) shows the effect of concentration of CMC on carbon soil deposition.

To augment claims made earlier (28), Bayley and Weatherburn (32) showed the influence of 0.005% CMC on built soap solutions. These results are shown in Fig. 8.12. In this figure they show that soap-sodium carbonate systems are improved most by CMC; the silicate built systems improved somewhat less. The same Micronex carbon soil was used as in (28).

Sanders and Lambert (33) deplored the use of carbon because it was not related to natural laundry soils. They compiled a complex mixture of

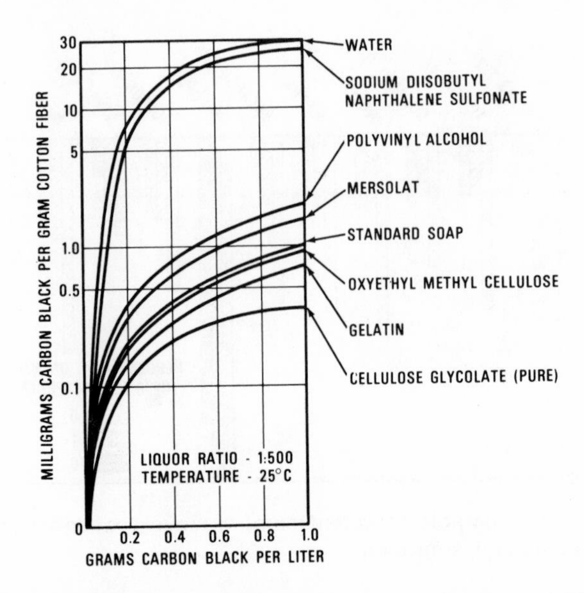

FIG. 8.10. Distribution of carbon soil between liquor and fiber with various additions to liquor (concentration of the liquor: 1 g/liter).

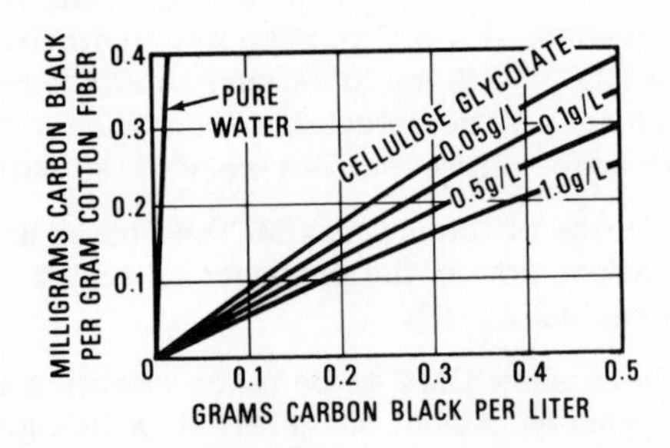

FIG. 8.11. Distribution of carbon soil between liquor and fiber (liquor being pure water and water solutions of cellulose glycolate).

naturally occurring soils and used this to test the efficiency of CMC as an antiredeposition agent. Adding CMC and tetrasodium pyrophosphate singly and together to an alkylaryl sulfonate solution, they showed the tertiary mixture to be better in all cases against redeposition. The same was found true for nonionic surfactant solutions; and additionally, the nonionic system somewhat better than the anionic in preventing whiteness loss on cotton cloth. They next took their experiments one step further and showed that carboxymethylation of the cotton cloth was an effective redeposition prevention method.

Jarrell and Trost (34) studied the use of CMC as a "size" for cotton cloth, and evaluated the resistance of this "size" to wet soiling with carbon soil. Their results indicated that an amount of CMC applied directly to cloth was more effective than the same amount of CMC added to the wash solution. They also investigated other polymeric materials for anti-wet-soiling properties. Polyvinyl acetate, polyvinyl alcohol, sodium alginate, and certain modified starches were found less effective than CMC.

Fong and Lundgren (35) researched a wide variety of polymeric materials in search of antiredeposition agents other than CMC. They found that certain proteins were quite effective in preventing carbon deposition on cotton fabric, and showed that proteins rich in the amino acid proline were particularly adept at whiteness maintenance. Proline itself had no such property.

In a second phase of these experiments, Fong and Lundgren (35) showed that certain uncharged synthetic polymers were effective in whiteness retention of cotton versus a carbon soil. Of particular interest was polyvinylpyrrolidone. Figure 8.13 shows the chemical structure of polyvinylpyrrolidone (36). A U.S. letter patent was subsequently issued to Fong et al. (37) covering the use of this compound for antiredeposition.

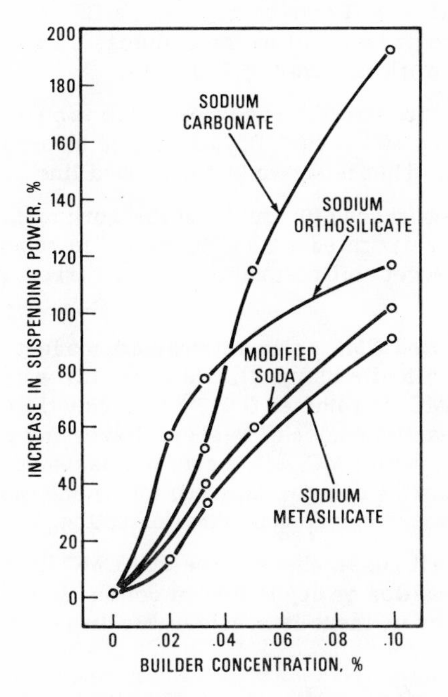

FIG. 8.12. Increase in the suspending power of built soap solutions due to the additon of 0.005% CMC.

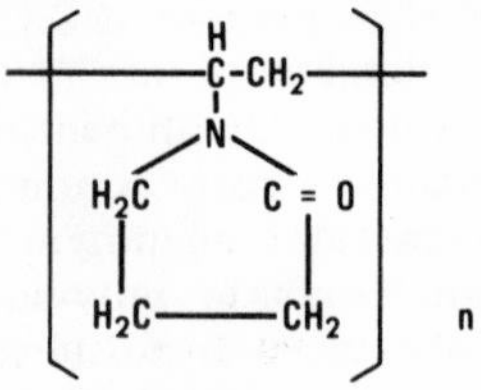

H
$-C-CH_2-$
N
H_2C $C = O$
H_2C — CH_2 n

PVP

FIG. 8.13. Chemical structure of polyvinylpyrrolidone (36).

Martin and Davis (38) researched the use of CMC in preventing deposition of clay onto cotton cloth. They showed that additon of 1% CMC, based on dry weight of detergent, showed a marked improvement over no CMC. Addition of 5% and 10% CMC again demonstrated improvements, but they concluded that 1-2% CMC in a detergent mix offered the best economic compromise.

Trost (20) showed the effect of degree of substitution (DS) of glycollic groups on the antiredeposition properties of CMC.

Figure 8.14, shows the rapid decline of effectiveness in prevention of deposition of a carbon soil onto cotton above a DS of about 0.7 glycollate groups per anhydroglucose unit on the cellulose. This DS is verified by Batdorf (39). His work is shown in Fig. 8.15.

In this case, Batdorf shows, in addition to the same curve shown in Fig. 8.14, the amount of CMC of each DS that can be incorporated into a liquid detergent formula. This is shown by the dashed line in Fig. 8.15.

Beninate and co-workers (40) emulated the work of Jarrell and Trost (34) in that they incorporated a 2% CMC "size" to resin-treated cotton fabrics to improve wet soil resistance when a carbon-laden soiling solution was employed.

Schott (41) evaluated CMC as an antideposition agent for cotton cloth against a montmorillonite clay soil. He found that when the cloth was pretreated with CMC, it retained 0.017% less clay than fabric not pre-treated after the same number of washes. Reversing the process, he pretreated the clay with CMC. This clay was taken up to a lesser extent by the fabric and that which was taken up was much easier to remove than the clay not pretreated before deposition on cotton.

As can be seen, all researchers agree that CMC is an effective agent for minimizing redeposition or deposition on cotton cloth. On the mechanism of this effect, however, there is considerable disagreement among researchers.

Mechanism of Action of Antiredeposition Agents. Since their introduction into mass usage during the late 1940s, the mode of operation of antiredeposition agents has been studied by various scientists. The methods employed by these researchers have generally been quite sophisticated techniques brought over from physical chemistry and applied to the problem. Usually, however, they have attempted to verify their conclusions through soil deposition or redeposition studies. Nieuwenhuis (26), in 1947, cited three circumstances which he held to be responsible for the actions of CMC:

1. The close molecular relationship between the cellulosic portions of the CMC molecule and the cotton fiber.

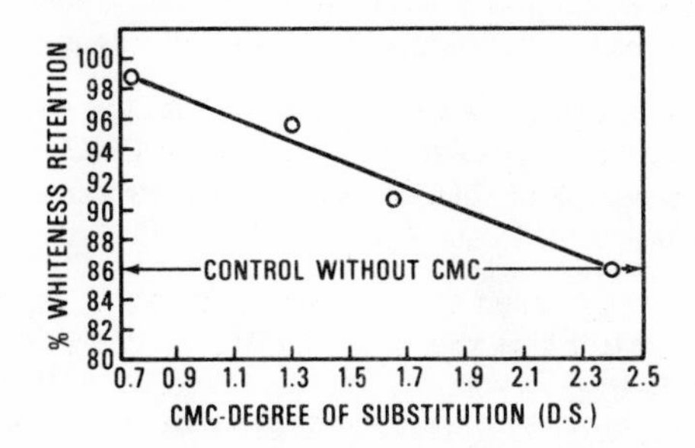

FIG. 8.14. Effect of degree of substitution on the antiredeposition properties of CMC. (20).

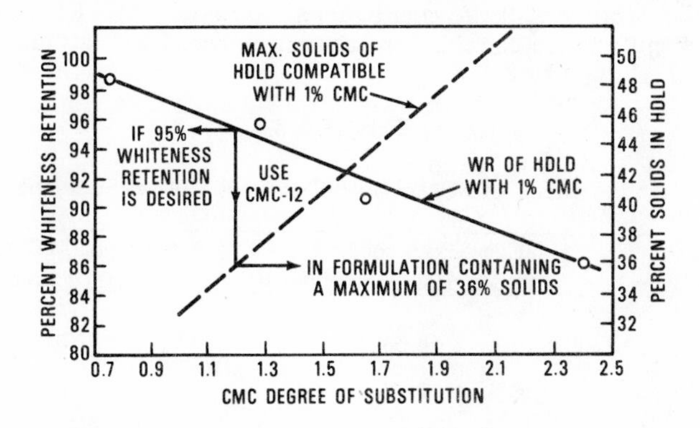

FIG. 8.15. Compatibility and whiteness retention of CMC with heavy-duty liquid detergents (39).

2. The presence of carboxyl groups which give negative charges to both the soil and to the fibers which adsorb CMC. The COOH groups also serve to bind a considerable amount of water through hydration.
3. The COOH groups, on CMC and on cotton cellulose, have fixed, similar positions in relation to one another.

He postulated that the CMC adsorbed onto both soil and cotton and created a repulsion between the two. He based these conclusions on deposition tests conducted with his standard soil of carbon, vegetable oil, and ground silica.

By 1954, Nieuwenhuis (42) had extended his work to other fibers and concluded that CMC was ineffective on noncellulosic fibers for the reason hydrogen bonds could not form between CMC and noncellulose molecules because of the adverse positioning of the COOH groups.

Johnson and Foster (43), however, showed that the degree of carboxylation of the cellulose substrate caused a reduction in CMC adsorption by the cotton. They demonstrated this on cotton fabrics variably bleached with sodium hypochlorite; see Table 8.3.

Stüpel and Rohrer (44) used a fluorescent dyestuff, Acridine Orange, in a series of investigations of the role of CMC in the wash bath. They first treated bleached cotton, rayon, ramie, and wool with the dyestuff and found no fluorescence on the fabrics. Next they added 0.0025-0.01% CMC to Acridine Orange solutions and could determine no fluorescence. When they treated fatty soils and pigments with CMC-dyestuff complex, they did find fluorescence. From these, Stüpel and Rohrer concluded that the CMC operated by adsorption onto soil but not onto fabric. Nieuwenhuis disputes the validity of these conclusions, however (42). Fong and Ward (45) also dispute these conclusions, asserting that CMC is "highly

TABLE 8.3

Adsorption of SCMC onto Oxidized Cotton Sheeting

Length of oxidation time	Carboxyl content, mmole/100 g cotton	Amount SCMC adsorbed, μg/g cotton
Control	1.55	70
6 hours	1.32	80
1 day	1.77	70
4 days	4.54	55
7 days	7.13	45

substantive" toward cotton. They used deposition of a channel black (carbon) soil in deriving data to back this conclusion.

In what must be ranked as one of the best studies of the redeposition process, Stillo and Kolat (46) outlined three primary factors which govern colloidal stability and showed that soil deposition was controlled by them. The factors were:

1. There is present an electrical force which may be either attractive or repulsive and is due to a double layer phenomenon.
2. There are present attractive forces which arise from London or dispersive forces.
3. There exists a nonelectrical repulsive force that is due to a nonelectrical interaction of the surface adsorption layer of the soil particles and fiber.

They agreed with Nieuwenhuis that CMC was adsorbed on both soil and fiber; these adsorptions serving to increase the electrical repulsive forces extant between soil and fiber. Through investigations of electrophoretic mobilities, they established that the zeta potentials of cotton and carbon soil were equalized in an alkaline bath when CMC was added. Rutkowski (47), though she did not determine zeta potentials, showed that the electrophoretic mobilities of cotton and clay indeed were equalized in the presence of CMC and anionic surfactant under all water conditions; but that with CMC-nonionic surfactant solution, the velocities were equalized in hard waters but not in deionized water. She correlated these studies with clay deposition tests on cotton cloth and concluded that double-layer forces of repulsion were not of major import in preventing clay deposition. She pointed out that some correlation seemed to exist between deposition and adsorption processes which occurred at soil-solution and fiber-solution interfaces.

Lange (48) pointed out the negativity of soils and fabrics in water and showed that anionic surfactants, certain alkaline builders, and CMC increased the negative interfacial potentials between soil and fiber. He hypothesized that this increased negativity accounted for the efficiency of these products in the washing process. Kling and Lange (49) also showed that anionic detergents, as well as many builders, caused increased repulsion between dirt and fiber. Von Stackelberg and co-workers (50) in 1954 showed that materials which influenced the charge on common textile fibers also influenced the charges on soil particles in the same bath. (If the reader desires to read further on the subject of the electrical phenomena at the soil-fiber interface, he can find an explanation in Chap. 4 of this volume.)

The data reported by the preceding authors establishes firmly that CMC is absorbed on both fiber and soil during the washing operation. This was not fully agreed upon in early studies of the role of CMC. Stawitz (51) in 1956 found no adsorption of CMC on highly cleansed, unbleached cotton. Bartholomé and Buschmann (30) reported in 1949 that they had found 0.0037 g of CMC per gram of cotton cloth washed in a 0.01% solution of CMC. Later, in 1957, Stawitz and co-workers (52) showed an "irreversible occlusion" of about 10 μg of CMC per gram of cotton. They found no adsorption on wool, acetate, or Perlon, however. In 1960, Stawitz and Höpfer (53) showed the action of CMC to be that outlined in Fig. 8.16.

This diagram demonstrates that a cloth without CMC adsorbs soil with CMC heavily; soil without CMC is adsorbed less markedly. Fabric with CMC adsorbs CMC-coated soil very little, if any, and soil with no CMC very lightly.

Schott (41) somewhat verified these results when he showed that cotton cloth pretreated with CMC before agitation in a clay deposition slurry retained 0.017% less clay, after the same number of washes, than an untreated fabric. Montmorillonite clay, pretreated with CMC, was taken up by cotton cloth to a lesser extent than untreated clay; and that clay which adhered was easier to remove. Schott postulated a weak reversible adsorption of CMC by cotton.

Johnson and Foster (43) also showed, using radiotracer techniques, that an exchange occurred between CMC in the detergent bath and CMC on cloth during each wash. They concluded that soil removal was facilitated by this partial desorption of CMC by the fabric.

Using CMC radioactively labeled with ^{14}C, Hensley and Inks (54) showed adsorption of 400-500 μg of CMC on a gram of cotton cloth. They calculated that this was the approximate amount required for a monomolecular layer. They found no adsorption on wool, Orlon, or acetate; but on nylon and rayon established the presence of some CMC.

Johnson and Foster (43) measured initial adsorption of CMC on cotton cambric fabric at 260 μg per gram of cotton. This figure is somewhat

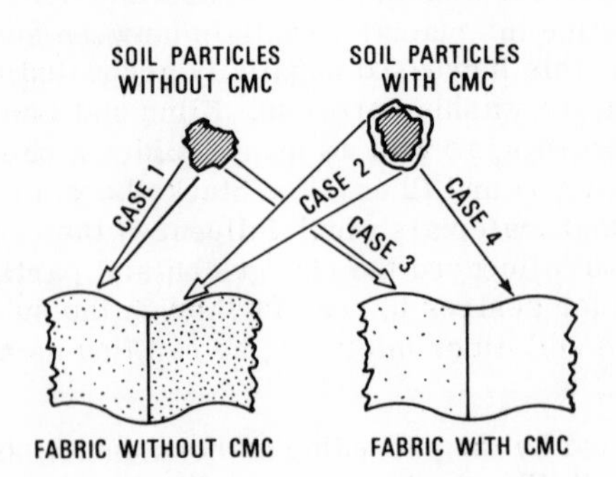

FIG. 8.16. Influence of CMC sorption on fabric and soil on graying.

lower than that reported by Hensley and Inks (54), but far in excess of that reported by Stawitz and co-workers. Johnson and Foster also reported that approximately 180 μg CMC per gram of cotton was sufficient to constitute a monomolecular layer; again lower than the results of Hensley and Inks. Batdorf (39) reported that the surface of cotton appeared to become saturated with CMC when the concentration of CMC solution reached approximately 0.006%.

As reported by Rutkowski (47), the presence or absence of polyvalent ions in the wash solution appear to influence sorption of CMC by cotton. Hensley and Inks (54) reported earlier that the adsorption of CMC by cotton depended on the presence of excess cations in solution, increasing in adsorption with cation concentration and the valency of the cations. Adsorption was insensitive to temperature and to pH within the alkaline range. Fong and Lundgren (35) also found pH no factor in adsorption between pH 3.9 and 10.2.

Ross et al. (22) demonstrated that CMC inhibited deposition in the presence of monovalent as well as polyvalent ions. They reported that CMC did not sequester or remove calcium ions from solution to any marked degree.

The theoretical and practical work of Stillo and Kolat (46) went beyond explaining the role of CMC in deposition minimization. It also established certain general properties that any compound should possess to exhibit deposition inhibition. Their work suggested that any ionogenic polymer capable of adsorption onto fabric or soil might have antiredepositional activity if the degree of substitution of ionic groups was modified so that zeta potentials of soil and substrate were equalized. They theorized that the activity of nonionogenic polymers and weakly ionogenic polymers was due to adsorption on the substrate, the result of such sorption serving to lower the nonelectrical attractive forces. The function of the additive was thought to be steric in nature.

Their work seems to verify some of the conclusions of Fong and Lundgren (35) which were reported in 1953. These investigators, working with nonionogenic and weakly ionogenic polymeric materials showed that an optimum molecular size existed for uncharged polymers used to minimize deposition of carbon soil. They established 40,000 as the optimum molecular weight for polyvinylpyrrolidone. They found polyvinylpyrrolidone insensitive to the presence of calcium ions in the bath, and found polyvinyl alcohol highly sensitive. In disagreement with other investigators, they found CMC to be moderately sensitive to the presence of calcium ions.

Extending the work of Fong and Lundgren (35), Fong and Ward (45) investigated CMC, gliadin (a protein), polyvinyl alcohol, polyvinylpyrrolidone, and polyethylene glycol. They used deposition of channel black carbon as a test method.

As a result of these studies, Fong and Ward (45) showed that CMC and gliadin increased the negative surface charge of cotton but did not alter the surface charge of the carbon soil. Polyvinyl alcohol and polyvinylpyrrolidone showed no effect on surface charge of either soil or substrate. This conclusion seems to be in agreement with that of Stillo and Kolat (46) in regard to substrate, but not completely so in regard to soil. They found polyvinylpyrrolidone substantive to carbon but not so to cotton.

They concluded that ionogenic agents influence redeposition (deposition) by their effect on the surface of the cotton, probably through an increase in the electrostatic repulsion between soil and fabric. Nonionogenics, they postulated, apparently act at the surface of the carbon soil, possibly through reduction in van der Waals attraction between fabric and soil. Another nonionogenic polymer recommended for antiredeposition properties is polyvinyloxazolidone (55).

From these studies, it seems clear that the effectiveness of CMC and other antiredeposition agents is due to the adsorption of the agent by soil, by fabric or by both soil and fabric. The last is probably the most plausible explanation. Once the agent is adsorbed, it causes a mutual repulsion to be set up between soil and substrate. The nature of this repulsion is conjectural at the moment, and one is led to agree with Trost (20) who said that the "mode of action of a polymeric additive is not limited to a single mechanism."

4. Evaluations of Substrates

The overwhelming majority of work done to evaluate soil deposition or redeposition onto fibrous substrates has been done on pure finish cotton-that is, cotton which, theoretically, has been freed of any surface additives and is presumed to present the pure cellulose molecules to the actions of the experiment. Typical prepreparation of the substrate has been repeated "strippings" of the surface with solutions of sodium hexametaphosphate (47). Such a treatment would probably rid the surface of anionic surfactant residues by conversion of those residues to soluble sodium salts, but cationic and nonionic materials would probably not be removed in this way except by dilution. Schott (56) has shown that the presence of nonionic molecules on the surface can materially change the energy system binding montmorillonite clay to the cotton substrate.

Schwarz et al. (57) after stripping with sodium hexametaphosphate, washed the cotton surface repeatedly with 0.1 N HCl to assure surface purity. They found that even momentary contact between this wet, acidified surface and blotting paper was sufficient to reestablish measurable quantities of calcium ion onto that surface. Lambert (58) accomplished this acidification with acetic acid, as did Weatherburn et al. (6).

Probably the best method of assuring surface purity is that used by Stawitz (51). This procedure involves extraction with organic solvents followed by washing with pure inorganic chemical agents in distilled water.

Another facet of substrate surface which seems to have had too little attention is the aging or chemically changing condition of the cellulose presented for activity during the experimentation. As far back as 1950, Lambert and Sanders (59) mentioned possible implications of the aging of cotton during a history of its total wash life. They deplored the drawing of wide conclusions from data based on a few washes of new cotton cloth. Johnson and Foster (43) showed that the degree of carboxylation, obtained by repeated bleachings in sodium hypochlorite, significantly changed the adsorption of CMC by cotton. They did not carry this experiment forward to report the effect of such change on deposition of soil, but the lowered pickup of CMC might be assumed to effect soil depositional properties.

Nuessle et al. (60) did show that the presence of oxycellulose at the surface of cotton fiber significantly reduced adsorption of carbon black by that surface. They also oxidized the surface by bleaching with sodium hypochlorite.

Another factor which seems to have had too little consideration in soil deposition studies is the fabric construction. Hart and Compton (7, 8) showed that macroocclusion, or interfiber and interyarn entrapment of soil particles, played a large role in both primary and secondary soil deposition. They showed that as the weight of the fabric and the complexity of fabric and yarn structures increased, macroocclusion of soil particles increased accordingly.

The above factors relate mostly to studies made on pure finish cotton fabrics. Such fabric is rapidly being replaced on world markets by blends of cellulosic fiber and synthetic polymeric fibers of various chemical species, particularly polyester. Further, these blends are almost always treated with one or more different polymeric materials to impart crease recovery or muss resistance. Such fabrics make up the great bulk of what today is called "permanent press" garments. A more recent complication has been the addition of still other polymers to the permanent press substrate to achieve "soil release" properties. All of these treatment have served to complicate still more the redeposition picture in the laundry process.

Pingree (61) asserts that all commercially used cross-linking agents (resins) for permanent press treatments are excellent scavengers for oily substances. Ghionis and Browne (62), on the other hand, show marked differences in the wet soiling behavior of certain resins against soil composed of dry-cleaner sludge. Their work, present in Table 8.4, shows distinct differences in the behavior of three such resins.

As early as 1958, Mazzeno and co-workers (63) reported wide changes in deposition properties of fabrics treated with various resins. Figure 8.17 shows the results of their work. Pure finish cotton was affected

TABLE 8.4

Wet Soiling Properties of Cross-Linking Agents

Cross-linking agents (used with catalyst and nonionic softener)	Hunter D-40 whiteness	
	Original	After one laboratory soiling
Dimethylol carbamate	95	73
Dimethyloldihydroxy ethylene urea	101	52
Dimethylol propylene urea	111	70

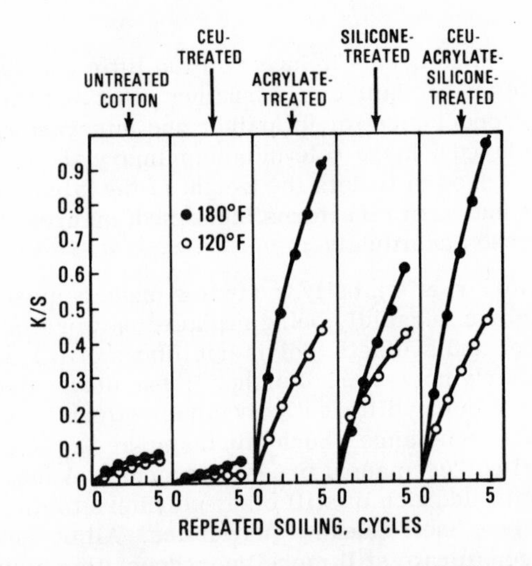

FIG. 8.17. Repeated soiling of cotton treated with the complete resin formulation and its components as compared to untreated control at 180 and 120°F.

very little by five washes insofar as deposition was concerned. The addition of cyclic ethylene urea (CEU) resin to the pure cotton seemed to effect deposition little at either 120 or 180°F. When unspecified acrylic resin and silicone resin, added singly and together, were included with the cross-linking resin, however, serious deposition occurred. This was greater at 180°F than at 120°F, but washing at both temperatures showed serious whiteness losses. The addition of CMC reduced but did not eliminate discolorations due to soil deposition. Mazzeno et al. blamed the tackiness of the added thermoplastic resins for the increased soil pickup.

Matsukawa (64), working with CEU resin, found that the addition of maleic anhydride-vinyl-methyl ether polymer was effective against soil deposition and facilitated release of adhered soil, possibly due to dissociation of the maleic residues of the polymer mixture.

Beninate et al. (40) combined CMC with dimethylol ethylene urea resin and found good resistance against wet soiling with carbon soil through 25 washes.

Somers (65) used carbon-black deposition tests to show that a series of cross-linking textile resins did not contribute to soil pickup. He confirmed the work of Mazzeno et al. (63) by showing that the addition of acrylic and silicone polymers increased deposition markedly.

The two most definitive studies of whiteness retention of resin-treated cotton fabrics appeared in 1963. Berch and Peper (66) did depositional studies using iron oxide, iron oxide-oleic acid-coated, carbon furnace black, and vacuum cleaner soils. They evaluated deposition of these soils on pure cotton surfaces, the B-aminoethyl ester of cellulose, phosphonomethylated cotton, hydroxyethylated cotton, cotton treated with acrylic resins of varying hardnesses, and cotton treated with two species of fluorocarbons.

Their results indicated that hydrophobic soils, carbon and oilborne iron oxide, showed little affinity for cotton in any hydrophilic state (pure finish, phosphonomethylated, hydroxyethylated); but that the addition of hydrophobic finishes (fluorocarbons) caused heavy soiling with such hydrophobic soils. Their results with fluorocarbons have been verified by Connick and Ellzey (67). Best resistance to wet soiling was obtained with treatments which contributed anionic groups to cellulose according to Berch and Peper (66). Carboxylates, such as CMC, and phosphonates were found to resist such deposition well. The presence of cationic amino groups on the cellulose caused heavy wet soiling with oilborne stains.

Hardness of the acrylic coatings did not appear to influence the deposition of the vacuum cleaner soil nearly as much as it did the carbon and oilborne iron oxide soils (66). Hardness of the acrylic coatings were found to be contributors to soil deposition of oilborne soils, however, confirming the results of Mazzeno et al. (63). Harder acrylics showed less soiling than softer coatings. Deposition onto such coatings was also shown to be directly proportional to temperature of the soiling bath. Silicones used in these experiments did not show temperature sensitivity because they did not soften, particularly in the range of temperatures employed in washing.

The acrylic coatings also invariably soiled more heavily than the untreated cotton.

In a later work, Peper and Berch (68) asserted that the ideal anti-wet-soiling finish for cotton, and one which would maximize release of soil, would be a thin, permanent, nonswelling coating on the fiber which possessed strong hydrophilic properties. They cited CMC as the agent which most nearly approached these criteria when the article was written in 1963.

In partial confirmation of and partial rebuttal to these works, Nuessle and co-workers (60) published the other definitive work on the role of resinous substrates in wet soiling. Their work indicated that the soft, thermoplastic acrylics did indeed contribute to wet soil pickup, but they hastened to add that other factors contributed equally as much in the process. The ionic nature of the polymer and the presence or absence of cross-linking agents and their catalysts affected greatly the ultimate degree of wet soiling. The soil employed was found to be most significant; see Fig. 8.18. This figure shows a complete lack of correlation between deposition of nine different carbon soils, manganese dioxide, and ferric chloride on two differnet resinous substrates.

Polymers containing anionic carboxyl groups tended to repel carbon soil, while cationic polymers containing amino groups tended to attract

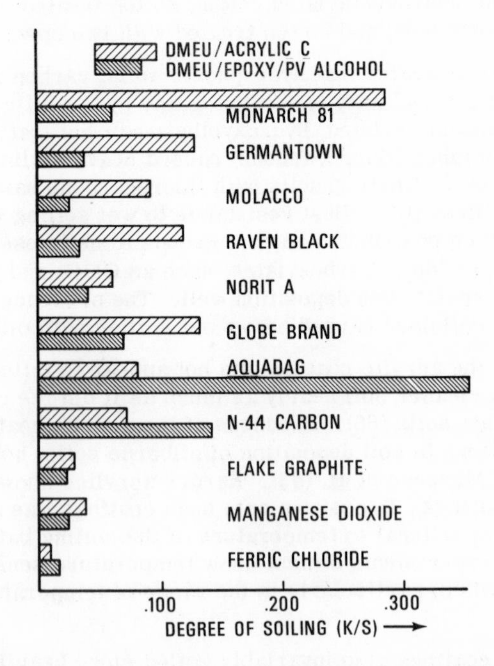

FIG. 8.18. Effect of eleven soils on two finishes.

carbon heavily in deposition tests. These results are in agreement with those of Berch and Peper (66). Nonionic polymers, made hydrophilic by virtue of the presence of many hydroxyl groups, were found to pick up carbon soil heavily. These polymers, dry, are quite hard; but when heated in a hot, alkaline bath, they tend to become hydroplastic and adsorb carbon. Polyvinyl acetate was cited as an example of such a hydroplastic polymer. When freed of vinyl alcohol, polyvinyl acetate would not wet soil in carbon-containing baths.

In summary, Nuessle et al. (60) concluded that the problem of wet soiling by thermoplastic polymers added to fibrous substrates, while not necessarily nonexistent, had been overemphasized. Other variables, they contended, were equally as important.

The most recent intensive research in the textile industry has been pointed toward the so-called soil release finishes. Actual work has been directed, advertently or inadvertently, in two directions: soil release and soil antiredeposition. Perry (69) points this out clearly, as shown in Fig. 8.19. Here Perry is showing the discrete paths soiling can take. Hydrophobic fibers are static prone and attract soil in air. They wet out poorly in water washing, hence decreased soil removal is realized. They tend also to adsorb soil from the wash bath. Hence, there are three distinct paths to soil buildup on hydrophobic fibers. The addition of a soil-release agent causes the fabric so treated to avoid all of these paths. The right-hand path is the one typically for antiredeposition finishes. The hydrophobic fibers attract oilborne material in the dirty wash water, and increased deposition is realized. The solution to this right-hand path, according to Perry, was to add to the surface of the hydrophobic (polyester) fiber a proprietary product—Cirrasol PT.

Cirrasol PT, according to Ghionis and Browne (62), renders the surface of the polyester fiber hydrophilic when deposited thereupon. This hydrophilicity is imparted by the presence in the molecule of ester-linked oxygen having two free electron pairs which bind water molecules to the surface. Oily materials present in the wash bath cannot deposit on the hydrophilic surface. As evidence of this, they present Table 8.5. Here they show serious wet soiling with high-density polyethylene softener and a "reactive" softener; intermediate soiling with a nonionic ester and cationic softeners; and good resistance to soil deposition by Cirrasol PT. The soil used in this case was sludge from a dry-cleaning still. They cited that Cirrasol PT was effective only on the polyester fiber and that this product was durable to repeated home launderings.

Aliman and co-workers (70), using a mixture of mineral oil, vegetable oil, clay, and carbon in depositional tests, showed the effectiveness of Cirrasol PT on 100% polyester, 65% polyester/35% cotton, and 50% polyester/50% cotton substrates. Figure 8.20 shows the results of their work.

Most works on soil redeposition in regard to the substrate have been

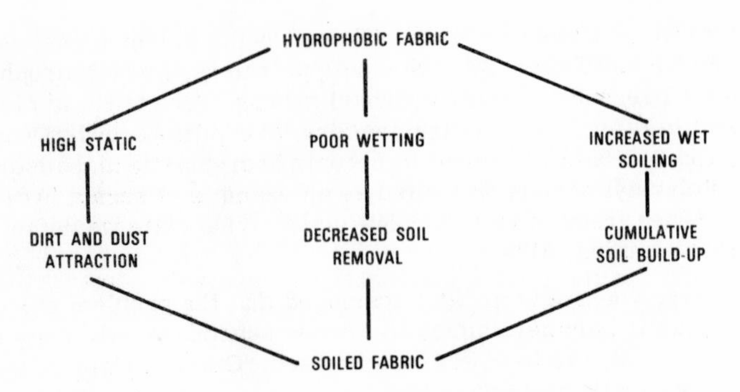

FIG. 8.19. Paths to soiling of hydrophobic fibers.

TABLE 8.5

Wet Soiling Properties of Softeners

Softeners (used with carbamate resin and catalyst)	Hunter D-40 whiteness	
	Original	After one laboratory soiling
High-density polyethylene	92	59
Nonionic ester	96	82
Cationic	98	85
Reactive	100	55
Cirrasol PT (hydrophilic finish)	105	94

done on cotton or on blends of cotton and polyester fiber with the presence of other resinous additives in the latter case. Work on the anti-wet-soiling properties of other fibers has lagged considerably, even though some work is now being published in the literature. Kennedy and Stout (71), though working on a somewhat different problem, were able to show depositional properties of numerous fibers in wet contact with a clay soil. Table 8.6 shows the approximate average Hunter reflectance readings they obtained after ten washes. Although these data were not presented in this way by Kennedy and Stout, the writer has taken the liberty of so arranging them.

Another problem in regard to deposition of soil onto synthetic fibers is presented by Oldenroth (72). Working with Perlon (caprolactam) nylon, he showed that the type of heat curing done to the nylon at the time of manufacture dictated to a marked degree what the wet soiling history of that nylon would be. Uncured nylon and that cured at 170°C in hot air showed distinctly heavy wet adsorption of carbon soil. Samples cured at 200°C in hot air of 125°C in saturated steam showed strikingly less adsorption of carbon in laundering. Nylon cured in hot air at 190°C wet soiled to a degree somewhat intermediate between the two extremes. Oldenroth cited these as evidence of the necessity for strict quality and process control during nylon manufacture. He also showed that the presence of a small amount of adsorbed detergent (surfactant) on the fiber decreased the carbon adsorption.

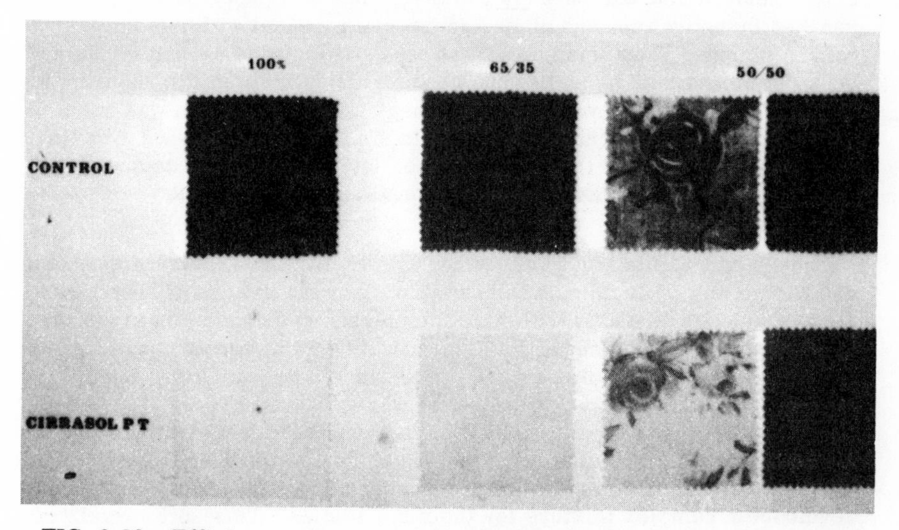

FIG. 8.20. Efficiency of antiwet soiling of Cirrasol PT, according to Aliman et al. (70).

TABLE 8.6

Hunter reflectance	Fiber
41-50	Wool, Orlon
51-55	Nylon, Dacron, Kodel, acetate
56-60	Viscose, Avril, Arnel, silk
Above 60	Cotton, Zefran

Wet soil adsorption experiments, too long confined to studies of the cotton fiber, are belatedly being extended to include the synthetic fibers and the newer resinated finishes being applied to cotton/synthetic blends. This experimentation will probably accelerate markedly in the future.

5. Evaluations of Various Soils

Detergency researchers have used a variety of materials as particulate soil during the history of their studies. This variety includes almost every colored compound, pigment, or mineral conceivable; but by and large the field has narrowed down recently to two materials: carbon and clay. Sanders and Lambert (73), in their classical paper of 1950, mildly rebuked investigators who dipped cotton cloth, coated with fat, into a carbon slurry. They pointed out that heavy deposits of carbon might occur in mechanics' overalls, but were highly unlikely in real laundry such as sheets, shirts, pillowcases, etc. They stated that carbon was not a realistic soiling material; and around this statement has raged a battle since that day. One of the chief needs of detergency research now is to find a realistic soil upon which all laboratories can agree and seek a common ground.

Sanders and Lambert (73) asserted, with good reason, that the researchers had no way of deciding whether laboratory evaluations of detergent performances correlated with practice unless performance in practice was known. They devised a practical use scheme with cotton roller towels within their laboratory and office washrooms, and proceeded to show that results obtained with these towels, washed in four different detergents, in no way correlated with end results of tests run with the carbon-vegetable oil soils then prevalent. To develop a meaningful soil, they analyzed street dust from six different cities about the country and developed a complex mixture for their lab studies. This soil correlated well with the towel test in multiple-cycle washing evaluations.

During the same year, 1950, Snell and associates (74) pointed out that natural soil was not a simple material; conversely, it was a complex mixture as found on clothing. They cautioned that for light soiling, the darkening of the fabric was roughly proportional to the amount of soil. As the degree of soiling increased, the amount of soil needed to give darkening increased greatly: "This means that the difference between good and bad wash results depends on removal of the last traces of soil." They might well have said, also, that good washing depends on successful prevention of redeposition of the removed soil. As one of their arguments against carbon, Sanders and Lambert (73) had shown that 10 mg of carbon was equal in tinctorial power to 1 g of the soil mixture which they developed. Thus, it can be seen that very small amounts of carbon can swamp out redeposition values in reflectance measurements.

Many researchers have assumed that carbon presented a neutral, nonionic surface to the surfactant and cloth in detergency tests. Doscher

(75) showed in 1960 that carbons of different types could be either positively or negatively charged in distilled water. He showed the electrophoretic mobility of Excelsior carbon to be -0.18 and that of Malacco to be +2.1.

Particle sizes of carbons used as test soils are frequently cited by researchers. These generally are figures based on manufacturers' specifications. Dannenberg and co-workers (77) destroyed the myth that these ultimate particle sizes were attained when they showed that 4 min in a Waring blender operating at 12,000-15,000 rpm was required to reduce the carbon to those sizes. They cited three separate points in the manufacture of carbon black where agglomeration of the ultimate particles occurred. In an exhaustive study in 1951, Dannenberg and Seltzer (76) uncovered other facts about carbon which had not been considered in detergency researches. The first was that there existed in carbon blacks a benzene-extractable hydrocarbon fraction which was formed by partial cracking of the raw material during the ignition and burning of that raw material. This fraction has generally not been accounted for in "grease-free" carbon systems.

Hart and Compton (8), working with grease-free carbon, reported that the suspending power of various surfactants were found to differ widely in respect to various fibers. One is tempted to speculate that they did not truly have a grease-free carbon.

In practice, clothes have been found cleaner after 1 min of washing than after 30 min. This was caused by the instability of the suspension of small particles after removal from the clothes (78). This observation was confirmed by Dannenberg et al., who showed that as the particle size is decreased, it becomes more difficult to disperse the pigment in any medium (77).

Second, Dannenberg and Seltzer (76) showed the presence of inorganic oxides on the surface of the carbon particles which could effect their polarities and suspendabilities in surfactant solutions. Third, they established that surface oxidation of the carbon took place, another factor which could affect their polarity. These factors may account for some of the differences in carbons found by Nuessle and co-workers (60), shown in Fig. 8.18. Dannenberg and Seltzer (76) presented Fig. 8.21 to show the effect of these factors on the dispersability of two different types of carbons—furnace blacks and channel blacks.

One predominant factor here is probably the difference in the amounts of metallic oxides in the two blacks. These oxides have many polyvalent cations. (77). That the suspendability of graphite was sensitive to its ash content and to its condition of oxidation had been shown earlier by Veselovski (79).

But these results have not halted carbon research. Weatherburn et al. (6) made carbon slurries at 105 cpm on a lab shaker device, running for

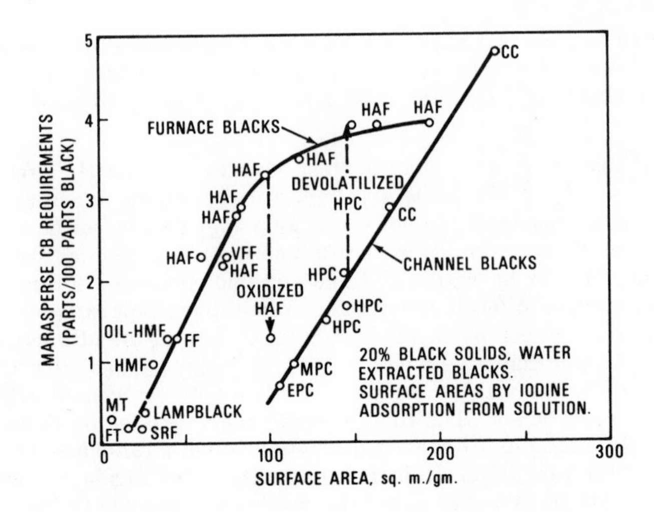

FIG. 8.21. Effect of surface area on dispersing agent requirements (76).

20 min. They also used a slurry so heavy with carbon that consecutively treated swatches showed no reflectance differences.

Compton and Hart (80) used 10% carbon soil based on the weight of cotton employed. In this series of experiments, they used a special method of deposition which will be discussed in the next portion of this chapter.

As further evidence of the unreliability of carbon soil cloths available on the market, Sanders and Lambert (33) published Fig. 8.22 in 1951.

In this series of experiments, they used four detergents developed in their natural soil simulation work (73). It will be noted that detergent No. IV, which had shown best on graphite-vegetable oil soils in earlier work, showed up as least effective on each of the commercial cloths. Cloth-to-cloth correlation was bad, as was detergent efficiency correlation.

Since the late 1950s, there has been a detectable shift toward using clays as model materials in soil redeposition studies. Hensley is one investigator who has worked widely in both carbon and clay testing, and in his work the two materials stand face to face for close examination. Working with Kramer, Ring, and Suter, Hensley in 1955 published results obtained using ^{14}C as a test soil (81). In this case, the elemental carbon was made by reduction of carbon dioxide. Carbon thus formed resembled graphite more than it did the amorphous carbons in that the individual particles were platelike structures having dimensions similar to graphite. Figure 8.23 shows results obtained with this soil.

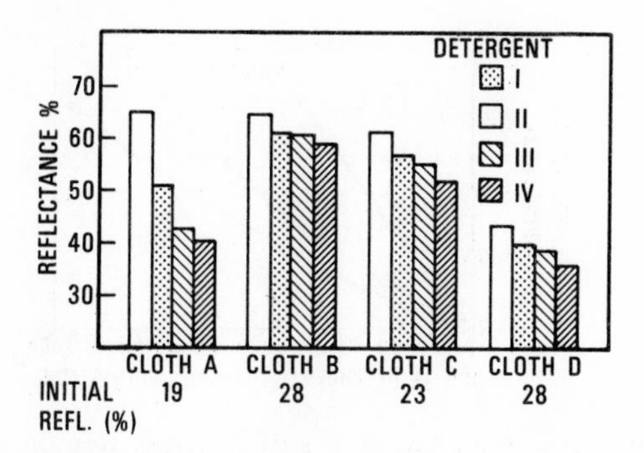

FIG. 8.22. Terg-O-Tometer wash tests with commercial test fabrics (33).

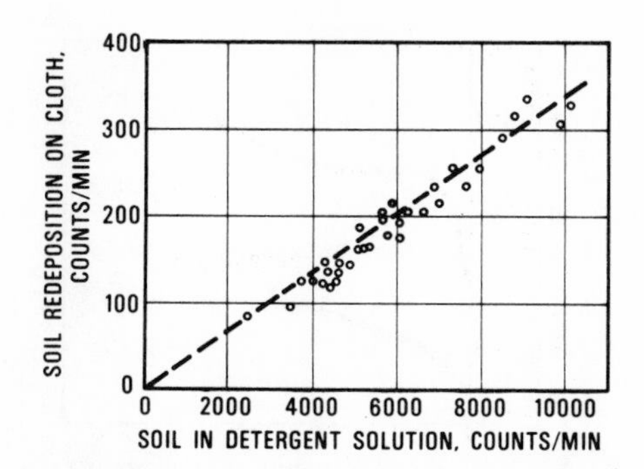

FIG. 8.23. Soil redepositions on individual cloth disks (initially clean) washed with two soiled disks.

This figure shows that graphite like-carbon deposition onto cotton cloth in the presence of an alkyl benzene sulfonate (ABS) surfactant is a direct function of the concentration of that carbon in the bath.

This figure should be compared to results obtained by Hensley and Inks (82) in working with ^{45}Ca tagged clay. Figure 8.24 shows this comparison.

The deposition of clay onto cotton cloth was found to be a straight line again in the presence of ABS-type surfactant; the amounts varying somewhat from the carbon curve of Fig. 8.23, but the trend identical. Also noted here were the influences of builders and the effectiveness of soap in clay suspension. In each case, another outstanding feature was the influence of soil in bath concentration on deposition.

Figure 8.25 shows the effect of CMC on the deposition of carbon and

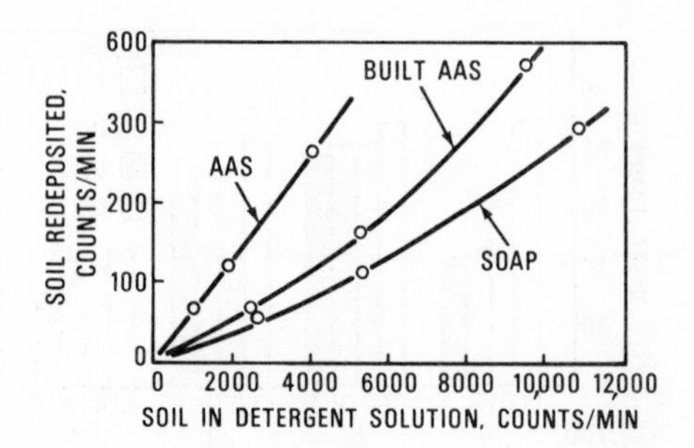

FIG. 8.24. Relation between clay soil redeposition on cloth and soil loading in solutions of alkylaryl sulfonate, built alkylaryl sulfonate, and high-titer soap.

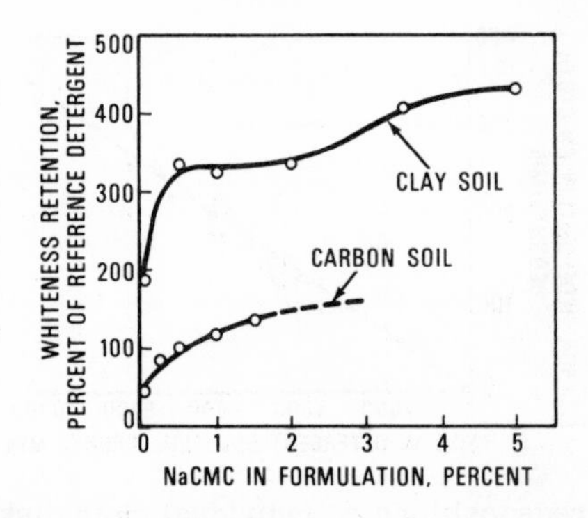

FIG. 8.25. Effect on whiteness retention of varying content of sodium carboxymethyl cellulose in household laundry detergent as measured with the ʻagged clay and carbon black soils.

clay onto cotton cloth (82). The deposition of graphitelike carbon seems to be more or less insensitive to CMC above about 2% CMC, whereas clay deposition is inhibited by higher concentrations of CMC. These latter results are in general agreement with those of Martin and Davis (38).

Powe (83) established that clay was the principal ingredient of "tenaciously bound" soils on cotton fabric. This was not surprising in view of the ubiquity of clay minerals. His electron migrographs of the soiled fibers established that the particles tenaciously bound were generally platelike in structure (see Fig. 8.26) with major dimensions ranging from 0.02-1.0 μ. Figure 8.27 shows an electron micrograph of graphite. The platelike structure is evident here also. One is tempted to speculate that this physical similarity between clay and graphite might account for the depositional similarity observed by Hensley and Inks (82).

Working with amorphous carbon soil, Hart and Compton (8) concluded that microocclusion of the soil in fiber irregularities accounted for soil adhesion. Powe (83) disputes this saying that the accumulation in fiber rugosities did not appear to be an important mechanism of soil retention since the majority of observed clay particles were 100-200 nm in major dimensions, and the upper limit of fiber rugosities was about 50 nm.

That the majority of soil tenaciously bound to fabric was clay, was established by Tsunoda (84). He showed, by mass spectroscopy and X-ray analysis, that the soil had the general formula $Mg_2Al_4Si_5O_{18}$. He also showed the analysis of metals percent to be that shown in Table 8.7. A later version of this table published by Tsunoda and Oba (110) showed some slight deviations from Table 8.7.

Using neutron activation techniques, Gordon and Bastin (85) have shown that a considerable portion of the residual clay on a washed swatch is due to redeposition rather than to retention. They showed that about half the residual clay on a cotton swatch was due to redeposition, while as much as 70% of the residual clay on nylon and polyester fibers was of the redeposited type.

Martin and Davis (38, 86) used Bandy Black clay, a kaolinite in

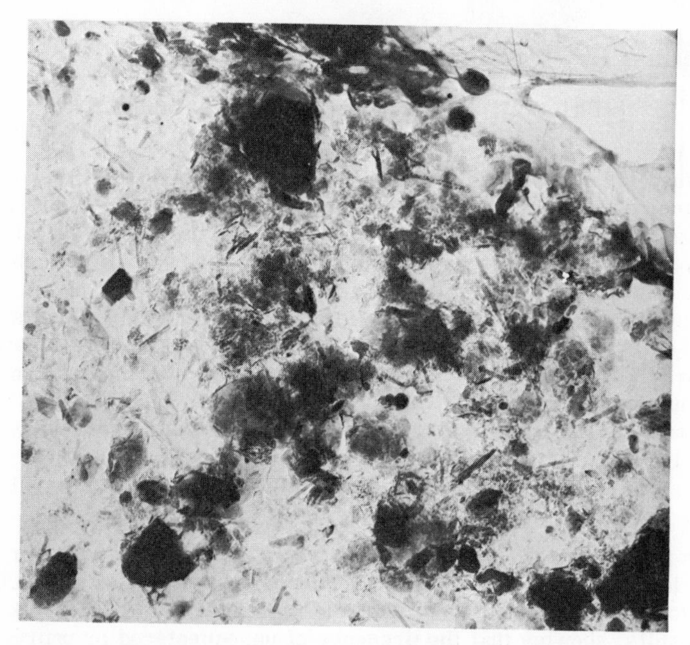

FIG. 8.26. Electron micrograph of tenaciously bound soil on cotton fiber.

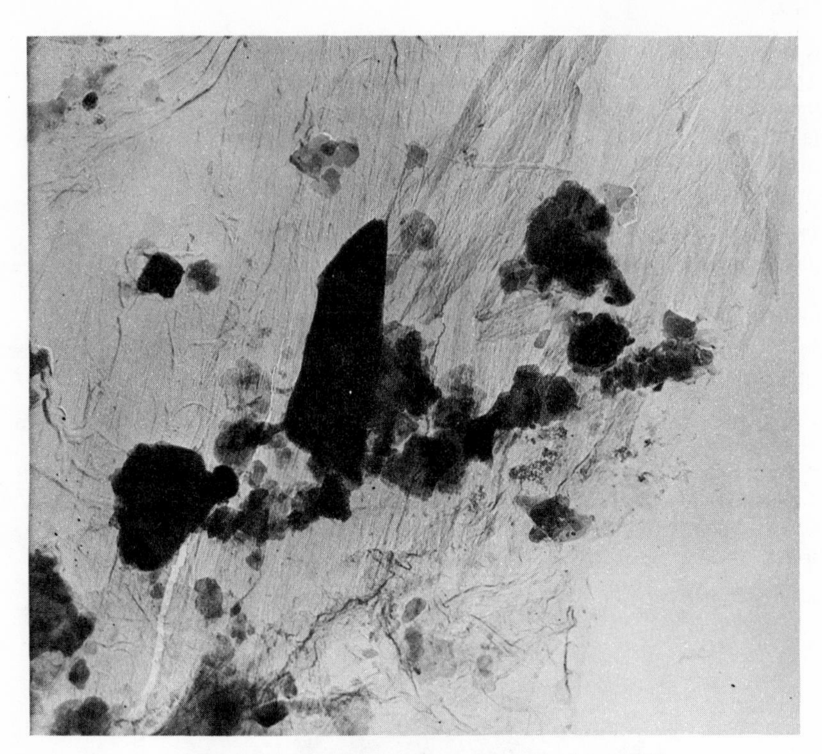

FIG. 8.27. Electron micrograph of graphite.

comparing clay deposition to that of natural soil. To do this, they obtained soil indices of loads of home-soiled clothing washed in 0.35% built nonionic detergent. This soil index was obtained by filtering a 250-ml aliquot of wash water through No. 5 Whatman filter paper in a Büchner funnel and measuring the dry reflectance of the soiled filter paper. Next they constructed a calibration curve for their clay in 0.35% of the same nonionic detergent, obtaining soil indices of varying concentrations of clay slurry. This done, they conducted deposition tests through ten washes using sufficient clay to achieve the same soil index as was realized in each wash of the home-soiled laundry. Whiteness-loss curves were essentially identical. Where reflectance values of clay versus the natural soil were plotted against each other, correlation was essentially +1.0 (see Fig. 8.28).

Martin and Davis (38) also showed that clay reacted to hard water in a manner identical to natural soil in that deposition of clay was directly proportional to water hardness. Rutkowski and Martin (23) corroborated these results, showing that the presence of unsequestered or unprecipitated calcium correlated directly with clay deposition. Clays used in these two separate experimental series were identical.

TABLE 8.7

Spectroscopic Analysis Data for
Ash Prepared by Heating Inorganic Soil

Element	Content, %	Element	Content, %
Al	Strong	Ni	0.01-0.05
Mg	Strong	Ti	0.01-0.05
Si	Strong	Cu	0.005-0.01
Ba	0.05-0.1	Sn	0.005-0.01
Zn	0.05-0.1	Sr	0.005-0.01
Cr	0.01-0.05	B	0.001-0.005
Fe	0.01-0.05	Pb	0.001-0.005
Mn	0.01-0.05	V	0.001-0.005
Na	0.01-0.05	Ag	0.0005-0.001

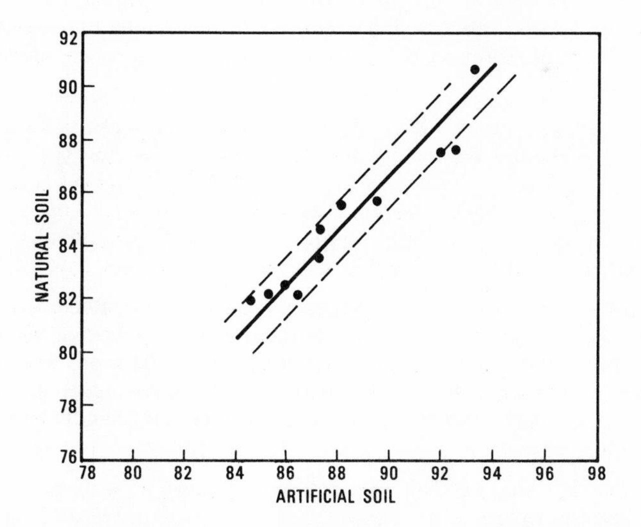

FIG. 8.28. Comparison of reflectances of swatches washed with naturally soiled laundry and swatches washed in clay suspensions.

Schott (41) showed that cotton fabric would adsorb about 0.12% clay from aqueous dispersion. He found this clay to be tenaciously bound to the cotton, repeated washings reducing it but little. He showed that nonionic surfactants prevent clay from reacting firmly to cotton by being adsorbed on both clay and cotton surfaces. Washing in nonionic surfactants left 0.04% clay (montmorillonite) firmly attached to the substrate. Schott (56) showed that the montmorillonite existed on unwashed cotton fabric in a four-layer agglomeration. The 0.04% unremoved by nonionic washing indicated that all but that layer in direct contact with the cotton was removed (41).

In some very recent work, Shimauchi and Mizushima (87) have cast doubts upon the efficacy of using clay alone as a soiling material. Using scanning electron microscope techniques, they found that clay and carbon soils reacted quite differently from natural soils in both home and commercial washings. They recommend the use of natural soils in all deposition and redeposition studies. Thus, with particulate soils, research has gone full cycle back to Saunders and Lambert (73). A large list of carbon soils used in detergency studies can be found in Chap. 9.

The subject of deposition of organic soil onto fibrous substrates is open to a lot of questions; the major query being: "Is residual soil on fabric redeposited or is it simply unremoved?" Powe and Marple (88) established that most of the organic soil present was traceable, in identity, back to body secretions. They did not, however, answer the basic question cited above. They showed the presence of rather large amounts of Ca^{2+} soaps on sheets and other fabrics, but these might have been deposited onto the fabric by Ca^{2+} interaction with laundry soap per the mechanism suggested by Albinson (89).

That surfactants react with polyvalent ions in the wash bath was shown early by Meader and Fries (90) and Boyd and Bernstein (91). Knowles and co-workers (92) went further, suggesting that oleyl methyl taurate or alkylbenzene sulfonate be incorporated in detergents as lime soap dispersants. Schwarz et al. (57) showed that even trace amounts of calcium can cause precipitation of insoluble detergent residues onto cotton cloth.

In recent years, evidence has begun to appear that indeed organic soils are redeposited. McLendon and Richardson (93) showed in 1963 that organic soil, of the body oil type, could be extracted from the folds of used, contour-fitted sheets. These extractions were made in areas not normally in contact with the human body during use, hence could only be redeposited.

Oldenroth (72) showed that very minute amounts of skin fat were retained by caprolactam nylon.

Much of the main evidence about retention of organic soils is negative in nature; that is, it has developed as a result of investigation of new anti-wet-soiling treatments for polyester fiber. Ghionis and Browne (62) used dry-cleaner still pot sludge to evaluate the efficacy of an antiredeposition treatment. This was not new, Ringeisen (94) having used such soil in 1943. Aliman and co-workers (70) simulated the dry-cleaner sludge for their evaluations. In either case, natural sludge or simulated, the textile treatment minimized deposition of the organic soil onto the treated polyester fabrics. Antideposition, and by analogy antiredeposition of organic soil had been accomplished. The fact that such soil deposited on untreated portions of the cloth indicates that redeposition of organic soil is a reality. Tsuzuki and Yabuuchi (111) emphasized the importance of fiber rugosities in the retention; and further demonstrated that multilobal fibers of polyester were more oil retentive than round cross-sectional fibers. In the future, it can be expected that continuations will be made on modified versions of the mixed organics formula used in deposition evaluations on polyester fibers. The era of simplistic testing, i.e., a single soil for all purposes, is passing. Complex organic mixtures and multicomponent particulate blends, similar to natural soils, will probably be subjects of exploration in the next decade.

6. Evaluations of Mechanical Action

The washing parameter which seems to have had the least amount of attention in soil deposition and redeposition studies is mechanical action. The necessity for mechanical action has been realized since the early days of the wash process. Powney and Noad (5) in 1939 condemned the use of static sedimentation or suspendability tests as criteria for final judgment in detergency evaluations. They affirmed that mechanical input was necessary in order to prevent readherence of removed soil to the fabric substrate. Bailey (95) established the fact that there was an upper limit to agitation speed, although he failed to define this maximum energy input clearly. He simply stated that excessive motion redeposited soil onto the fabric, and emphasized that laundry machinery in use in 1943 was geared to perform at optimum speeds of agitation. Snell and co-workers attempted to "put numbers" into the mechanical action phase of detergency testing by asserting that in a 0.3% soap solution in 15 grains/gal hard water, 40% of the brightness recovery realized was due to the mechanical input and 60% was due to the soap (74).

Vaeck and De Ville-Boorsma (96) showed that the deposition of carbon soil was due to the concentration of the soil, duration of the test, and, to a lesser extent, to the intensity of mechanical agitation. The importance of the duration of the wash cycle was shown by Rhodes and Brainard (97). These researchers established 7.5 min as the optimum time for washing. Shorter times gave inferior soil removal, and longer times gave increased redeposition. Compton and Hart (8, 80) showed that the violence of mechanical action had a larger effect on the stability of the soil-fiber complex than did thermal energy. The addition of a small amount of mechanical energy was more effective in dispersing soil than the input of relatively large amounts of heat. Durham (78) showed that increasing the temperature of the bath tended to increase redeposition.

Probably the most definitive study on the role of mechanical action in deposition testing was that of Tuzson and Short (98). These investigators showed that, with the modified Terg-O-Tometer they used, 36 strokes/min represented a critical mechanical input value. Clay soils applied below that energy level were believed to be in a high state of agglomeration, hence such soils were easily removed. Clay soil applied at or above the 36-strokes/min energy level were reasoned to be very finely divided and were thus very difficult to remove.

This parameter of mechanical action is covered more thoroughly in Chap. 7. It is a parameter which needs far more study. Additional investigation is especially urgent today when highly resinated polyester-cotton blend fabrics are becoming dominant in the home laundry operation. No successful method has as yet been devised by the textile industry to alter the great losses in tensile and tear strengths imparted to the cotton fiber by application of these treatments to achieve "permanent press" features. A result of these strength losses has been serious abrasion damage during the washing process. An objective of investigations into the role of mechanical action during laundering should be the definition of the correct mechanical input to achieve good cleaning, good antiredepositional properties, with minimal damage to the fabric substrate.

7. Evaluations of Test Methods

Just as soil depositional tests have been used to investigate each facet of the overall detergency processes, so have these procedures been employed to evaluate the total process through studies of special techniques devised to offer improvements to the washing procedure or to the understanding of the wash process. The purpose of all the work reported by Lambert and Saunders (59) was to bring out the questionability of conclusions based on what they considered incomplete evidence. They proceeded to show that laboratory tests with carbon soils failed to correlate in any way with practical laundry tests run on shop towels used in their laboratories and offices. They then compared these roller-towel tests to actual home washes and found good correlation. From these tests, Lambert and Saunders concluded that natural, or naturallike, soils must be used in multicycle procedures to achieve meaningful results in detergency studies.

Vaughn and co-workers (10), in studying the system of washing in sea water, first used carbon soil deposition tests for laboratory screening of surfactants and other adjuncts to the washing process. Then they carried these tests forward to actual redeposition studies on naturally soiled Navy uniforms to prove their method. They had, essentially, verified a test method by actual laundry practice—utilizing multiple wash procedures as recommended by Lambert and Saunders. A year later, 1950, Vaughn and Suter (99) emphasized, however, that it was unnecessary and undesirable to attempt close simulation of practical washing conditions in laboratory testing. They recommended soil deposition tests for measurement of whiteness retention and emphasized that soil removal and soil redeposition tests should be measured separately. Merrill and Getty (21) disagreed, saying "that for practical studies, the soil and testing techniques used should closely approximate conditions under which the product under test was to be used." It is apparent that little agreement between researchers was developed on the point of realistic testing during the early and middle 1950s.

By 1960, this picture began to change. Weatherburn and Bayley (12), who had previously worked with heavy loadings of carbon in depositional testing, reported that in order to simulate soil concentration found in practical conditions and possibly to improve accuracy of tests, it was better to employ low concentrations of soil with a number of multiple washings. In essence, they advocated that each laboratory should define its own tests and test conditions to accomplish the research underway.

Martin and Davis (86) compared deposition of clay soil to natural soil through multiple cycles. They cited the extreme variability of soil loadings on naturally soiled garments and recommended the use of only those loads with moderately heavy soil loadings. All of these test methods pointed toward realistic, end-use-type tests which would gain in prominence.

Probably the most extensive use of depositional tests to evaluate a test method can be found in the works of Compton and Hart (7, 8, 80, 100, 101). These researchers studied the deposition of carbon soils onto cotton fiber using the "chopped fiber" technique. This technique was first used by Kornreich (102) in 1946 to evaluate the removal of sericin gum from silk. The method was brought over into detergency research by Powney and Feuell (103). The method was described by Compton and Hart (80), who cited as their reason for using the chopped fiber technique, that this method eliminated fabric geometry and yarn geometry from soil retention consideration. Hence, they were able to eliminate macroocclusion as a soiling mechanism. The method consisted first of agitating cotton fiber, which had been ground in a Wiley mill, in a suspension of carbon black. Two grams of carbon were used to soil 5 g of ground cotton—an unusually heavy soiling. These ingredients were stirred for 30 min, then the mixture was "classified" with tap water until clear water resulted. The volume was then reduced to 500 ml, and this loading was milled in a Waring blender at 10,000-12,000 rpm for a total of 30 sec. The chopped fiber was filtered out in a Buchner funnel and the resultant mat was dried and reflectance was measured.

Using variants in soil, fiber, detergent, and other parameters, Compton and Hart formed the following general conclusions:

1. Most soiling of fibers was caused by microocclusion of soil particles in the surface rugosities of the fibers (8).
2. Sorption was shown to be a minor factor in soiling. This conclusion was based on the observation that large particle carbon was insensitive to the presence of surfactant and salts; and change in pH had no effect (80).
3. Fibers were shown to have varying surface irregularity profiles (8).
4. In a grease-soil-fiber system, soil may be attached to fiber by sorption or microocclusion, or both. Sorption occurs when soil contacts fiber as water is removed from the system (101).
5. The suspending power of various surfactants for a given soil system will vary widely with different fibers (8). Detergent formulations must be carefully tested or it is possible that redeposition values might rise to objectionable levels in multicycle washings (7).
6. Temperature reactions of deposition are dependent on the system under study (7, 8).
7. Number of particles adhering to fiber increases with decreasing size of particles (80). The chief factors controlling formation and stability of the soil fiber complex are the geometric relationships between size and shape of the soil particles, the size of the functional rugosities on the fiber, and the probability of close approach of soil and fiber (100).
8. Bonding energies vary from near zero to a magnitude requiring disruption of the fiber to effect removal (100).
9. Anionic cellulose derivatives and polyvinyl alcohol are best for minimization of deposition; but when surfactants are added, CMC or polyvinyl alcohol reacts either synergistically or antagonistically to their presence.
10. The proper water-to-fabric ratio, to minimize deposition on cotton, should be 20:1 or above. Below this ratio, serious deposition was encountered (100).

These conclusions have been subjected to inspection by numerous authors since they were published between 1951 and 1953. Phansalkar and Vold (104) showed that the state of aggregation of carbon had little effect on deposition onto cotton. Weatherburn and Bayley (12) verified these results.

But there have been other evidences brought forward which do not agree with the finding of Compton and Hart. Phansalkar and Vold (104) showed that reflectance measurements gave inaccurate results due to changes in particle size distribution of the suspended carbons and with changing concentrations of detergents. They used radiotracer methods to arrive at this conclusion. Nuessle et al. (60) found that the type of soil and age of soil after suspension in water and before use can effect results of tests. Powe (83) asserted that sorption was more important than microocclusion; showing clay particles to be the principal cause of soiling and citing that the major dimensions of these particles were such that macroocclusion would be precluded.

Dannenberg and Seltzer (76) had shown that 4 min of Waring blender agitation was required to reduce carbon to the ultimate particle size. Weatherburn and Bayley (12) more nearly approached proper dispersion of carbon soil by agitating in a Waring blender for 1-2 min at about 15,000 rpm. This was still not sufficient to achieve complete dispersion as recommended by Dannenberg et al. (77), however.

The similarities between deposition of model soils and redeposition of natural soils have been explored widely since about 1960. The use of natural soils, by definition, means end-use testing or testing with home-soiled laundry. Recommendations for this procedure go back at least to Saunders and Lambert (73) and Vitale (11). Their work was strengthened by the work of Marple (105); said work leading to the issuance of U.S. Patent 3,197,980, which covers reduction of redeposition of natural soil by use of two washes as opposed to one wash. Rhodes and Brainard (97) had earlier shown the efficacy of using a multiple bath system to give increased soil removal and better redeposition protection.

Martin and Davis (38) compared deposition of clay to redeposition of natural soil from dirty laundry. They showed that, within limits set by the variability of soil loadings, redeposition of natural soil and deposition of clay were equal in results over a ten-wash history in 0.25% built soap. An identical test in 0.35% built nonionic detergent gave excellent correlation, as shown by Fig. 8.29.

Hunter and colleagues (106) showed good correlation between laboratory-controlled practical washings and home laundering on a series of fabrics. They showed that fabrics containing polyester fiber tend to gray by redeposition more than other fabrics. Further, they showed that permanent press fabrics, polyester and resin-treated cotton, were more susceptible to graying than an untreated counterpart. They concluded that under-usage of detergent, over a history of ten washes, was primarily responsible for most unsatisfactory launderings on fabrics of these types.

In recent years, however, another facet of the deposition-redeposition problem has begun to emerge: to wit, that certain results obtained by deposition testing cannot be verified by studies of redeposition. Indeed, certain researchers have reported cases of essentially inverse behavior between the two methods. Nuessle and co-workers (60) appear to have been the first to allude to this phenomenon. They cited the reason for homogenizing their soils was "to yield more uniformity because soil removed from fabric by a detergent during normal washing procedure

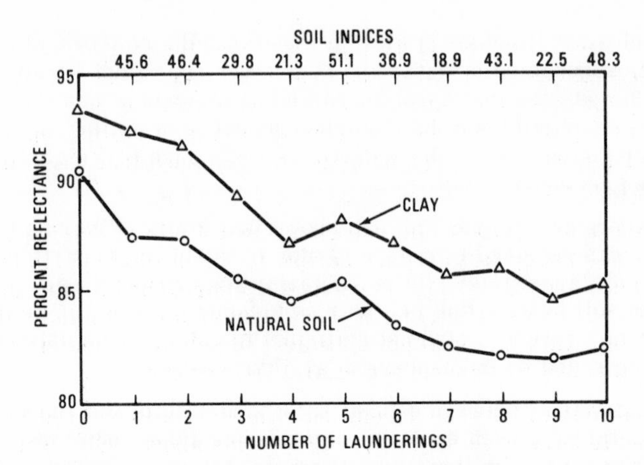

FIG. 8.29. Comparison of redeposition curves of natural soil and clay, 0.35% built nonionic detergent, 120°F.

is usually well dispersed." They had not, at the time of publication, studied the role of homogenization, however. Snell et al. (74) reported that redeposited soils were much harder to remove than deposited soils. They theorized that the redeposited soils had achieved direct contact with the fiber after both fiber and soil had been stripped of any oil layers which might have prevented tenacious attachment.

Hensley (107) reported some dramatic differences in the nature of behavior of soils washed from fabrics and soils added directly to the detergent bath. He found that certain nonionic polymers were effective in reducing deposition of soil added to the detergent bath, but were ineffectual in preventing redeposition of soil washed from fabric. He found polyvinyl alcohol, polyvinylpyrrolidone, and polyethylene glycol superior to carboxymethyl cellulose against depositional soils, whereas only CMC was effective against redeposition of clay and carbon soils. He theorized that these results might be related to particle size distribution, degree of agglomeration, or presence on the soil surface of adsorbed materials.

Smith and colleagues (108), working in nonaqueous systems, also found anomalous behaviors in deposited versus redeposited soils. They used "rug-beater soil," soil obtained from mechanical cleaning of numerous carpets. They designed a detergent system which would, intentionally, produce bad results. Using this detergent as their standard, they evaluated good systems against it. They applied the rug-beater soil to cotton fabric and ran tests using both soiled and unsoiled swatches in the same bath. They also tried adding soil directly to the wash bath with white deposition swatches. They compared good detergent systems to their standard system, using a "graying index" and a "cleaning index." The former they defined as the ratio of percentage graying of the test detergent to the standard, or poor, detergent. The latter was defined as the ratio of the increase in reflectance of the soiled cloth in test detergent compared to the increase in reflectance in the reference detergent.

These investigators concluded first that graying was more severe with redeposited soil than with deposited soil despite there being far more actual soil in the deposition-type bath. The reasons for this, they concluded, are: first, the great difficulty in reproducing the exact type of soil suspension present in actual cleaning baths of their laundering or dry-cleaning systems; and second, the difference in degree of dispersion of model soils causes problems, that is, soils removed from fabric are in a different degree of dispersion than those added directly to the bath.

An intriguing approach to the soil-detergent interaction process is that taken by Hock (109). In this work, he froze small droplets of a 0.15% sodium oleate solution in liquid nitrogen in both the presence and the absence of graphite soil. Water was removed in vacuo at -40°C or lower. These slides were then shadowed with chromium and photographed under the electron microscope.

In systems containing only soap, the particles of soap were evenly distributed; but in the system which contained soil, uneven distribution was noted; the soap particles appearing to aggregate at the edges of the graphite particles. The aggregations at the edges of the graphite were thought to be larger than soap micelles. Hock suggested that redeposition of graphite might be prevented by surrounding the soil particles by the soap. Unfortunately, it does not appear that Hock or anyone else has followed up this method of measurement.

It is apparent that the whole question of test methodology, deposition versus redeposition, model soil versus natural soils, multiple washing versus single washes, etc., remains a moot question. Where will test methods go in the future? It seems now that much evidence is piling up to indicate that a single-bath, simple-soil test will not be accepted widely by detergency researchers. Model soils will have to be devised, as recommended 20 years ago by Lambert and Saunders (59), which proximate closely those found in naturally soiled clothing. Further multiple washings, probably ten, will be demanded before acceptance of a given methodology. In all probability, most researchers will in addition require comparison of given model systems to actual soiled clothing tests before wide acceptance is granted.

IV. CONCLUSION

The author of this chapter has attempted to define as closely as possible the "state of the art" in regard to soil deposition and redeposition studies as of the current time. He has presented what he considers to be the best possible picture of the overall field; presenting those data he considers most germane to the subject and to the development of that subject over the years. He has intentionally refrained from presenting too much of the intricate detail to each relevant experiment; electing to leave to the reader the decision about which works he should pursue more thoroughly.

The author has also in certain cases attempted to point out areas which, in his opinion, need further elucidation by meaningful research. Such areas are these:

1. The study of fibers other than cotton in regard to their redepositional properties. In particular, the polyester fiber, and blends of polyester-cellulosic fibers, should be subjected to investigations.

2. The study of the role of resinous substrates in redeposition should be continued with main emphasis being put on those fabrics commonly sold today as "permanent press."

3. Meaningful soiling mixtures typical of those soils which are encountered on naturally soiled clothing should be investigated more closely and adopted for general use in the entire technology.

4. Test work should be expanded to include whiteness retention under realistic end-use conditions.

5. The search for and evaluation of antiredeposition agents specifically for the synthetic fibers should be pursued.

6. The role of mechanical action should be studied more deeply in order to assure maximum cleaning with minimum mechanical damage to the substrate.

7. New surfactants and new builders should be investigated. These new materials should in some manner compensate for certain inherently bad properties of some of the new synthetic fibers. These materials should also be completely degradable in sewerage disposal systems.

These are some of the frontiers for researchers in the field of detergency for the next several years.

REFERENCES

1. J. W. McBain, R. S. Harborne, and A. M. King, J. Soc. Chem. Ind., 42, 373T (1923).

2. F. D. Snell and I. Reich, ibid., 68, 98 (1949).

3. A. M. Mankowich, Ind. Eng. Chem., 44, 1151 (1952).

4. J. W. McBain, W. W. Lee, R. C. Merrill, and J. J. O'Connor, Chem. Prod., 19, Jan. to Feb., 1941.

5. J. Powney and R. W. Noad, J. Text. Inst., 30, T157 (1939).

6. A. S. Weatherburn, G. R. F. Rose, and C. H. Bayley, Can. Res. J., 28F, 213 (1950).

7. J. Compton and W. J. Hart, Ind. Eng. Chem., 40, 597 (1953).

8. W. J. Hart and J. Compton, Text. Res. J., 42, 164 (1953).

9. C. R. Caryl, Ind. Eng. Chem., 33, 731 (1941).

10. T. H. Vaughn, E. F. Hill, C. E. Smith, and L. R. McCoy, ibid., 41, 112 (1949).

11. P. T. Vitale, J. Amer. Oil Chem. Soc., 31, 341 (1954).

12. A. S. Weatherburn and C. H. Bayley, ibid., 37, 20 (1960).

13. W. Strauss, Kolloid-Z., Band 158, Heft 1, 30 (1958).

14. J. D. Carter, Ind. Eng. Chem., 23, 1389 (1931).

15. J. D. Carter and W. Stericker, ibid., 26, 277 (1934).

16. A. S. Weatherburn, G. R. F. Rose, and C. H. Bayley, Can. Res. J., 27F, 363 (1949).

17. F. D. Snell, Ind. Eng. Chem., 25, 162 (1933).

18. D. M. Morgan, Can. J. Res., 8, 429 (1933).

19. T. H. Vaughn and A. Vittone, Jr., Ind. Eng. Chem., 35, 1094 (1943).

20. H. B. Trost, J. Amer. Oil Chem. Soc., 40, 669 (1963).

21. R. C. Merrill and R. Getty, Ind. Eng. Chem., 42, 856 (1950).

22. J. Ross, P. T. Vitale, and A. M. Schwartz, J. Amer. Oil Chem. Soc., 32, 200 (1955).

23. B. J. Rutkowski and A. R. Martin, Text. Res. J., 31, 892 (1961).

24. Kalle and Cie., "Tylose HBR für Waschmittel," Wiesbaden-Biebrich, Germany, 1940.

25. L. J. Armstrong, B. C. Sher, E. W. Scott, Jr., L. O. Meyer, P. Soderdahl, A. J. Feit, and J. Kritchevsky, Amer. Dyst. Rep., 37, P569 (1948).

26. K. J. Nieuwenhuis, Chem. Weekbl., 43, 510 (1947).

27. French Pat. 805718, Kalle and Cie (1936).

28. C. H. Bayley, A. S. Weatherburn, and G. R. F. Rose, Laundry and Drycleaning J., Dec., 1948.

29. P. B. Law, Text. J. Aust., 189, April, 1950.

30. E. Bartholomé and K.-F. Buschmann, Melliand Textilber., 30, 249 (1949).

31. J. Inamorato (to Colgate-Palmolive), U.S. Pat. 3,144,412 (1964).

32. C. H. Bayley and A. S. Weatherburn, Text. Res. J., 20, 510 (1950).

33. H. L. Sanders and J. M. Lambert, ibid., 21, 680 (1951).

34. J. G. Jarrell and H. B. Trost, Soap Sanit. Chem., 28, 40 (1952).

35. W. Fong and H. P. Lundgren, Text. Res. J., 23, 769 (1953).

36. J. L. Azorlosa, Soap Chem. Spec., 38, 58 (1962).

37. W. Fong, W. H. Ward, H. P. Lundgren, U.S. Pat. 3,000,830 (1961).

38. A. R. Martin and R. C. Davis, Soap Chem. Spec., 36, 74 (1960).

39. J. B. Batdorf, ibid., 38, 58 (1962).

40. J. V. Beninate, E. L. Kelly, G. L. Drake, Jr., and W. A. Reeves, Amer. Dyst. Rep., 55, 25 (1966).

41. H. Schott, Text. Res. J., 36, 226 (1966).

42. K. J. Nieuwenhuis, J. Polym. Sci., XII, 237 (1954).

43. G. A. Johnson and F. G. Foster, J. Appl. Chem., 18, 235 (1968).

44. H. Stüpel and E. Rohrer, Fette, Seifen, Anstrichtm., 56, 588 (1954).

45. W. Fong and W. H. Ward, Text. Res. J., 24, 881 (1954).

46. H. S. Stillo and R. S. Kolat, ibid., 27, 949 (1957).

47. B. J. Rutkowski, J. Amer. Oil Chem. Soc., 45, 266 (1968).

48. H. Lange, Amer. Dyst. Rep., 50, 25 (1961).

49. W. Kling and H. Lange, Kolloid-Z., 127, 19 (1952).

50. M. von Stackelberg, W. Kling, W. Benzel, and F. Wilke, ibid., 135, 67 (1954).

51. J. Stawitz, Fette, Seifen, Anstrichtm., 58, 736 (1956).

52. J. Stawitz, W. Klaus, and H. Krämer, Kolloid-Z., 150, 39 (1957).

53. J. Stawitz and P. Höpfer, Seifen, Ole, Fette, Wachse, Nr. 3, 1 Februarheft, 51 (1960).

54. J. W. Hensley and C. G. Inks, Text. Res. J., 29, 505 (1959).

55. E. A. Vitalis (to American Cyanamid), U.S. Pat. 2,874,124 (1959).

56. H. Schott, J. Colloid Interface Sci., 26, 133 (1968).

57. W. J. Schwarz, A. R. Martin, and R. C. Davis, Text. Res. J., 32, 1 (1962).

58. J. M. Lambert, Ind. Eng. Chem., 42, 1394 (1950).

59. J. M. Lambert and H. L. Sanders, ibid., 42, 1388 (1950).

60. A. C. Nuessle, L. M. Nageley, and E. O. J. Heiges, Text. Res. J., 33, 146 (1963).

61. R. A. Pingree, Mod. Text. Mag., 49, 61 (1968).

62. C. A. Ghionis and C. L. Browne, Amer. Dyst. Rep., 57, 28 (1968).

63. L. W. Mazzeno, Jr., R. M. H. Kullman, R. M. Reinhardt, H. B. Moore, and J. D. Reid, ibid., 47, 299 (1958).

64. T. Matsukawa, Nat. Sci. Rep. Ochanomizu Univ., 10, 570 (1959).

65. J. A. Somers, Text. Recorder, 63, May 1959.

66. J. Berch and H. Peper, Text. Res. J., 33, 137 (1963).

67. W. J. Connick, Jr., and S. E. Ellzey, Jr., Amer. Dyst. Rep., 57, 17 (1968).

68. H. Peper and J. Berch, ibid., 54, 36 (1965).

69. E. M. Perry, ibid., 57, 42 (1968).

70. W. T. Aliman, R. K. Dunlap, and W. C. Zybko, ibid., 57, 28 (1968).

71. J. M. Kennedy and E. E. Stout, ibid., 57, 11 (1968).

72. O. Oldenroth, Textilind., 68, 413 (1966).

73. H. L. Sanders and J. M. Lambert, J. Amer. Oil Chem. Soc., 27, 153 (1950).

74. F. D. Snell, C. T. Snell, and I. Reich, ibid., 27, 62 (1950).

75. W. L. Marple (to Whirlpool Corp.), U.S. Pat. 3,197,980 (1965).

76. E. M. Dannenberg and K. P. Seltzer, Ind. Eng. Chem., 43, 1389 (1951).

77. E. M. Dannenberg, M. E. Jordan, and C. A. Stokes, India Rubber World, 22, 662 (1950).

78. K. Durham, J. Appl. Chem., 6, 153 (1956).

79. V. S. Veselovski, Issled. Fiz.-Khim. Tekh. Suspensii, 83 (1933).

80. J. Compton and W. J. Hart, Ind. Eng. Chem., 43, 1564 (1951).

81. J. W. Hensley, M. G. Kramer, R. D. Ring, and H. R. Suter, J. Amer. Oil Chem. Soc., 32, 138 (1955).

82. J. W. Hensley and C. G. Inks, Special Technical Publication No. 268, Symposium on Applied Radiation and Radioisotope Test Methods, American Society for Testing Materials, 1959.

83. W.C. Powe, Text. Res. J., 29, 879 (1959).

84. T. Tsunoda, unpublished work.

85. B. E. Gordon and E. L. Bastin, J. Amer. Oil Chem. Soc., 45, 754 (1968).

86. A. R. Martin and R. C. Davis, Soap Chem. Spec., 36, 49 (1960).

87. S. Shimauchi and H. Mizushima, Amer. Dyst. Rep., 57, 35 (1968).

88. W. C. Powe and W. L. Marple, J. Amer. Oil Chem. Soc., 37, 136 (1960).

89. E. Albinson, Text. Rec., 68, May 1958.

90. A. L. Meader, Jr., and B. A. Fries, Ind. Eng. Chem., 44, 1636 (1952).

91. T. F. Boyd and R. Bernstein, J. Amer. Oil Chem. Soc., 33, 614 (1956).

92. D. C. Knowles, Jr., J. Berch, and A. M. Schwartz, ibid., 29, 158 (1952).

93. V. McLendon and F. Richardson, Amer. Dyst. Rep., 52, 27 (1963).

94. M. Ringeisen, Teintex, 8, 31 (1943).

95. A. E. Bailey, J. and Proc., Roy. Inst. Chem. Great Britain and Ireland, 105 (1943).

96. S. V. Vaeck and G. DeVille-Boorsma, Ann. Text., No. 4, 44, Dec. 1958.

97. F. H. Rhodes and S. W. Brainard, Ind. Eng. Chem., 21, 60 (1929).

98. J. Tuzson and B. A. Short, Text. Res. J., 32, 111 (1962).

99. T. H. Vaughn and H. R. Suter, J. Amer. Oil Chem. Soc., 27, 249 (1950).

100. W. J. Hart and J. Compton, Ind. Eng. Chem., 44, 1135 (1952).

101. J. Compton and W. J. Hart, Text. Res. J., 23, 158 (1953).

102. E. Kornreich, Text. Mfr., Manchester, 72, 271 (1946).

103. J. Powney and A. J. Feuell, Res., 2, 331 (1949).

104. A. K. Phansalkar and R. D. Vold, J. Phys. Chem., 59, 885 (1955).

105. W. L. Marple, Whirlpool Corp. Research Report 1824-4, unpublished.

106. R. T. Hunter, C. R. Kurgan, and H. L. Marder, J. Amer. Oil Chem. Soc., 44, 494 (1967).

107. J. W. Hensley, ibid., 42, 993 (1965).

108. W. H. Smith, M. Wentz, and A. R. Martin, ibid., 45, 83 (1968).

109. C. W. Hock, Text. Res. J., 25, 682 (1955).

110. T. Tsunoda and Y. Oba, Yokagaku, 17, 475 (1968).

111. R. Tsuzuki and N. Yabuuchi, Amer. Dyst. Rep., 57, 472 (1968).

Chapter 9

EVALUATION METHODS FOR SOIL REMOVAL AND SOIL REDEPOSITION

J. J. Cramer
BASF Wyandotte Corporation
Chemical Specialties Division
Wyandotte, Michigan

I. INTRODUCTION

This chapter is a state-of-the-art report on test methods for evaluation of the degree of cleanliness of textile articles with respect to both optical assessments and actual soil content. Schwartz et al. (1) have stated that: "A measurement of detergency implies simply an estimate of the amounts of soil present on the substrate before and after the cleaning operation," and ". . . we are usually interested in the whiteness of the fabric rather than in its actual soil content."

Reviews of methods for measuring detergency have been presented by a number of writers (1-7). These methods find most frequent use in evaluation of the soil removal and antiredeposition properties of aqueous solutions of detergent compositions which normally comprise blends of surfactants, an antiredeposition agent, alkaline salts, water softening agents, and fluorescent whitening agents. The methods are also used for studying both dry and wet soiling characteristics of fabrics, carpeting, and other textile articles. Further use is in connection with developing theories and elucidating phenomena of detergency.

Optical assessments relate to the photometric and colorimetric attributes of a textile article such as cloth. Is the shirt white, gray, yellowed, overblued, "soiled," or flourescent near-white? This type of appearance can be rated subjectively by visual inspection, and it can also be evaluated quantitatively with suitable types of photometers and reflectometers. A great amount of color physics is present in the sciences underlying these assessments.

Achromatic grayness of fabric can be measured with relatively simple types of instruments (1, 8). The degree of gray may approach black, as with fabric artificially soiled with relatively large amounts of graphite and carbon, or may be near-white, as with fabric onto which a small amount of neutral pigment has redeposited during wash operations. Grayness is conventionally expressed in terms of percent of magnesium oxide, complete blackness being 0 and perfect brightness being 100% of MgO, using 45° incident light of 5500 ± 1000 Å units and 0° reflection, or vice versa, 0°/45°.

Yellowing and other chromatic effects on textiles are measured with instruments such as tristimulus reflectometers, trichromatic colorimeters, or spectrophotometers. Chapter 10 deals with these devices. Assessments of chromaticity have become increasingly important during the past two decades which have seen artificial soils of unrealistic dark color and chemical composition progressively replaced with both simulated and actual naturally occurring soils comprising clays plus fatty and oily organic substances. Almost universal use of fluorescent whitening agents during the past decade has brought about further sophistication of reflectance measuring devices. In addition to previous features, an ultraviolet filter is needed and a special uv light is sometimes provided.

Actual soil content of the fabric has also received a large amount of attention, both qualitatively and quantitatively, during recent years. Some simulations of natural soil provide ready means of measurement such as radiotracers in clay and in organic constituents, including artificial sebum. Solvent-soluble soils such as wool grease may be extracted and determined gravimetrically. Soils such as iron oxide can be determined by chemical analysis.

Much concern has been devoted to need of good correlation among the various methods used to evaluate the degree of cleanliness of fabrics (9). With respect to optical assessments, subjective evaluation by the eye is all important and reflectance methods are expected to provide the same relative ranking as arrived at visually by the customer (10). An understandable relationship should also exist between optical ratings and actual soil content of fabric.

Good correlation is also necessary between laboratory test data and those obtained in practical wash programs (1). One rather fundamental difference between the two approaches might well be pointed out: In lab work, contents and tenacities of artificial soils are generally greater than those of natural soils in practical programs, as the detection of relative performance differences among detergent formulations usually requires that even the most effective composition not remove 100% of the soil from the fabric. In contrast, practical washes are supposed to free the fabric of all soils and stains (11). It should also be noted that artificially soiled swatches are frequently included in test loads of practical washes.

It is obvious that the ultimate objective in washing fabric is to eliminate all soils from the substrate and to leave the fabric with a most pleasing color. Currently, this means sufficient detergent action to remove soils from a wide variety of fabrics, made from both natural and man-made fibers, plus prevention of redeposition of soils back onto the fabric during wash operations, removal of stains when needed by oxidative bleaching without damage to the fabric, and color correction or enhancement by use of blueing and/or fluorescent whitening agents. It also means that the successful detergent chemist has had to master knowledge in the areas of color science and textile technology.

Information on selection of types of washing machines and soiled fabrics to be used, as well as other detergency test conditions, will be found in other chapters of this book. Here, we are concerned only with cleanliness and appearance of the fabric before and after washing.

II. OPTICAL ASSESSMENTS

A. Color Concepts

In order to gain an appreciation of technical problems in optical assessments of washing results on fabric, and to obtain a proper understanding of the color physics involved, the reader is directed to standard texts (12, 13). Admittedly, assessments of degree of grayness, as with dark artificial soil based on carbon black, may pose no special problems to the detergent chemist, during either visual ranking or by reflectance readings. But evaluation of wash quality with more realistic soils, such as clays plus artificial sebum, as well as presence of color-corrective blueing and/or fluorescent whitening agent should be approached with adequate understanding of underlying sciences.

A relatively simple color diagram has been used by Hemmendinger and Lambert (14) and others (15, 16) to illustrate chromaticity aspects encountered both with the concept of whiteness, a quality to which the eye is highly sensitive with consequent preference tendency, and with the concept of brightness, an entity measured by reflectance of light in the green portion of the spectrum. The diagram, which also includes concepts of yellowness, blueness, and grayness, is shown in Fig. 9.1. Perfect whiteness of 1.00, same as reflectance of a pure surface of magnesium oxide, is shown at the top of the diagram. Moving down along the achromatic scale from this point, brightness decreases and grayness increases. Typical white fabric of commerce falls near 0.90 brightness, and below this, the reflectance

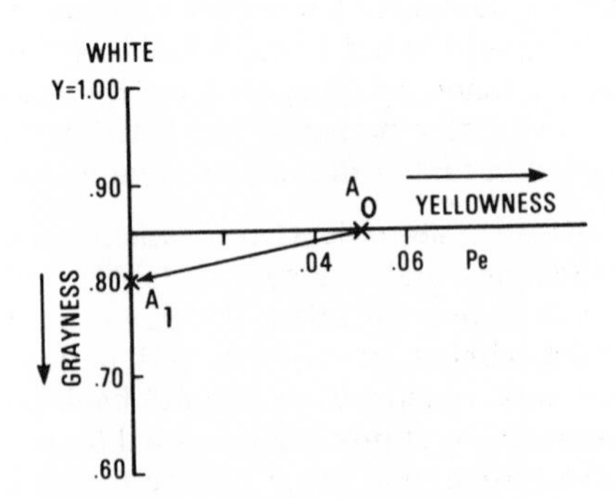

FIG. 9.1. Diagram of gray and yellow near-whites. Reprinted from (14, p. 164) by courtesy of Journal of the American Oil Chemists' Society.

range of carbon-soiled cloth. On the horizontal coordinate to the right of the vertical brightness scale is shown increasing amounts of instrumentally measured yellowness. Hemmendinger and Lambert (14) expressed yellowness in terms of excitation purity, P_e (12) at 575 nm. Thus, point A_0 on the diagram has a green filter reflectance of 85% of magnesium oxide and P_e = 0.05.

Addition of the proper amount of blueing dye to the fabric will just neutralize the initial yellowness, producing a fabric of tinctorial properties described by point A_1 on the vertical, achromatic scale. As can be grasped from the color diagram, the color-corrected point A_1 is achromatic but less bright than the yellowed fabric of point A_0. As pointed out by several authors (16, 17, 18), the eye prefers whiteness to yellowness, and hence rates point A_1 higher than point A_0. In contrast, simple green filter reflectance shows the greater brightness of point A_0. It is evident that elucidation of this apparent contradiction is provided by tristimulus colorimetry plus relevant color science.

Several comprehensive treatments of near-white chromaticity are found in the literature (14, 18, 19, 20). Of these, Goldwasser (18) presented the equal-whiteness diagram shown in Fig. 9.2. This diagram shows the relationship of reflectance and purity with 572 nm illumination for constant whiteness of laundered cloth as judged by a panel of observers. No fluorescent whitening agents were present. The interesting relationship is that the observers gave equal whiteness ranking to all fabrics whose properties fell on the same line, for example the curve which begins on the left side of the diagram at about 70% reflectance-zero purity (achromatic) and ends at 92% reflectance-3.5% purity (yellowed).

Another well-known means of overcoming yellowness of fabric is addition of fluorescent whitening agents, which are present as an ingredient in all of today's leading detergent compositions. Chapter 12 is devoted to the

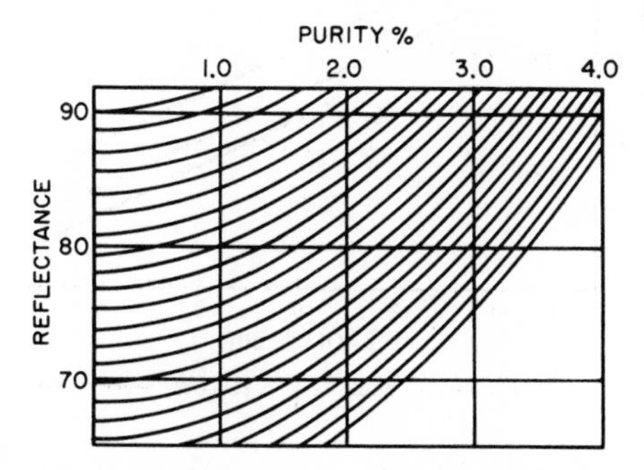

FIG. 9.2. Equal-whiteness diagram. Reprinted from (18, p. 50) by courtesy of *Soap and Chemical Specialties* and the author.

chemistry and application of fluorescent whitening agents. Such agents have the ability to absorb energy from the near-ultraviolet region (360-400 nm) of the spectrum and to reemit this energy in the visible-blue region (425-475 nm), thus producing the desired whiteness. Referring again to the Goldwasser equal-whiteness diagram, addition of a properly chosen fluorescent whitening agent will cause the actual light emanating from the cloth to change in purity in a direction which is nearly horizontal, for example from 92% reflectance-3% purity (yellowed) to 92% reflectance-zero purity (achromatic).

Robert Hunter (17) depicted visual preference as a function of yellowness versus blueness of optically whitened fabric at various levels of brightness, as shown in Fig. 9.3. This source (17) reported that "fabrics of a yellow hue were subjectively unsatisfactory to a panel of judges whereas fabrics with a blue-white hue caused by fluorescent whitening agents tended to be acceptable irrespective of their grayness within reasonable limits. Thus, FWA effects appeared to be the single most important detergent-related factor in determining the laundered appearance of these fabrics." Additional discussion on whiteness and brightness is given by Evans (21) and R. S. Hunter (22).

Another aspect of chromaticity is the effect of type of illumination on judging of near-whites. Textile chemists refer to this general phenomenon as metamerism. It arises from the fact that two objects have slightly different spectral reflectance curves (23, 24). Two pieces of fabric match when viewed in daylight, but not under tungsten light (or vice versa).

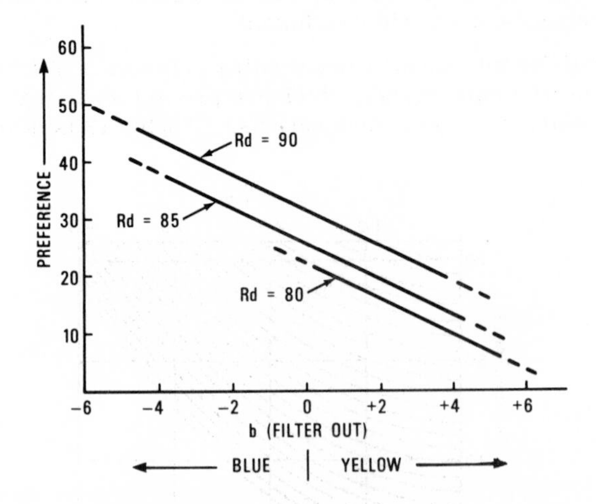

FIG. 9.3. General effect of reflectance values of Rd and b on preference rankings of panels of observers. Reprinted from (17, p. 365) by courtesy of Journal of the American Oil Chemists' Society.

Goldwasser (18) dealt with similar concepts in connection with various color types of fluorescent whitening agents to provide subjective appearances that are preferred by housewives. His findings, which resulted from a statistically designed and evaluated, paired-comparison test, are seen in Fig. 9.4. The four fluorescent whitening agents were applied to both new and aged cotton fabric to give equal energy of emission in all cases. These agents gave slight tinges of red, blue, none (white), and green under 365 nm excitation. The order of preference of each tinge is shown under three types of viewing conditions. Overall conclusions are difficult to find. However, "green" is either most preferred or most disliked; on the other hand, "blue" seems to elicit no strong feeling either way.

B. Subjective Visual Appearance

The human eye is obviously the ultimate judge of wash fabric appearance (10, 14) in spite of its subjective and sometimes emotional nature.

1. Evaluations

Visual evaluation of wash quality must surely have originated during

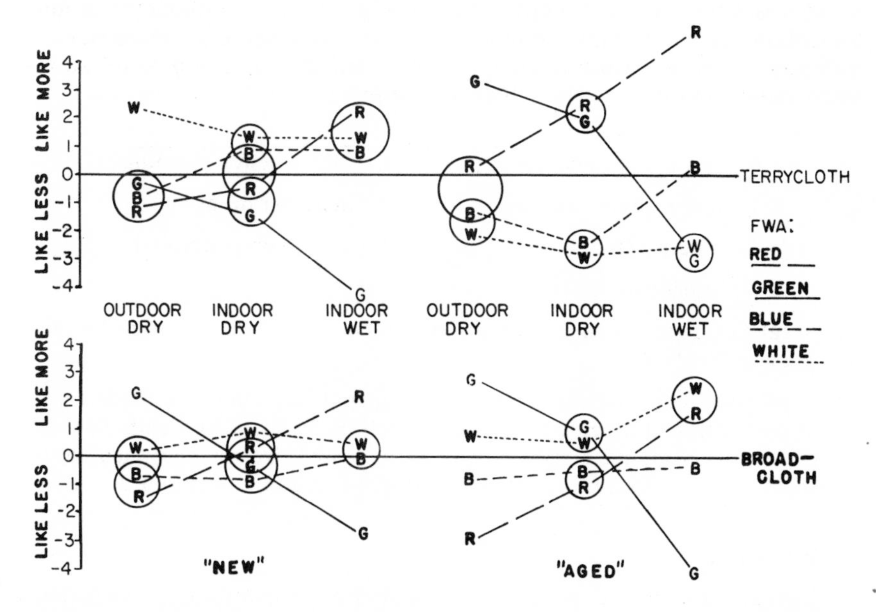

FIG. 9.4. Visual order of preference for various tinted or white fluorescent agents on "new" or "aged" cotton fabric as a function of type of "normal" viewing condition. Reprinted from (18, p. 121) by courtesy of Soap and Chemical Specialties.

prehistoric times (25). Attainment of white linens was probably an important criterion in the laundering activities practiced in the ancient Egyptian court (26). The practice of visual assessment of all types of wash continues in the twentieth century. Two early scientific reports with artificial soil washing were by Spring in 1909 (27), using cotton fabrics soiled with carbon black, and by Shukov and Shestakow in 1911 (28) on lanolin-lampblack soil. In 1923 Stericker (29) reported visual examination of mechanics' coveralls after practical wash tests using silicate with and without soap. Two recent patents report rating of detergency by visual evaluation (30, 31).

Viertel (32) reported visual assessments of wash quality in the washing and drying of easy-care shirts.

Brown et al. (33) visually assessed residual amounts of oily soils after hand washing of polyester, polyester-cotton (PEC) blend, and cotton fabrics with a built alkylbenzene detergent at 0.36% concentration for 3 min at 48°C. Seven oil stains that were applied separately to fabric panels were heavy and light lubricating oils, olive oil, vegetable cooking oil, sebum-type oil alone or mixed with linseed oil. A trace of oil-soluble, Sudan black dye was added to each oil to act as an indicator when the panels were assessed for oil residues. The black dye was essentially nonsubstantive to the test fabrics. Staining was with a standard amount of each oil dispensed from a buret onto each 2ft × 1 ft fabric test piece giving a stain diameter of about 2 in. One series of stained cloths was aged 2 days while a second series was aged 3 weeks. After washing, rinsing, and drying, residual oil stains were rated according to the following scale:

0 = stain undetectable	4 = fairly heavy staining
1 = very slight staining	5 = heavy staining
2 = slight staining	6 = very heavy staining
3 = moderate staining	

The total gradings for the seven oil stains on each of the five test fabrics are seen in Table 9.1.

In a second study, the same writers subjected five lab coats made from the aforementioned fabrics to 1 week of soiling in a fish and chip establishment. Degree of visual dirtiness was assessed before washing, and degree of cleanliness was observed after washing, as shown in Table 9.2.

2. Photography

A logical extension of visual assessment of wash quality is photographing the test fabric before and after washing. Examples of this were first noted in the silicate washing studies of Carter (34), and have been continued by other workers at various times (35-37). Carter's work and the last-named

TABLE 9.1

Total Gradings for Seven Oil Stains[a]

	100% Filament polyester	100% Staple polyester	67/33 Staple polyester/ cotton blend	100% Cotton	100% Filament nylon
Aged 2 days	$14\frac{1}{2}$	24	23	$20\frac{1}{2}$	$5\frac{1}{2}$
Aged 3 weeks	$16\frac{1}{2}$	$28\frac{1}{2}$	$28\frac{1}{2}$	$28\frac{1}{2}$	$8\frac{1}{2}$

[a]Reprinted from (33, p. 736) by courtesy of Textile Research Journal.

reference include reflectance data in addition to visual assessment and photographs.

Photographic reproduction of visual wash qualities seems to find greatest application with relatively dark soils. Lambert and Sanders (11) have observed that, in contrast to response of reflectometers, the eye is quite sensitive to differences in appearance among relatively dark gray test swatches. They correlated the sensitivity of the eye with advantages in presenting photographs of such test swatches. Common experience indicates that realism in prints and reproductions with both high whiteness and optically whitened fabrics is restricted in practice by the quality of printing paper and pigments used in reproduction. However, differences among test panels generally persist to some extent. Electrostatic copying machines can be used to record reasonable degrees of quantitative differences from actual test swatches. Color photography is sometimes used to illustrate yellowing of fabric and unremoved, colored stains (26).

Deering Milliken (38) has photographic standards for visually gauging the intensities of stain retention, from class 1 (greatest degree of staining) to class 5 (complete removal of stain). Use of these standards was reported by Tsuzuki and Yabuuchi (39) who used an alkylaryl sulfonate detergent to wash 11 types of fabrics. Of these, two were cottons, with or without resin treatment; three were PEC, with or without resin treatment; three were straight polyester, and three were of different types of rayon. Finishes included soil release.

Kennedy and Stout (40) used the Munsell color chip system in addition to reflectance measurements for color evaluation of various fabrics soiled with clay-oleic acid and wash with synthetic detergents. Their lack of enthusiasm with this approach was attributed to relative difficulty to use, absence of very light cream yellow to blue chips, and due to its subjective nature, of doubtful value except with experienced observers.

TABLE 9.2

Oil Soiling and Washability of Filament and Staple Polyester, Polyester/Cotton Blend, Filament Nylon, and Cotton Fabrics[a,b]

Dirtiness scale		Cleanness scale	
Degree of soiling	Grading	Degree of cleanness	Grading
Trace	1	Clean	0
Slight	2	Slight residual soiling	1
Medium	3	Moderate residual soiling	2
Heavy	4	Fairly heavy residual soiling	3
Very heavy	5	Heavy residual soiling	4
		Very heavy residual soiling	5

Gradings before and after washing on the above scales

Wear/wash cycle No.	100% Filament polyester		100% Staple polyester		67/33 Staple polyester/cotton blend		100% Cotton		100% Filament nylon	
	Before wash	After wash at 48°C	Before wash	After wash at 48°C	Before wash	After wash at 48°C	Before wash	After wash at 85°C	Before wash	After wash at 48°C
1	3-4	1	3	1	3-4	2-3	5	2	2	0-1
2	3	0-1	3-4	4	5	3	4	1-2	3	0-1
3	5	2	3-4	5	2-3	2	5	2	3-4	1-2
4	5	0-1	5	4-5	4	2	3	2	5	1
5	—	—	—	—	3-4	3	5	1-2	5	1-2

[a]Wear/wash tests on coats worn in fish and chip establishment.

[b]Reprinted from (33, p. 737) by courtesy of Textile Research Journal.

3. Viewing Conditions

A number of workers have been concerned with establishing optimized conditions for viewing fabric. Some relevant factors of lighting with respect to chromaticity were given by Reese (41) who also showed a schematic diagram covering the relationship between overhead light source, surface of the viewing table, and position of the observer. Standard viewing conditions were likewise described by Rizzo (42), Coppock (43), Anders (44), and Rauchle and Schramm (45).

For visually examining laundry goods, Furry and co-workers (46) have used special lighting of daylight plus ultraviolet, in addition to north daylight. McLaren (47) graded yellowed near-whites as a function of 5000-20,000°K color temperature of the illuminant. Luechauer (48) mentioned use of an inexpensive titration lamp obtainable from a lab supply house. This lamp, or a daylight fluorescent tube, as well as a small ultraviolet lamp, would normally be used in a darkened room. MacBeth and Reese (49) reviewed the history of color matching in the United States with respect to standard illumination practices.

Benischeck et al. (10) reported the visual whiteness ranking of wool under six types of illumination: incandescent light, cool fluorescent light, northern light with and without intervening glass, daylight, and sunlight, as well as twilight. The bleached or unbleached wool had no treatment (yellowish), was blued, brightened, or was treated with blue plus fluorescent whitening agent (FWA). Whiteness after simple bleaching ranked below simple blueing in all lights except for a near tie in north light through glass, this distinction having been more apparent under artificial and natural twilight. Under artificial lights, blueing with bleaching exhibited better whiteness than FWA alone, but in light with more ultraviolet, the relation was unmistakably reversed. In sunlight and daylight, the off-white yellow was completely masked in the samples with blue and FWA.

4. Test Methods

The American Association of Textile Chemists (50) has recently published a test method for the visual rating of carpet soiling. The degree of difference in cleanliness between an original or clean area and the area under examination is determined by visual matching with a stepwise series of differences in gray chips selected to form a geometric scale of differences on a dark-light axis.

The American Society of Testing and Materials Standard Method D-1729-64 (51), Visual Evaluation of Color Differences on Opaque Materials, covers the spectral, photometric, and geometric characteristics of light sources, illumination, and viewing conditions, sizing of specimens, and general procedure to be used in visual evaluation of color differences among opaque materials.

The ASTM D-12 Committee has recently drafted new performance test methods for home laundry detergents (305). It includes subjective criteria and home performance as test standards, using naturally soiled household laundry bundles.

5. Fluorescent Soils

Soils derived from petroleum oils and other fluorescent materials lend themselves to enhanced visual detection when viewed under ultraviolet light. This is a very simple way to visually assess removal of industrial laundry, oily soils. However, the intensity of such fluorescence diminishes with age (52).

Morgan and Lankler (53) have pointed out that in the evaluation of metal cleaning, animal and vegetable oils which do not fluoresce in their own right can be made to show this attribute by adding an oil-soluble dyestuff, Fluorescent Oil Green HW. It seems logical that this technique could be applied to fabric detergency. The two cited authors also presented luminograms, which are photographs of the fluorescence under ultraviolet light of soil on metal panels, thereby providing a permanent record of cleaning ability of detergents.

Reinhardt *et al.* (54) showed photographs taken under ultraviolet radiation of fabrics treated with dimethylol ethylene urea resin plus blue fluorescent whitening agents. The uv was provided by GE fluorescent BLB-type lamps with uv filters. Differences in fluorescence under uv radiation are clearly seen in the photographs.

C. Instrumental Measurements

1. Introduction

Although the eye is quite adept in detecting relative visual differences among a few pieces of fabric at a given point in time and space, it cannot store absolute types of information (25). Consequently, visual assessments performed on a given date may not have continuity with those made either before or afterwards. Even when the eye is assisted by reference to established visual standards, these materials are subject to change in appearance with time. Another deficiency of the eye is its inability to handle relatively large numbers of assessments.

Instrumental means of measuring whiteness can furnish recordable and repeatable numbers rather than mere mental impressions (25). Use of photocells in instrumental approaches to the evaluation of wash results was a natural outgrowth of assessment by the eye. Rhodes and Brainard (55) are credited (56) with starting the use of photocolorimetric systems for making quantitative tests on soiled and washed fabrics. They reported a Taylor-type integrating photometer equipped with a 32-cp incandescent lamp and MacBeth illuminometer.

R. S. Hunter (57) cited three advantages of barrier-layer photocells (58) for colorimetric and photometric apparatus: the cells are physically simple and rugged; they generate a voltage directly from the light flux incident upon them, requiring no externally applied potential; they respond more strongly to the energy of the visible spectrum than to other energies.

A number of commercial instruments are available for measuring relative brightness, chromatic properties, and fluorescent characteristics of soiled and washed fabrics. Chapter 10 deals with such devices. R. S. Hunter (59) has recently reviewed the history of the development of photoelectric tristimulus colorimeters. Harris and Brown (60) compared use of the Hunter Multipurpose (H. A. Gardner Laboratory) and the Photovolt Model 610 reflectometers and found no significant differences due to instrument. Kennedy and Stout (40) reported on a similar comparison. Machemer (61) reviewed construction features and whiteness data with three types of German reflectometers and the Hunter Multipurpose reflectometer. Some variations in the data among instruments did not greatly affect their practical applications.

A 1961 AATCC report (62) was concerned with a cooperative study of the characteristics and performance of one dozen instruments representing six different types of color and color-difference meters used in the Washington, D.C. area. Conclusions were that careful work is needed both in instrument calibration and operation in order to obtain repeatable measurements of reflectance, yellowness, and whiteness of textile fabrics. Instruments may vary as to make and model, but standards must have the same basis of calibration, and the technique of presenting the specimen for measurements must be standardized.

Reflectance meters vary with respect to ability to assess fluorescent effects of fluorescent whitening agents. Meters with filters between the fabrics and the photoelectric cells are practically insensitive to these agents. A partial sensitivity is found when the tristimulus filters are located between the light source and the fabric sample (63). Meters suitable for properly assessing fluorescent effects are described in Chap. 10.

2. Reflectance Meter Evaluations

Perhaps the most frequently used type of detergency test is the washing of artificially soiled fabric swatches plus some white swatches, followed by reflectance readings to assess soil removal and soil redeposition. Alternately, the clean swatches may be washed in a separate detergent solution containing added amounts of soil. This latter approach toward measuring the soil suspension properties of a detergent solution is called soil deposition. Stüpel (5) reviewed redeposition (backwashing) and deposition studies.

Two workers have demonstrated the differences between soil redeposition and soil deposition testing. Hensley (64) found that the two approaches may give contradictory results. In soil deposition tests both polyvinyl alcohol and polyvinyl pyrrolidone were indicated as superior to sodium carboxymethyl cellulose in suspending carbon black soil. In redeposition tests, only the extensively used sodium carboxymethyl cellulose was found to be really effective in prevention of redeposition of carbon black or radioactive tagged clay soils. Hensley's conclusion was that his data cast doubt on the validity of deposition tests.

Smith et al. (65) described a dry-cleaning detergent evaluation method of measuring soil redeposition. They found invalid past arguments that redeposition methods were basically unsound due to unequal soil concentration in baths when two detergents of unequal soil removal properties were compared. They found that detergents of low soil removal also gave high graying, and vice versa. They ascribed this to either correlation of the two qualities or to two aspects of the same quality. A possible explanation of differences between deposition and redeposition tests is that degree of soil dispersion is much greater in redeposition methods. Two excellent bibliographies in these areas are: "Bibliographical Abstracts on Redeposition of Soil on Cotton Fabric," Kramer et al. (66), and "Bibliography and Abstracts on Primary Soil Deposition, 1931-1961," Schwartz et al. (67).

The need for satisfactorily differentiating between the separate effects of soil removal and suspending power has been cited by Vaughn and Suter (68), Martin and Davis (69), and Ford (70). Ford proposed a theory and procedure designed to allow quasiqualitatively to distinguish and separate the effects of soil removal and soil suspension properties of a detergent.

a. Soils. Soiled fabric can be either prepared in the detergency testing lab or purchased commercially. Lab production typically commences with agitation of fabric in a bath consisting of pigment, fatty substances, emulsifier, and solvent. The last may be either water or volatile organic solvent. The soiled fabric is separated from the bath, and excess solvent removed by draining, wringing, or centrifugal extraction. After drying by suitable means, the soiled fabric is aged and tested for desired degree of soil removal by washing in solutions of selected detergents.

The production and standardization of a detergent soil based on colloidal graphite has been described by Harris and Brown (71). In 1950, Sanders and Lambert (72) reported the development and properties of a synthetic soil based on the chemical analysis of street dust from six American cities. The ingredients of their soil are shown in Table 9.3. Airborne dust has been used as a fabric-soiling ingredient. The dust may be deposited directly onto wash test fabric by sucking dust-laden air through the cloth with a vacuum cleaner for a suitable duration of time (73, 74). More often, the dust is collected by means of conventional vacuum cleaning of rugs

TABLE 9.3

Sanders and Lambert Synthetic Soil[a]

Humus	35%	Stearic acid	1.6%
Cement	15%	Oleic acid	1.6%
Silica	15%	Palm oil fatty acids	3.0%
Clay	15%	Lanolin	1.0%
Sodium chloride	5%	n-octadecane	1.0%
Gelatin	3.5%	1-octadecene	1.0%
Carbon black	1.5%	Lauryl alcohol	0.5%
Iron oxide	0.25%		

[a] Reprinted from (72, p. 156) by courtesy of Journal of the American Oil Chemists' Society.

(65, 75) or from air conditioners and electrostatic precipitators (76, 77) followed by removal of oils by solvent washing, drying, and screening to remove lint; then the particulate dust is applied to cloth, either with (78) or without mixing with fat by agitation in a suitable type of liquid solvent bath. Alternately, dry tumble solvent-extracted yarn with screened vacuum cleaner dirt and steel balls, followed by separation of loose soil by air blasting or screening (79). The composition of a sample of screened and ether-extracted vacuum cleaner dust from English city offices was said by Jones et al. (80) to be similar to that of the American Sanders and Lambert artificial road dust (72).

Powe (81) found that clay rather than carbon black was the typical pigment in natural soil. Also, he and Marple (82) found that the organic phase of natural soils was rich in fatty acids and fatty acid esters. These and other workers at the Whirlpool Corp. (69, 83-85) subsequently developed artificial soils based on a kaolinite type of clay, Spinks "Bandy Black," with or without organic constituents such as triolein and oleic acid.

In 1965, Spangler et al. (76) reported the development of a synthetic soil based on use of natural airborne particulate material collected from air conditioners and similar equipment, plus the synthetic sebum moiety described in Table 9.4. A year later Spangler et al. (86) reported a detergency test based on rapid aging of unremoved sebum. The advantage of this approach is that hard-to-detect residues of oily material may be heat developed into sufficient yellowness which can be readily measured instrumentally.

TABLE 9.4

Spangler Synthetic Soil[a]

Palmitic acid	10%	Olive oil	20%
Stearic acid	5%	Squalene	5%
Coconut oil	15%	Cholesterol	5%
Paraffin	10%	Oleic acid	10%
Spermaceti	15%	Linoleic acid	5%

[a]Reprinted from (76, p. 725) by courtesy of Journal of the American Oil Chemists' Society.

Gustafson (87) reported the composition of typical carpet soils to be that described in Table 9.5. Nieuwenhuis (88) stained fabrics with four different types of mixed artificial soils: (a) peanut oil, tallow, petroleum jelly, quartz powder, and iron oxide; (b) fats, carbohydrates, proteins, and organic pigments; (c) natural dyes, acids, and sugar; and (d) blood. Commercially available soiled test fabrics were reviewed by Harris (3), Weder (89), and Stüpel (90). Current information can be obtained from a number of suppliers (91, 92, 93).

With respect to fabrics artificially soiled with carbon and graphite, literature references cite specific sensitivities toward detergent ingredients. For example, Warren (94) reported that F. D. Snell soil gave relatively high soil removal and relatively low redeposition results with mixtures of silicates and phosphates. U.S. Testing soil was found to be relatively insensitive to temperature in soil removal tests, while detergents showing best soil removal properties gave the poorest redeposition results. Testfabrics, Inc. soil appeared to show inverted temperature effects in soil removal. Ginn et al. (95) reported that compositionwise among these same three soils, the U.S. Testing cloth would appear to be the most hydrophobic, favoring nonionic detergents, the Testfabrics, Inc. cloth less hydrophobic, while the F. D. Snell soil furnishes a different chemical type. It was proposed that the latter two soils favor anionic detergents.

Other criticism with respect to carbon-based soils has not been lacking. In 1954, Diehl and Crowe (96) reported using four types of artificially soiled cloths to evaluate cleaning by representative detergent products. Three of these fabrics were commercially available while a fourth was from a private laboratory. Indications were that artificially soiled fabric must be used with caution. No substitute was seen for actual performance tests for detergent products under practical conditions.

TABLE 9.5

Gustafson Typical Carpet Soil Composition[a]

Moisture	2-4 wt.%
Silica and silicates	30-40
Oxides, carbonates, and phosphates	6-24
Carbon	0-3
Animal fibers	10-12
Cellulose fibers	10-12
Resins and gums	6-10
Fats, oils, and tars	3-8
Miscellaneous	1-3

[a]Reprinted from (87, p. 195) by courtesy of Proceedings of Chemical Specialties Manufacturers' Association, and the author.

Hensley et al. (97) have pointed out that soiled test fabrics must be designed to give light reflectances in ranges most suitable for accurate measurement. This has placed rather severe limitations on standard soils, particularly with respect to the soil's loading range, and has resulted in fabric loaded heavily with carbon black, a condition representative neither quantitatively nor qualitatively of the most commonly encountered natural soils.

Lambert and Sanders (11) have commented similarly, pointing out that photometry at the high reflectance levels is of great importance because the task of a detergent is to wash the fabric as clean as possible. Although reflectometers are very sensitive to near-white fabrics, experimental errors in such washings are very high, with consequent need to run a large number of individual washings. Hence, this may also explain why very dark swatches have been used in wash testing.

Shimauchi and Mizushima (98) concluded that carbon black was not satisfactory for use in soil deposition studies because deposition with this artificial pigment was far different from that with natural soils in both lab tests and commercial laundering. They also concluded that their artificial particulate soil which was based on clay was inferior to natural dirt. Nuessle et al. (306) discussed anomalous results from soil redeposition testing, using carbon black soil, reporting that various carbon blacks and other test soils gave soiling properties differing from those of other test soils or natural soils. Vaeck and Maes (77) feel that many experimental facts contradict the concept that clay minerals are a major cause of soil buildup and redeposition on cotton fibers. Schoenberg (7) reviewed the composition of both lab-prepared and commercially available soils from 1911 to 1961. He also reviewed reported compositions of natural soils.

b. Evaluations. Typical examples of assessment of washing effectiveness with respect to soil removal, soil redeposition, and soil deposition are shown in Tables 9.6-9.13. Nomenclature and abbreviations used in these tables are shown in the table on page 397.

Considerable attention during recent years has been devoted to soiling, soil removal, and soil redeposition problems basically unique to polyester cotton blend fabrics including those specially finished to provide durable press and soil release features (311-318).

Attention has also been given during the late '60s and early '70s to measuring the soil removal properties of detergent formulations containing enzymes. The test methods may include reflectance measurements, particularly on washed test panels originally soiled with such organic materials as blood, gravy, chocolate and grass stains. Wash operations of 10 to 20 minutes duration under conventional conditions of temperature, mechanical action, and detergent concentration may be preceeded by soaking the soiled test panels in the detergent solution for one hour to overnight at room temperature.

It should be noted that wash tests may be run on either a single or multiple wash basis. The repetitive approach is probably closer to practice than that of the single test. In the repetitive approach the original test fabric is transferred through the successive fresh detergent solutions, with rinses interspersed between; sometimes the test swatches are dried and read for reflectance between washes.

Along these same general lines Bernstein and Sosson (133), developed a multicycle alternate soil-wash test for evaluation of fabric detergents for use in sea water aboard naval vessels. In their tests white cotton fabric was given $3\frac{1}{3}$ cycles of treatment in a Terg-O-Tometer, a complete cycle consisting of a graphite soil deposition-detergent operation plus two rinses, and then washed in a fresh solution of the detergent without soil present. Reflectance was read after each half-cycle.

In 1963, Schwartz and Berch (134) reported the development of a laboratory soil accumulation test method utilizing reflectance values as detergency measure. The flow chart for swatches used in comparing two detergents is shown in Fig. 9.5. Use of natural soils was encouraged, and there were indications that field performance of laundry detergents could be reliably predicted by the Schwartz-Berch method.

Assessment of detergency characteristics of wash formulas, as well as evaluation of tensile strength losses on the fabric after 20 to 50 repetitive

TABLE 9.6

Assessment of Removal of Colloidal Carbon- and Graphite-Based Soil by Reflectance

Soiled fabric						
Pigment	Fabric	Prep.	Detergent	Washer	Reflectometer	Literature Source
Lampblack	Cotton	Lab.	Soap, synthetic	LOM	Lange	Bacon '45 (99)
Graphite	Cotton, syn.	Lab.	Sodium oleate, synthetic	LOM	HMP	Clark & Holland '47 (100)
Graphite	Cotton	Lab.	Nonionic, anionic, cationic; soap	LOM	PV610	Armstrong et al. '48 (101)
C black	Cotton	ACH	Built soap, nonionic	DM	HMP	Wollner & Freeman '51 (102)
C black	Cotton	ACH	Nonionic and anionic	TOT	HMP	Barker & Ranauto '55 (103)
C black	Cotton	Lab.	Alkylbenzene sulfonates	Ultrasonic	HMP	Ludeman et al. '58 (104)
C black	Cotton	UST	Heavy duty anionic	TOT	HMP	Linfield et al. '62 (105)
C black, clay Lampblack C black	Cotton Cotton Cotton	FDS TFI UST	Silicate built, anionic and nonionic detergents	TOT	HMP	Warren '63 (94)
Various	Cotton	Comm.	Nonionic detergents	TOT	HMP	Schmolka & Hensley '65 (106)
C black	Cotton	UST	Built sucrose esters; anionic	TOT	—	Schwartz & Rader '65 (107)
C black, clay Lampblack C black	Cotton Cotton Cotton	FDS TFI UST	Commercial; built anionic and nonionic	TOT	—	Ginn et al. '66 (95)
C black	Cotton	EMPA	Sucrose derivative	—	Zeiss	Gerhardt '66 (108)
C black C black	Cotton Cotton	ACH UST	Built anionic containing NTA	TOT	PV	Pollard '66 (109)
C black C black	Cotton Cotton	EMPA UST	Tallow-derived	TOT	D-40	Rader & Schwartz '66 (110)
C black, clay C black	Cotton Cotton	FDS UST	Alcohol sulfates, ethoxysulfates Ethoxylates; built and unbuilt	TOT	—	Finger et al. '67 (111)

TABLE 9.7

Redeposition of Colloidal Carbon and Graphite by Reflectance Assessment

Type	Fabric	Detergent	Washer	Reflectometer	Literature source
Graphite	Cotton	Surfactants, NaCMC, soap	LOM	PV610	Armstrong et al. '48 (101)
C black	Cotton	Anionic syndets, NaCMC	LOM	PV	Griesinger & Nevison '50 (36)
Lamp-black	Cotton	Built soap, NaCMC	Experimental, tumbler	—	Nieuwenhuis '54 (112)
Lamp-black	Silk, rayon, nylon	Soap, phosphate	Automatic home washer	GE spectrophotometer, integrator	Lindsey '54 (113)
C black	Cotton synthetics	Soap, syndets	LOM	HMP PV 525 nm	Bacon et al. '56 (114)
C black, clay	Cotton	Silicate-built detergents	TOT	HMP	Warren '63 (94)
C black	Cotton	Built alkylbenzene sulfonate, NaCMC, PVP, PVA, etc.	Miniwasher	HMP	Hensley '65 (64)

TABLE 9.8

Deposition of Colloidal Carbon and Graphite by Reflectance Assessment

Soil					
Type	Fabric	Detergent	Washer	Reflec-tometer	Literature Source
C black	Cotton	Nonionic	Manual	—	Gruntfest & Young '49 (115)
C black	Cotton	Various	LOM	HMP	Vaughn & Suter '50 (68)
C black	Cotton	Various	LOM	—	Weatherburn & Bailey '60 (116)
C black	Cotton	Built alkylbenzene sulfonate	Mini-washer	HMP	Hensley '65 (64)
C black	Cotton	H. D. detergent	LOM	Hunter	Johnston & Bretland '67 (117)
C black	Cotton	Syndet containing NaCMC, etc.	LOM	Hunter D-40	Morris '67 (118)
C black, 9 types	Polyester, w/wo finishes	Anionic detergent	LOM	Photo-meter	Shimauchi & Mizushima '68 (98)

TABLE 9.9

Reflectance Assessments of Removal of Non-Carbon-Based Soils

Soiled fabric						
Soil	Fabric	Detergent	Washer	Meter	Filter	Literature source
-80 Mesh vacuum cleaner dirt	Cotton	Nonionic, anionic	TOT	PV610	Green	Ferris & Leenerts '56 (75)
Vacuum cleaner dirt-oil, 5:1	Cotton, synthetics	Built soap, syndets	Mechanical	Hunter	—	Jones et al. '60 (80)
Spinks Bandy Black clay-oleic acid	Cotton	Built anionic, built nonionic	TOT	Gardner CD	—	Davis '63 (83)
Face and neck natural soil	Cotton	Built LAS	TOT	Gardner AC-2 CD	Rd	Trowbridge & Rubinfeld '67 (119)
Airborne dust-synthetic sebum	Cotton	H.D. anionic syndet, high foaming	TOT	CD	Rd	Spangler et al. '65 (76)
Various	Cotton	Nonionic detergents	TOT	HMP	Green	Schmolka & Hensley '65 (106)
Airborne dust-synthetic sebum	Cotton, PEC, polyester	Built anionic, built nonionic	TOT	D-25	Rd	Hunter et al. '67 (120)
Clay	Cotton	Nonionic, anionic, cationic	TOT	—	—	Vandegrift & Rutkowski '67 (121)
EMPA	Wool	Anionic-nonionic surfactants	—	Elrepho	7	Hoff '67 (122)
Bandy Black clay-oleic acid	3 Natural, 9 synthetic	Neutral detergent, water softener	LOM	PV Hunter	—	Kennedy & Stout '68 (40)

TABLE 9.10

Redeposition of Noncarbon Soils by Reflectance Assessment

Type	Fabric	Detergent	Washer	Reflectivity		Literature Source
				Meter	Filter	
Family, Spangler	Cotton w/wo finish, PEC, 4 synthetics	Commercial, built anionics, built nonionics	Practical TOT	D-25	w/wo uv	Hunter et al. '67 (120)
Bandy Black clay-oleic acid	3 natural, 9 synthetics	Neutral detergent, water softener	LOM	PV Hunter	— —	Kennedy & Stout '68 (40)

TABLE 9.11

Deposition of Noncarbon Soils by Reflectance Assessment

Type	Fabric	Detergent	Washer	Reflectivity Meter	Reflectivity Filter	Literature Source
Spinks Bandy Black clay	Cotton	Soap, built anionic-built nonionic	TOT	Gardner CD	—	Martin & Davis '60 (69)
Spinks Bandy Black clay	Cotton	Anionic, nonionic, NaCMC	TOT	Gardner CD	—	Rutkowski '68 (85)
Artificial particulate and fatty soils	Polyester w/wo finishes	Anionic syndet	LOM	Photometer	—	Shimauchi & Mizushima '68 (98)

TABLE 9.12

Color and Color Difference Assessment of Wash Results

Soiled fabrics					
Type of soil	Fabric	Detergent	Washer	Reflectivity meter/readings/ reporting	Literature source
Natural	Cotton, silk, rayon, nylon	Neutral soap, phosphate	Household	GE spectrophotometer and integrator/chromaticity coordinates/NBS CD units	Lindsey '54 (113)
None, naturally soiled pillow cases	Bleached and unbleached cotton sheeting	Built syndet containing FWA unbuilt soap wo FWA, Na perborate, Cl bleaches	LOM, agitator household washer	Hunter color and CD modified to use or reject uv by Noviol filter; fluorescence-sensitive meter w/wo uv/Rd, "a", "b"/ ΔRd, Δ"a", Δ"b"; blue refl.; CD, NBS units = $\underline{E} = (\underline{L}^2 - \underline{a}^2 - \underline{b}^2)^{\frac{1}{2}}$, where $\underline{L} = 10(Rd)^{\frac{1}{2}}$	Furry et al. '59 (46, 123), '61 (124)
Natural, N compounds of perspiration	Cotton sheets and T-shirts	Built syndet, unbuilt soap, Na perborate	Household	Gardner Automatic CD/ "b", with or without uv on sample	McLendon & Richardson '63 (125), '65 (126)
Unsaturated fatty compds	Artificial, cotton, nylon	None	–	Hunter/G, B/yellow G-B	Wagg et al. '67 (127)
Natural	Cotton	Various	Household	Gardner large area/Rd, "b" with and without Noviol filter, Rd	Leigh '67 (128)
Mechanics	Cotton huck towels	Heavy duty det.	5-gal. semi-automatic	Gardner large area/Rd, "b" with and without uv	Mihalik & Cross '65 (129)
(Heating yellowing of cotton fabric)		–	–	GMP/yellow (G-B)/G	Loeb et al. '66 (130)
None	25 Types of fibers, finishes, etc.	Built synthetic	6 household types	Gardner Automatic CD/Rd, "b"	Taube & Poole '67 (131)
Artificial sebum	Cotton, synthetics	Built anionic	TOT	CD/L, "b"	Spangler et al. '67 (86)

TABLE 9.13

Whiteness Assessment of Wash Results

Soil						
Type	Fabric	Detergent	Washer	Reflectivity meter	Mode of reporting	Literature source
Worn by women	Cotton, blend, nylon, Dacron; resin-treated blouses	Built soap Built liquid syndet	Household	Gardner automatic CD	Whiteness = $10\,\overline{Rd} - 3b$	Poole et al. '62 (132)
Natural	Cotton without FWA	Representative detergents, without FWA	Household	Hunter C & CD	100 - whiteness = $[(100 - \underline{L})^2 + 2.3^2(\Delta\underline{S})^2]^{\frac{1}{2}}$	Diehl '67 (25)
Synthetic sebum	Cotton polyamide polyester	4 comm.	TOT	CD	Whiteness = $\underline{L} - 3b$	Spangler et al. '67 (86)

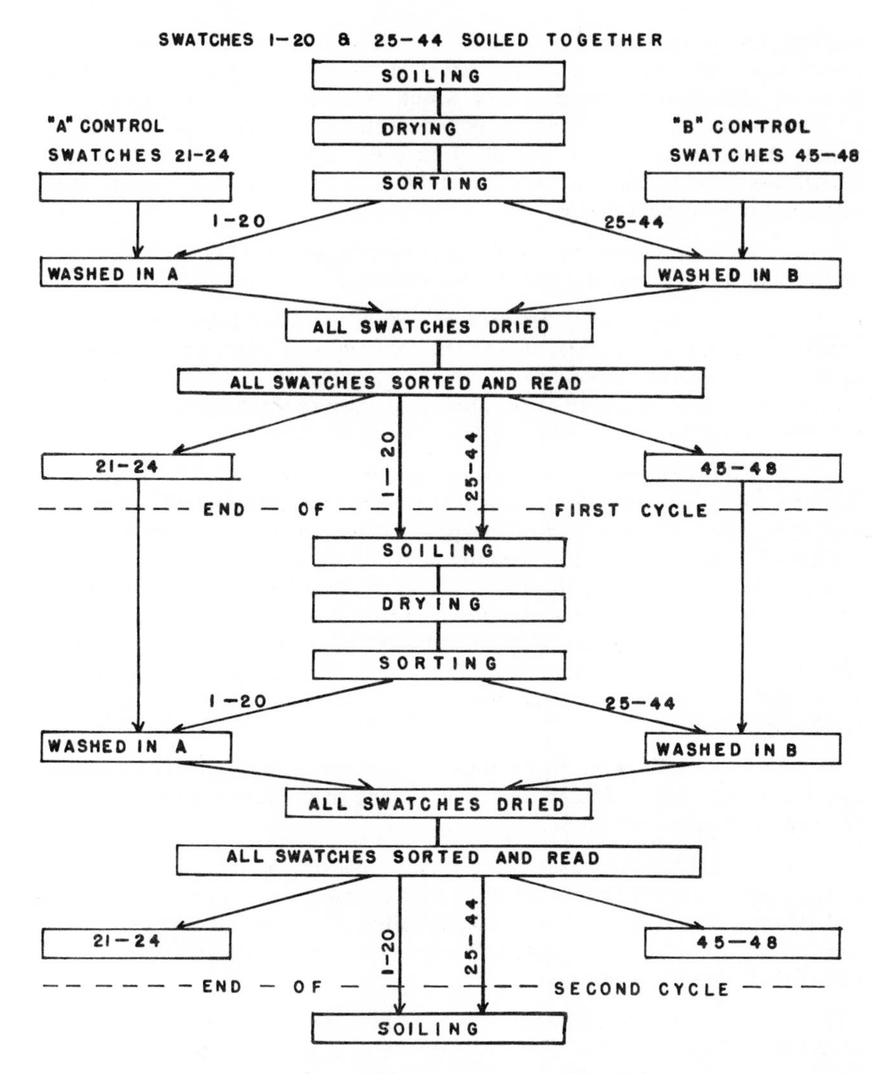

FIG. 9.5. Flow chart for handling of test swatches in Schwartz and Berch soil accumulation test method. Reprinted from (134, p. 79) by courtesy of Soap and Chemical Specialties.

washes, has been handled by test bundles in commercial, institutional, and rental laundries for many years. Most of these test bundles are simply a piece of white cotton sheeting upon which reflectance measurements can be made. Recently, the Institute of Industrial Launderers and the Linen Supply Association of America (319) have provided a relatively elaborate bundle also comprising fabrics for measuring soil removal, bleach intensity, and mechanical action.

With respect to fluorescent whitening agents, an excellent bibliography on their detergent application has been published by ASTM (135).

c. Reproducibility. In the interest of improved precision of reflectance readings, a number of variables in the technique of presenting cloth swatches to the reflectometer have been recognized and studied. These factors comprise orientation of fabric, number of thicknesses of the fabric, and type of background plaque or sample holder.

Diaz et al. (8) reported a change in reflectance from 67.8 to 65.6% of magnesium oxide when the weave of the cloth was oriented 90° on the exposure head of a Hunter Multipurpose reflectometer (incident light 45°, reflected light 0°; viewed perpendicular to cloth surface). This change was ascribed to variation in light-scattering qualities from the light source due to difference in surface structure in the warp and fill directions of the cloth. Stephens and Brown (136) noted similar rotational effects under generally similar test conditions. In Tables 9.14 and 9.15, R. S. Hunter et al. (62) showed percent changes in reflectance due to 90 rotating of cloth specimen.

Johnson (320) measured directional reflectance properties from acrylic pile fabric of various luster levels by means of a Hunter Model D-10 Recording Goniophotometer.

Means of avoiding errors arising from these differences lie in prescribing a fixed orientation (Diaz) or reading both in warp and fill directions (Stephens and Brown, Hunter et al.). The construction features of one type of reflectometer avoids cloth orientation effects by using 0° incident light and integrated circular pickup at 45° (52).

Diaz et al. (8) showed data on effect of thickness of cloth and reflectance of backing plaque: In Fig. 9.6, reflectance of white cloth increases when backed up by multiple thicknesses of the same fabric; in the case of black backing, seven thicknesses were required to reach the maximum value. Oldenroth (137) reported increase in whiteness degree with eight thicknesses over four thicknesses of cloth using an Elrepho instrument.

In Fig. 9.7, Diaz et al. (8) showed the increase in reflectance with a single thickness of the white cloth when backed up with plaques of increasing whiteness; this increase was attributed to partial transmission of light through the cloth fabric, which was not completely opaque. Three advantages were recognized when using a black backing over that of multiple

TABLE 9.14

Changes in Reflectance Measured by One Laboratory as Each Specimen Was Rotated 90°[a,b]

Specimen	Reflectance change with rotation (%)	
	Green	Blue
Cover glass omitted		
Light gray panel	0.1	0.4
Greige 80 x 80 cotton	0.2	0.6
Wool flannel	0.9	1.5
Bleached 80 x 80 cotton	3.0	4.4
Cotton carpet	3.1	4.0
Cover glass installed		
Light gray panel	0.0	0.3
Greige 80 x 80 cotton	0.4	0.3
Wool flannel	0.3	0.4
Bleached 80 x 80	3.1	4.9
Cotton carpet	0.4	0.8
Bulky Specimens behind glass		
Nonwoven	0.5	0.7
Terry	1.1	1.4
Nylon fibers	2.4	1.9
Brightened specimens behind glass		
80 x 80 cotton: brtnr included	2.6	3.1
brtnr excluded	2.6	3.7
Knit cotton: brtnr included	5.0	4.6
brtnr excluded	5.0	5.1
Nylon organdy: brtnr included	1.9	1.5
brtnr excluded	1.9	1.5

[a]Hunterlab D40 Reflectometer.

[b]Reprinted from (62, p. 819) by courtesy of American Dyestuff Reporter.

cloth thicknesses or a white background: neater operation, higher precision, and more accuracy. In accordance with this concept, Ashcraft (138) reported use of a black backing cloth of approximatley zero reflectance.

Stephens and Brown (136) recommended use of a white background as providing maximum differentiation in relative detersive efficiency on the basis of data shown in Table 9.16. Their recommendation was supplemented by visual examination of washed cloths: against a black background, fabric

TABLE 9.15

Changes in Reflectance of One Specimen[a] as Specimen Was Rotated 90° in Each Laboratory[b]

Lab no.	Instrument	Unmounted specimen		Specimen behind glass	
		Green	Blue	Green	Blue
2a	Multipurpose Reflectometer	4.8	5.1	0.2	0.2
2b	Multipurpose Reflectometer	4.9	0.7	1.6	0.9
6	Hunterlab D40 Reflectometer	3.0	4.4	3.1	4.9
13a	Hunterlab D40 Reflectometer	3.1	4.3	2.6	4.0
13b	Hunterlab D40 Reflectometer	3.0	0.8	4.1	5.1
9	Colormaster Diff Colorimeter	7.1	5.2	2.9	4.8
8	Color Eye	2.2	2.6	1.6	1.7
10	Gardner Color-Diff Meter	4.1	4.6	3.4	3.4
4	Gardner Color-Diff Meter	2.9	4.0	3.3	3.2
5	Gardner Color-Diff Meter	4.9	5.4	3.3	3.6
1	Hunterlab Color-Diff Meter	4.0	3.7	3.4	3.7
12	Hunterlab Color-Diff Meter	0.6	1.1	2.4	2.3

[a] 80 × 80 bl. cotton.

[b] Reprinted from (62, p. 819) by courtesy of American Dyestuff Reporter.

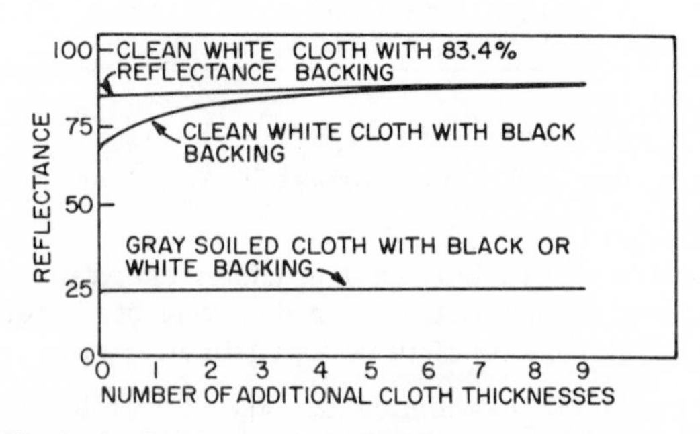

FIG. 9.6. Effect of additional cloth background on reflectance. Reprinted from (8, p. TP243) by courtesy of the American Society for Testing and Materials.

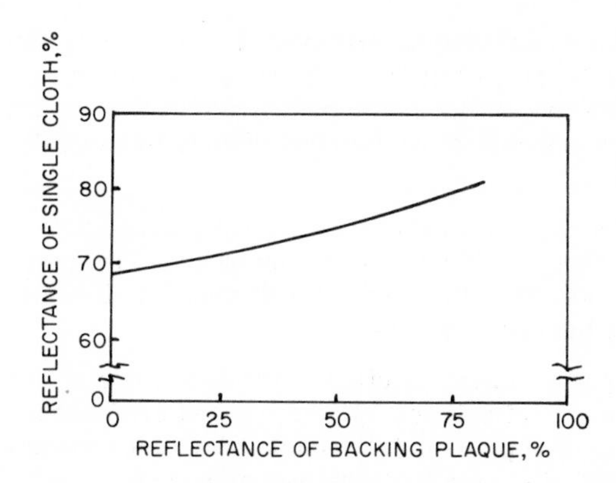

FIG. 9.7. Effect of reflectance of backing plaque on reflectance value obtained on single thickness of cloth. Reprinted from (8, p. TP243) by courtesy of the American Society for Testing and Materials.

TABLE 9.16

Relative Detersive Efficiency, Percent, Using Five Different Backgrounds[a]

Detergent solution	Black	Xylonite	White	Nearest	Special
D	100	100	100	100	100
C	88	84	81	86	84
B	37	33	30	32	33
A	27	24	23	22	24

[a]Reprinted from (136, p. TP104) by courtesy of American Society for Testing and Materials.

washed with detergent D appeared to be slightly lighter than that washed with detergent C; much greater lightness difference with detergents D versus C was reported with cloths against a white background.

Beninate et al. (139) reported use of a white background consisting of four layers of larger-diameter filter papers. Rees (73) studied the effect of white versus black background and amount of openness of fabric in connection with assessment of degree of soiling of fabrics. Experimental findings were that measurement made against a black background show that the extent of soiling of a fabric by airborne dirt increases with increasing openness of the fabric, while corresponding data obtained using a white background gave the erroneous impression that the very open fabrics soil less than the reasonably close fabrics. Hence Rees preferred a background of black velvet.

In general, detergency workers have used a wide variety of approaches to the cloth thickness and background question. Thicknesses vary from 2 (140, 141, 60) to 4 (114, 105) and to "sufficient to eliminate the effect of any extraneous background" (142).

Harris (3) noted frequent mention in the literature of need to use background of constant reflectance. He expressed preference for sufficient layers of fabric of approximately equal reflectance mounted by stretching with small sharp pins on a battleship-gray board.

A comprehensive report on basic information and means of achieving accuracy in reflectance, yellowness, and whiteness measurements on textile specimens has been presented by the Washington Section of the AATCC (62).

d. Special Techniques. In order to expedite reading and statistical averaging of results, Thompson (143) reported a simple method of measuring the average reflectance of 20 swatches using only two reflectance measurements. This was accomplished by mounting the swatches on a phonograph turntable which was rotated in the field of a Hunter Multipurpose reflectance meter. This technique has been used by other Colgate workers (144, 145).

Oldenroth (146, 147) has reported use of the Krefelder hydrosulfite treatment, a dye stripper, for removing grayness due to pickup of fugitive organic dyes. Optical brightness was evaluated by using a xenon lamp with and without ultraviolet barrier filter; assessment was made of the effects of both pigment and fugitive dye graying by making a reading after use of the hydrosulfite.

In England, Powney and Feuell (148) replaced the orthodox fabric of detergency tests with chopped fibers in the interest of improving reproducibility of wash results by eliminating nonuniformity of soiling and geometric problems associated with the fabric. They used a soil of graphite plus liquid paraffin. After washing and rinsing in a lab washer, the fibers were formed into a pad by filtration through a Büchner funnel equipped with a paper filter. The dried pad was examined visually and also by reflectance.

The chopped fiber technique was used in the United States by Compton and Hart (149, 150) for studying the interaction between carbon black and textile fibers.

Lambert (151) commented that the chopped fiber approach is too artificial in that the filter pad of disintegrated fibers retains irrelevant carbonaceous material and prevents measurement of trace amounts of ingrained soil which is most important in practical washing. Neither did Lambert accept the claim of improved reproducibility, mainly because the Powney and Feuell evidence was obtained with relatively dark pads with consequent photometric insensitivity.

e. Calculations Associated with Reflectance Methods for Detergency Assessment. As pointed out by Snell (152), the only uncontroversial way to handle detergency reflectance data is to report as "percent of magnesium oxide." Alternately, the reflectance of a swatch before washing may be subtracted from that of the swatch after washing, and the difference reported as "change in reflectance", or as some (153) prefer, ΔR.

An equation frequently used for reporting reflectance readings (11):

$$\% \text{ Detergency} = \left[(R_w - R_s)/(R_o - R_s)\right] \times 100$$

where R_w is the reflectance of the washed swatches; R_s is the reflectance of the soiled swatch; and R_0 is the reflectance of the swatch before soiling. If all swatches have the same initial reflectance, $D = K_1R_w - K_2$. Both K_1 and K_2 are constants.

Harris (3) has provided an expression for calculating washing effectiveness on the basis of the power laundering of loads containing both soiled and white test swatches:

$$\% \text{ Detergency} = \{[(A - E) - B]/(C - B)\} \times 100$$

where A is the reflectance of soiled swatch after washing; B is the reflectance of soil swatch before washing; C is the reflectance of redeposition swatch before washing; and E is equal to the net reduction in reflectivity of redeposition swatch (before washing-after washing). Bacon et al. (114) washed soiled and clean cloth in a Launder-Ometer jar, rinsed, blotted, and ironed the swatches dry. Results of reflectance measurements were handled as follows:

$$\% \text{ Return to original} = \left[(S_1 - S_0)/(W_0 - S_0)\right] \times 100$$

where W_0 is the reflectance of white swatch before washing; S_0 is the reflectance of soiled swatch before washing; W_1 is the reflectance of white swatch after washing; and S_1 is the reflectance of soiled swatch after washing.

Rutkowski (84) evaluated proprietary detergents using test fabrics soiled with a kaolinite-type clay plus triolein, and expressed results as:

$$\text{Performance} = \text{Soil removal} - \text{Soil redeposition} + (\tfrac{1}{2})\text{Optical brightener}$$

where soil removal values were obtained from Hunter D-40 reflectance readings made of the original and washed soiled swatches and application of the Kubelka-Munk equation; soil redeposition was reported as reflectance units lost by the clean swatches in three Terg-O-Tometer washings; and optical brightener effectiveness was reported as percent of the total reflectance due to fluorescence.

Several workers have investigated the kinetics of soil removal, usually as a means of elucidating the detergency phenomena. Schoenberg (7) has reviewed many of the interesting mathematical expressions generated by various workers. Bourne and Jennings (154) have developed a concept of two types of soil which vary greatly in rate of removal. Loeb et al. (155) studied soil removal as a rate process using three artificially soiled cotton test fabrics.

Numerous mathematical expressions have appeared for the purpose of calculating values of whiteness from reflectance readings using tristimulus or more sophisticated types of reflectance meters. The reader is referred to Judd (12), Nickerson (156), Oldenroth (147, 157), Diehl (25), Nieuwenhuis (158), Stensby (159), and Berger (160). Possibly typical of these formulas are that of R. S. Hunter (62):

$$\underline{W} = 4 \times \underline{B} - 3 \times \underline{G}$$

where $\underline{B}$ is blue filter reflectance with ultraviolet light included and $\underline{G}$ is green filter reflectance without ultraviolet light, and that of Stephenson (161):

$$\underline{W} = 2\underline{B} - \underline{A}$$

where $\underline{B}$ is blue filter reflectance and $\underline{A}$ is amber filter reflectance.

The handling of large amounts of data may be expedited by use of computers. A program of laboratory determination of fabric detergency frequently involves replicated wash tests made under a variety of conditions. Measurements before and after washing may reach hundreds of thousands of reflectance readings.

Illman et al. (162) converted the electric signal from the reflectometer to digital form, and fed these results to a card punch in order to record before and after washing values on punched cards. An appropriate computer program calculated and tabulated the reflectance changes for each test. Much operator time is saved. Statistical examination of the data can be provided in the program.

Gordon et al. (163) have handled data from radioactive tracer tests in a similar manner. Results were expressed in parts per million activity, basis fabric.

3. Light Transmission

a. Fabric. An alternate to reflectance on fabrics is determination of amount of light transmitted through the fabric. A number of German workers have been active in this approach (61, 164, 165).

b. Wash Liquor. Another type of optical measurement is assessment of amount of soil removed by running light transmission on the dirty wash liquor. Vaughn and co-workers (68, 166-168) pioneered this approach. They used a Fisher electrophotometer and a Lumetron colorimeter. Calibration curves were prepared using suspensions of known carbon concentration.

Wollner and Freeman (102) washed an experimental turbidity-type cloth with surfactant solutions in a Deter-Meter and used a Lumetron colorimeter to make turbidity measurements. Results were reported as "light transmission of solution."

Barker and Kern (169) compared the light transmission method to a light reflectance method, using cotton fabrics soiled with carbon black. Detergents consisted of a nonionic and an anionic surfactant, and tallow soap. Conclusions were that both methods lead to essentially the same conclusions with the synthetic detergents, but fail to give comparable results with the soap.

The soap turbidity difficulty as well as questions of degree of agglomeration of the soil as caused by the presence of surfactants or other detergents ingredients has been commented on by Lambert and Sanders (11), Vold and Phansalker (170), Martin and Fulton (4), Nieuwenhuis (171), Vaeck and Maes (77), and Evans and Camp (172).

D. Correlation between Subjective and Instrumental Evaluations

In a number of papers Furry and co-workers (46, 124, 173, 174) have reported good correlation between the visual ranking of wash quality and that obtained by use of reflectivity instruments. Observers were able to distinguish visually differences of only one NBS color difference unit on either woolen (173) or cotton (174) fabrics. The NBS value range is shown in Table 9.17 (113).

Wixon (175) reported that in connection with treatment of terrycloth hand towels with fabric softener plus sodium carboxymethyl cellulose, an instrumental difference of 0.5 + "b" in yellowing was visible to the eye.

In tests without soil, Furry et al. (46) assessed the fluorescent and nonfluorescent whitening effects on white cotton percale sheeting laundered repeatedly with different fluorescent whitening agents plus peroxygen or chlorine-containing bleaches. That reflectance results conform well with visual ratings of whiteness under a daylight lamp with added ultraviolet

TABLE 9.17

NBS Value Range[a]

Textile terms	NBS (Judd) units
Trace	0.0-0.5
Slight	0.5-1.5
Noticeable	1.5-3.0
Appreciable	3.0-6.0
Much	6.0-12.0
Very much	More than 12.0

[a]Reprinted from (113, p. 761) by courtesy of American Dyestuff Reporter.

light is shown by the data in Table 9.18. Satisfactory correlation was also obtained when polyester and nylon fabrics were used in similar tests (124).

Wagg et al. (127) found that the intensity of yellow color produced by repeated use in laundering of cotton fabric artificially soiled with squalene, chloresterol, and linolenic acid, and measured by a tristimulus reflectometer using a simple difference of reading with green and blue filters correlated with visual observation.

McCabe (176) obtained good correlation between visual ranking and reflectance readings on naturally soiled fabric washed with detergents with and without fluorescent whitening agent content. Vaeck (177) reported a 0.93 correlation coefficient between visual observations and tristimulus reflectance without fluorescence.

Nieuwenhuis (158) reported excellent agreement between results obtained with a reflectance method for measuring the whiteness of fabrics containing fluorescent whitening agents and those resulting from visual appraisal by 27 observers. The fabrics contained various amounts of several blue dyes and fluorescence, brightness, yellowness, or blueness and redness, or greeness.

The Rhode Island section of the AATCC (63) has reported on instrumental evaluation of color fluorescent textiles in terms of average observer response. Precautionary measures which are necessary to allow proper comparison of visual and instrumental assessments were listed. An experimental study was provided comparing samples of just noticeable difference by instrumental and visual methods of three areas of interest to practical workers: (1) fluorescent whiteners on bleached fabrics; (2) nonfluorescent

TABLE 9.18

Relationship between Visual Rankings and Instrument Measurements of Yellow-Blue ("b") and Blue Reflectance (uv Included) in Bleached Fabric Laundered 50 Times, Arranged in Decreasing Order of Whiteness According to Visual Ranking[a]

Composition and concentration of washing solution	Ranking[b]	Laundered swatches stored for two years		Laundered swatches not stored	
		Yellow-blue ("b") (uv incl)	Blue reflectance (uv incl)	Yellow-blue ("b") (uv incl)	Blue reflectance (uv incl)
			Percent		Percent
Syndet (with whitener), 0.10%; sodium perborate, 0.005% available 0 (0.05%); sodium sesquicarbonate, 0.05%	1	-0.8	89.1	-2.4	91.0
Syndet (with whitener), 0.10%; sodium perborate, 0.005% available 0 (0.05%)	2	-0.7	88.8	-2.4	91.0
Syndet (with whitener), 0.10%; hydrogen peroxide, 0.015% available 0	3	-0.5	89.2	-2.0	91.6
Syndet (with whitener), 0.10%; hydrogen peroxide, 0.005% available 0	4	-0.2	87.8	-1.9	91.4
Syndet (with whitener), 0.10%	5	+0.2	86.4	-1.4	89.3
Syndet (with whitener), 0.10%; sodium perborate, 0.015% available 0 (0.15%); sodium sesquicarbonate, 0.15%	6	0	88.1	-2.3	91.4
Average[c]		-0.3	88.2	—	—
Soap, 0.10%; whitener I, 0.00006%; sodium perborate, 0.005% available 0 (0.05%)	7	+1.2	86.7	-0.9	90.7

TABLE 9.18 (Cont'd)

Composition and concentration of washing solution	Ranking[b]	Laundered swatches stored for two years: Yellow-blue ("b") (uv incl)	Laundered swatches stored for two years: Blue reflectance (uv incl)	Laundered swatches not stored: Yellow-blue ("b") (uv incl)	Laundered swatches not stored: Blue reflectance (uv incl)
			Percent		Percent
Soap, 0.10%; whitener I, 0.00006%; hydrogen peroxide, 0.005% available O . .	8	+1.4	86.4	-0.6	90.0
Soap, 0.10%; whitener I, 0.00006%	9	+2.2	83.6	-0.3	88.7
Syndet (with whitener), 0.10%; dichlorodimethylhydantoin 0.022%, available Cl	10	+1.4	86.5	-0.6	89.6
Average[c]		+1.5	85.8	—	—
Soap, 0.10%; whitener II, 0.00006%; sodium perborate, 0.005% available O (0.05%)	11	+2.1	85.8	+0.4	89.4
Soap, 0.10%; whitener II, 0.00006%; hydrogen peroxide, 0.005% available O . .	12	+2.4	85.2	+0.6	87.6
Soap, 0.10%; whitener II, 0.00006%	13	+2.7	84.1	+0.8	87.8
Soap, 0.10%; whitener II, 0.00006%; sodium hypochlorite, 0.022% available Cl	14	+2.9	84.6	-0.1	89.4
Average[c]		+2.5	84.9	—	—
Syndet (with whitener), 0.10%; dichlorodimethylhydantoin, 0.066% available Cl	15	+3.3	83.2	+0.6	88.5
Syndet (with whitener), 0.10%; sodium hypochlorite, 0.022% available Cl . . .	16	+3.8	83.4	0	89.2

Soap, 0.10%; whitener I, 0.00006%; sodium hypochlorite, 0.022% available Cl	17	+3.3	83.7	+0.9	88.6
Soap, 0.10%	18	+3.5	81.5	+2.0	85.7
Average[c]		+3.5	83.0	—	—
Syndet (with whitener), 0.10%; calcium hypochlorite, 0.022% available Cl	19	+5.0	80.3	-0.5	90.0
Syndet (with whitener), 0.10%; sodium hypochlorite, 0.066% available Cl	20	+6.0	80.0	+0.3	89.0
Average[c]		+5.5	80.2	—	—
Soap, 0.10%; whitener I, 0.00006%; dichlorodimethylhydantoin, 0.022% available Cl	21	+6.8	78.8	+1.4	87.4
Soap, 0.10%; whitener II, 0.00006%; dichlorodimethylhydantoin, 0.022% available Cl	22	+7.0	79.0	+0.4	88.9
Average[c]		+6.9	78.9	—	—
Syndet (with whitener), 0.10%; calcium hypochlorite, 0.066% available Cl	23	+7.9	72.5	+1.4	86.4

[a]Reprinted from (46, p. 65) by courtesy of American Dyestuff Reporter.

[b]Visual ranking of samples was the same whether viewed in north daylight or under Macbeth Examolite, Type C-4D-UV.

[c]Laundered swatches in the groups averaged appeared visually to be much alike and different from those in the next group.

coloration with or without optical brightener; (3) fluorescent coloration other than fluorescent brightness. Good correlation was found in all areas examined.

Gailey (178) studied extensively both visual and instrumental color matching with respect to suitable means of correlating test data. He described at length a number of relevant details with both approaches.

Leigh (128) conducted a program wherein household laundering was evaluated subjectively in the home and by reflectivity using a Gardner Large Area reflectometer to determine the R_d values and "b" values with Noviol filter for ultraviolet in or out. The home rankings supported lab data which indicated greater loss in whiteness in items laundered in detergents without antiredeposition agent.

Robert Hunter (17) found blue-yellow scale, "b" readings to be germane with respect to visual preference of fabrics, a unit change in yellowness being more than four times as important as a unit change in grayness. He (120) compared the detergency and redeposition properties of several new fabrics under lab and home laundry conditions. Good correlation was found between data from lab-controlled practical washing and home laundry, although the latter was of poorer wash quality. Paitchel and R. S. Hunter (179) reported (see Table 9.19) typical comparisons between visual ratings and data obtained by using a Gardner large-aperature color difference meter.

Diaz et al. (8) noted that in the presence of achromatic soils and absence of the fluorescent whitening agents, visual preference of whiteness correlates with the widely used reflectance results expressed in terms of percent of magnesium oxide measured on such instruments as the Hunter Multipurpose reflectometer, Photovolt Reflectance Meter, Color Eye, Hunter Color and Color Difference Meter, and spectrophotometers. The reflectance readings referred to are those which are generally made with the green tristimulus filter (Y value), or at about 550 nm wavelength.

Mann and Morton (180) noted that the apparent degree of soiling of a colored object is a matter of subjective assessment, but as they demonstrated with a group of observers, is closely paralleled by a physical quantity, H_s, the diminution (%) in the photometric brightness coefficient of the surface:

$$H_s = \left[(Y_0 - Y_s)/Y_0\right] \times 100$$

where Y_0 and Y_s are the brightness of fabric before and after soiling, respectively.

Diehl (25) reported that optimum correlation with subjective visual evaluations of a large number of washed fabrics was obtained when the chromaticity vector was weighted 2.3 times the lightness vector. Thus,

TABLE 9.19

Typical Results from New Instrument[a]

	Specimen	Observed values R_d	a	b	Computer color difference (ΔE)[b]	Visual preference
Case I	Original	72.7	0.0	-0.6	—	
	A	70.4	-0.4	+1.4	2.3	A > B
	B	69.8	-0.4	+2.8	3.7	
Case II	Original	72.7	0.0	-0.6	—	
	C	69.4	-0.5	+4.2	5.1	C = D
	D	67.8	-0.5	+4.9	6.0	

[a] Reprinted from Ref. (179, Table I) by courtesy of author.

[b] $\Delta E = \Delta Rd^2 + \overline{\Delta} a^2 + \overline{\Delta} b^2$

$$100 - W = [(100 - L)^2 + 2.3^2(\Delta S)^2]^{\frac{1}{2}}$$

where W is whiteness; L is lightness in NBS units; and ΔS is chromaticity difference in NBS units with reference to MgO standard as 100.

Shanley (181) compared visual grading and spectrophotometric examination of white cotton fabric samples collected both at the white bin stage and after bleaching and finishing. The instrumental results, which were obtained in three different laboratories, were converted to CIE coordinates. The three different meters were found to be in only fair agreement with each other. All three failed to distinguish the smallest difference visible to the eye. As might be expected, spectrophotometry was very effective in analyzing the nature of the color differences among samples, and also provided a permanent record of color attributes.

Blakey (20) described adaptation of the Harrison colorimeter to overcome inadequacies of the CIE (1931) standard observer in the comparison of near-whites, anatase, and rutile types of titanium dioxide pigments. Also discussed were personal preferences of near-white tones of blueness relevant to the question of whiteness and brightness.

Several instances of development of interlocked systems of visual assessment standards and reflectance readings have been reported in the literature. Anders (44) reported a Cibanoid white scale made of light-fast and washable plastic materials for use in visually evaluating the degree of whiteness of fluorescent-brightened textiles and other materials. He discussed the color physics involved, and development of the scale on the basis of visual and reflectance data plus calculations.

Rauchle and Schramm (45) reviewed methods of estimating whiteness of textile fabrics, and described colorimetric investigations of the Geigy whiteness standards. The latter are 22 samples of bleached cotton fabric dyed with 0-35.3 mg/liter of Tinopal DMS, providing a linear measure of whiteness independent of uv in the incident light. Chromaticity coordinates were determined using the Color-Eye LS colorimeter, and comparisons made of color differences calculated from the Adam-Nickerson, MacAdam, and Wyszecki formulas. Whiteness scales based on equations derived by Hunter and Berger were also compared with the Geigy standards.

Coppock (43) described the establishment of a Chemstrand whiteness scale based on the visual grading of whiteness of nonfluorescent, near-white fiber samples, and is correlated with instrumental measurements. Samples are limited to those having a dominant wavelength of 560-580 nm. Percent whiteness can either be read from a purity-brightness chart or calculated according to:

$$W = 10 \times (Y - 2p^2)^{\frac{1}{2}}$$

where Y is % brightness; W is % whiteness; and p is % purity.

III. ACTUAL SOIL CONTENT OF FABRICS

The object of this approach is to estimate detergent action by performing a quantitative determination and/or analysis of amounts of soil associated with fabrics before and after washing. In contrast to optical appearance methods, this approach directly measures foreign materials actually present in the fabric.

Frequently, development of a quantitative analysis of soil constituents has been preceded by separation of soil from the fabric either by solvent extraction or ashing techniques, and qualitative analysis of resultant material. Modern techniques of identification include gas liquid chromatography (82) and electron microscopy (81), as well as infrared spectroscopy, gas chromatography, thin layer chromatography, and gel filtration (321).

Quantitative methods used to date include gravimetric, chemical analysis, radiotracers, neutron activation, X-ray fluorescence, and microscopic examination.

A. Gravimetric

Under gravimetric is classed all methods of assessing soil content of cloth whereby the weight of the soil can be determined directly or indirectly by appropriate procedures. The usual method is separation from cloth by solvent extraction of fatty soils such as sebum ingredients and wool fat. Ordinary particulate dirt, unless present in relatively gross amounts, cannot be separated quantitatively from fabric. Such amounts of either particulate or oily material can often be determined by weight gain or loss measurements on the fabric-soil system. Some soils, such as clays, can be determined by ashing the fabric samples. Examinations have been made of weight of dirt dispersed into solution by filtration, and of the total number of dirt particles per unit volume of solution.

Sisley and Wood (56) reviewed gravimetric methods employed prior to 1947.

1. Solvent Extraction

Activity with this approach formerly centered on determining amounts of wool grease before and after scouring operations. Soxhlet-type extractions were performed using suitable, volatile organic solvents (182). References to more rapid determinations and/or use of simpler equipment are also found (183, 184). Colorimetric analysis after coloring oily soils with oil-soluble dye is also noted (1).

Emphasis during recent years on use of natural soils such as sebum, plus attention to oily soil problems peculiar to relatively lipophilic synthetic fibers, has stimulated use of solvent extraction methods for purposes of quantitatively and qualitatively assessing soiling materials. Powe (185) extracted sebum from naturally soiled T-shirts using a 1:1 v/v chloroform-methanol mixture. Subsequent fat solutions were 6.6% w/v in 3:1 benzene-methanol. Cotton swatches were soiled with 1% of these sebum solids and used in Launder-Ometer or Terg-O-Tometer wash tests. After aqueous rinsing, the swatches were air dried and extracted overnight in Soxhlet apparatus using 1:1 v/v chloroform-methanol. The solvent was evaporated from the fat in a rotary evaporating device under vacuum and the fat then placed in a vacuum oven for 1 hour at 160°C., 30 psi, allowed to cool, and weighed.

Lyness and O'Connor (186) determined residues of lipid soil after wash tests by extraction with organic solvents. Arai et al. (187) used solvent extraction in Soxhlet apparatus to determine oily soil (beef tallow, liquid paraffin, and a mixture thereof) content of soiled cotton fabric test pieces to be used in detergency testing by reflectance assessment after washing.

Byrne et al. (188) extracted natural soil from nylon underskirts with petroleum and benzene-methanol solvents. Chemical analysis of soil constituents followed. Hoff (122) used a Tempo-Extraktor for quick determination of residual aliphatic mineral soil in wool washed with mixtures of anionic-nonionic surfactants.

Oldenroth (147) soiled cotton and polyester fabrics with skin surface fat using carbon tetrachloride solvent. After evaporating the solvent and storing for 24 hours, the soiled fabric swatches were washed in a built soap solution at 60 or 95°C. The same swatches were subjected to ten such cycles of soiling and washing. The residual fat was then extracted for 5 hours with 9:1 benzene-methanol. Less fat remained in the test series washes at a higher temperature. Residues in cotton yarn were greater than in polyester yarns. Possibilities that not all residual fat was extracted from the polyester were eliminated by detection of no fat by infrared spectra examination of the petroleum ether extract of polyester yarn samples dissolved in di- and trichloroacetic acid. Undershirts of PEC were worn by men for one week, washed, and residual soil was determined gravimetrically as above.

Brown et al. (33) soiled 100% filament and 100% staple polyester fabrics in sebum-type oil plus a small amount of nonionic surfactant in an aqueous bath. After rinsing and drip drying, oil content was determined by Soxhlet extraction of the cloths with diethyl ether. Residual oil was similarly determined after one, three, and five washes with synthetic detergent. The data obtained are shown in Table 9.20.

Tachibana et al. (189) extracted highly hydrogenated beef tallow residues from cotton fabrics after washing in sodium dodecyl sulfate solutions. The concentration of fat in the resultant benzene solution was measured by the equilibrium lens method. To do this, the benzene solution of the extract was added dropwise on a known area of a substrate of tap water in a tray until the first stable lens was formed. At this point the surface of the water is covered with a monolayer at a definite surface pressure and therefore the value V of volume of the solution added to the end point is in inverse proportion to the concentration. By means of a calibration curve, fat on the originally soiled cloth was estimated as 0.003 grams per 100 cm^2 of cloth. Washing efficiency W was expressed by:

$$\% W = 100 \times (V_w - V_s)/V_w$$

where V_w and V_s are V obtained from washed cloths and from soiled cloths, respectively.

Kantor (190) analyzed household cotton rags gravimetrically before and after washing, using water and carbon tetrachloride extractions. Tomiyama and Iimori (191) examined natural soil on cotton collar fabric worn for 3 days in August and then aged for 20 days at room temperature. Extraction was first with petroleum ether. The fabric was then soaked sequentially in water, 2% sodium chloride, 70% ethanol, and 0.2% sodium hydroxide. After the water step, extraction was repeated three times for each solvent at 25°C for 20 min. Nitrogen was determined before and after subsequent dialysis. Results are shown in Table 9.21.

Wash test data on the naturally soiled collar fabric suggested some beneficial interactions between sodium dodecyl benzene sulfonate and

TABLE 9.20

Oil Contents[a]

Experiment No.	Original oil, %	Residual oil, %, after			Oil removal, %, after		
		1 Wash	3 Washes	5 Washes	1 Wash	3 Washes	5 Washes
		Staple polyester fabric					
1	18.4	13.0	2.7	2.6	29.4	85.2	85.7
2	19.6	12.4	1.5	1.0	36.7	92.8	94.8
3	21.2	13.2	6.9	6.0	37.7	67.5	71.4
Mean	19.7	12.9	3.7	3.2	34.6	81.8	84.0
		Filament polyester fabric					
1	7.6	1.5	0.5	0.1	80.2	93.4	98.8
2	7.9	1.2	0.2	0.1	85.0	97.5	98.7
3	7.4	1.7	0.9	0.7	77.0	88.1	90.8
Mean	7.6	1.5	0.5	0.3	80.7	93.0	96.1

[a] Reprinted from (33, p. 737) by courtesy of Textile Research Journal.

TABLE 9.21

Nitrogen in Natural Soils[a]

Stage	Solvent	Nitrogen/mg % (wt. %) Before dialysis	(A)	After dialysis	(B)	(B) / (A) (%)
1	Pet. ether	0.0	(0.0)	—	(—)	—
2	Dist. water	58.8	(78.2)	6.8	(9.0)	12
3	2% NaCl	3.3	(4.4)	2.2	(2.9)	67
4	70% Ethanol	3.0	(4.0)	2.6	(3.5)	87
5	0.2% NaOH	5.7	(7.6)	5.3	(7.0)	93
	Total	70.8	(94.1)	16.9	(22.5)	24
	Natural soils	75.2	(100)	—	(—)	—

[a]Reprinted from (191, p. 449) by courtesy of Journal of the American Oil Chemists' Society.

nitrogen compounds of the soil. It was suggested that addition of proteins to artificial soil compositions might improve correlation between artificial and natural soil detergency results.

Bubl (322) extracted synthetic sebum from cotton fabric with perchlorethylene, using this plus ash content and reflectance readings to follow accumulation of oily soil and minerals during 24 soil-wash cycles.

2. Weighing the Dirt

Holland and Petrea (192) dried cloth to constant weight at 105°C, applied soil to the cloth in the Launder-Ometer, drained, air dried, brushed off excess, dried to constant weight, and reported as percent soil on the basis of the fabric.

Smith and Martin (193) applied a mixture of 99.5% "Air Cleaner Test Dust" plus 0.5% "Gas Black" carbon, and estimated the extent of soiling of cotton fabric by the increase in oven-dry weight of the fabric, to about 12% soiling, basis fabric weight.

3. Ashing Fabric

Schott (194) determined amounts of calcium and sodium montmorillonite as well as kaolinite retained in cotton fabric by drying a test swatch at

105°C, weighing, incinerating in a muffle furnace at 600°C for 1 hour, and weighing the ash on a micro balance. Corrections were made both for the ash of clay-free fabric and loss of water of constitution from clay at 600°C.

AATCC Standard Test Method 78-1961 (50), Ash Content of Bleached Woven Cotton Cloth, calls for conditioning cloth to constant weight at 105-110°C, charring fabric samples in a procelain crucible using a Meker burner, followed by heating to constant weight in a muffle furnace at 800°C for approximately 3 hours.

Stüpel (5) noted that ash contents of repetitively washed test fabrics provide indications of the hard-water stability of heavy-duty synthetic detergents, the silicate ash being traceable to sensitivity of the wash system to alkaline earth ion.

Bubl (322) ashed fabric in a muffle furnace at 480°C., finding that solvent extraction of synthetic sebum followed by ashing gave a better indication of total residual soil than reflectance measurement.

4. Millipore Filter

Smith and Martin and co-workers at the National Institute of Drycleaning (193, 195) have determined the quantity of soil suspended in dry-cleaning systems by using Millipore filtration. In the continuous flow system of dry cleaning, the concentration of insolubles suspended in the solvent in the washer cylinder increases from zero to a maximum and then decreases with time. The soil concentration approaches zero if a sufficient number of solvent changes occur to pump out the washer. Data for the concentration-time curve were obtained by removing samples from the washer at various intervals and determining the amount of soil in suspension by Millipore filtration. The total quantity of soil removed from the load was computed by integrating the curves graphically, and comparing the area with that of curves obtained with known quantities of soil. The percent soil removal can be computed when a known quantity of insoluble soil is placed on a load, thereby allowing measurement of a relative effectiveness of various detergents.

In the first report (195) test soil used for dry-soiling nylon tricot at the NID was General Motors Air Cleaner Dust. Filtration of solvent samples was through a 0.20 μ Gelman filter. In a second report (193), the soil was a mixture of 99.5% by weight of the Air Cleaner Dust and 0.5% of "Gas Black" carbon from the Fisher Scientific Company. The mix was applied to cotton fabric, to the extent of about 12%.

In related work (196), pressure data were determined as a function of grams of soil on a diatomite filter through which perchlorethylene was flowing at the rate of 2.5 gal/min. Soils included the Air Cleaner Dust. This method was used to measure the relative soil dispersion properties of several chemical types of surfactants used under a variety of practical conditions of dry-cleaning operations.

5. Coulter Counter

Tuzson and Short (197, 198) have used the Coulter Counter (199) to determine how much soil particles—individual plus agglomerates—were present at a given time in a detergent bath. This instrument is limited to particles larger than about 1 μ. Soil was kaolin clay particles of which about 70% were between 2 and 10 μ in diameter. The processes were conducted in a modified Terg-O-Tometer with variable agitation speed using a constant 360° stroke angle. Two-millimeter wash bath samples were diluted in 50 ml of 0.5% sodium tripolyphosphate, as use of the Coulter Counter required an electrolyte.

Agglomeration of the clay in deionized water was studied as a function of time and agitation speed. Results are seen in Fig. 9.8. Variation in clay concentration from 0.01 to 0.2% by weight had negligible effect on the agglomeration rate. Almost complete redispersion of the clay resulted from 2-min agitation at 96 cpm.

Deposition was studied by adding two 3" × 5" swatches of AHLMA cloth (edges taped to minimize linting) into a 300-mm, freshly prepared clay suspension and agitating at fixed speed for 5 min. Particle count data are given in Fig. 9.9. The declining count during the first 5 min is due to joint effects of agglomeration and deposition. The latter can be estimated from the deficiency of count found after reagitation at 96 cpm. See Fig. 9.10. Soil removal as a function of time, agitation speed, deposition speed, and soiling concentration was studied using swatches from the deposition work.

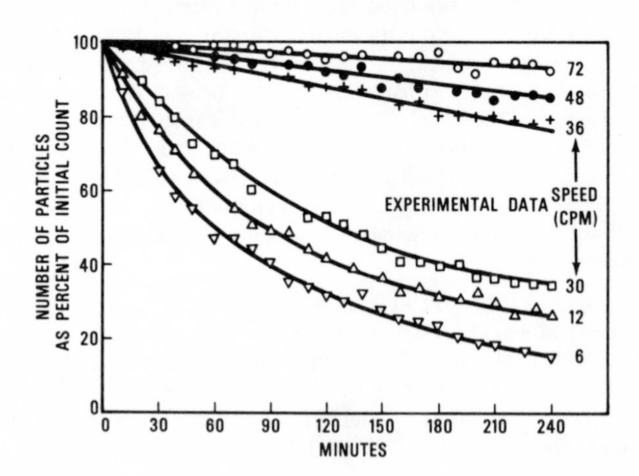

FIG. 9.8. The time variation of the total number of particles per unit volume in suspension as a percent of initial count. Mean diameter of counted particles is greater than 3 μ. No cloth present. Reprinted from (197, p.112) by courtesy of Textile Research Journal.

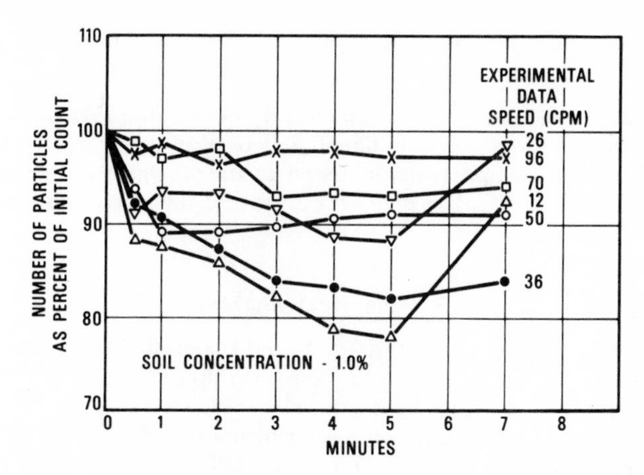

FIG. 9.9. The time variation of the total number of particles per unit volume in suspension as a percent of initial count. Mean diameter of counted particles is greater than 3 μ. Two cloth swatches are present during the first 5 min. After the first 5 min, the solution is reagitated for 2 min at 96 cpm. Reprinted from (197, p. 113) by courtesy of Textile Research Journal.

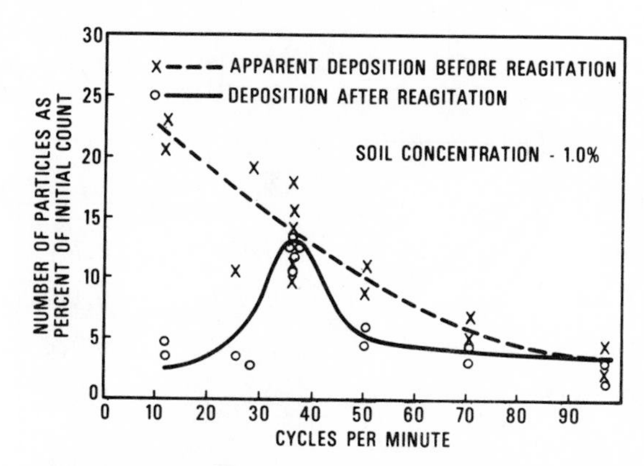

FIG. 9.10. Percentage deposition before and after a 2-min reagitation of the deposition solution vs. deposition speed. The two cloth swatches a are removed before reagitation. Reagitation speed is 96 cpm. Reprinted from (197, p. 114) by courtesy of Textile Research Journal.

In general, the data showed very rapid soil removal during the first few minutes and a subsequent very slow increase or decrease.

Krieger (200) used the Coulter Counter to determine that particulate soils found in laundry water ranged 0.65-50.9 μ in diameter. Approximately 45% of the natural soil particles present in the wash liquor were in the redeposition range of 1 μ or less, while 85% of the particles were less than 2 μ.

B. Chemical Analysis

Utermohlen and Wallace (142) determined magnetic iron oxide by ashing the soiled cloth at 600°C, fusing the ash with potassium pyrophosphate, dissolving the melt in dilute hydrochloric acid, reducing the iron to the ferrous state with hydroxylamine hydrochloride, adding orthophenanthroline, and comparing visually the intensity of the colored solution with standards prepared with magnetic iron oxide similarly treated. Additional work with iron oxide was reported by Gotte (201) and by Sauerwein (202).

Wagg and Kevan (203) reported the constitution of soiling matter of bed sheets to be as shown in Table 9.22. Matlin (204) presented a system of

TABLE 9.22

Constituents of the Soiling Matter of Bed Sheets, Expressed as Percentages of the Weight of the Fabric[a]

Component	Sheet 1	Sheet 2	Purified fabric
Acetone extract	0.93	0.70	0.06
Ash	0.137	0.170	0.018
Calcium as CaO	0.032	0.028	0.006
Magnesium as MgO	0.029	0.014	0.001
Other alkaline matter (other than CaO and MgO) as Na_2CO_3	0.081	0.059	0.014
Iron as Fe_2O_3	0.003	0.003	0.002
Aluminum as Al_2O_3	0.0006	0.0004	0.00025
Phosphorus as P_2O_5	0.007	0.008	0.004
Silicon as SiO_2	0.031	0.035	0.004

[a]Reprinted from (203, p. T102) by courtesy of The Textile Research Institute and the authors.

spot tests for analyzing the ash of the cellulosic fibers. Schaeffer (205) reported microanalytical methods for determining organic compounds and textile materials.

Much attention has been given to the determination of natural organic soil constituents after separation from fabric by extraction with organic solvents. Powe and Marple (82) used gas chromatographic analysis to determine the data given in Tables 9.23 and 9.24.

Byrne et al. (188) gave qualitative and quantitative data on fatty soil extracted with petroleum or benzene-methanol solvent from nylon undershirts after wear trials. The extracts were analyzed by thin layer chromatography, gas liquid chromatography, and infrared spectroscopy. Petroleum extracts from garments were analyzed as shown in Table 9.25.

C. Radiotracers

The availability of radioactive compounds, which began in the late 1940s, made possible a most unique and valuable tool for use in measuring amounts of soils and surfactants associated with both fabric and wash water before and after washing operations. The excellent sensitivity of these methods and the wide variety of available radioisotopes permitted accurate measurements at very low levels of soil, amounts which were generally much more realistic than the relatively heavy loads of colloidal carbon black and graphite needed for reflectance readings on artificial

TABLE 9.23

Accumulated Organic Soil[a]

Sample	Garment	g	%[b]	Free fatty acids	Esterified fatty acids[c]	Lime soaps	Unsap.
A	Sheet	15.1	2.2	3.2	49.8	23.4	23.4
B	Sheet	31.7	4.4	2.9	58.2	20.1	18.7
C	Pooled sample[d]	15.4	—	—	—	12.5	—
D	T-shirt	8.4	7.6	—	—	38.0	—

[a]Reprinted from (82, p. 137) by courtesy of Journal of the American Oil Chemists' Society.

[b]Based on weight of garment.

[c]By difference.

[d]4 T-shirts, 1 pillow case, 4 dress shirts.

TABLE 9.24

Fatty Acids in Unremoved Clothes Soil[a]

Carbon atoms	Percentage						
	A			B	C		D[b]
	Total[c]	Free	Lime soap	Total[c]	Total[c]	Free	Total
<C_{12}	1	2	—	1	1	1	6
C_{12}	1	6	—	2	2	6	1
C_{13}	2	3	—	1	1	2	—
C_{14} myristic	6	24	6	7	9	24	7
C_{14} unsat.	1	1	—	1	1	1	5
C_{15} total	11	12	5	10	9	13	10
C_{16} palmitic	33	30	38	36	33	29	30
C_{16} branched	2	—	—	2	1	—	—
C_{16} unsat.	10	7	16	9	11	10	9
C_{17} total	5	2	3	6	5	4	5
C_{18} stearic	18	6	26	18	15	5	9
C_{18} oleic	6	7	6	5	10	5	10
>C_{18}	2	—	1	2	3	—	—

[a]Reprinted from (82, p. 138) by courtesy of Journal of the American Oil Chemists' Society.

[b]Extracted with ethanol.

[c]Excluding lime soaps.

TABLE 9.25

Scheme of Chemical Analysis[a]

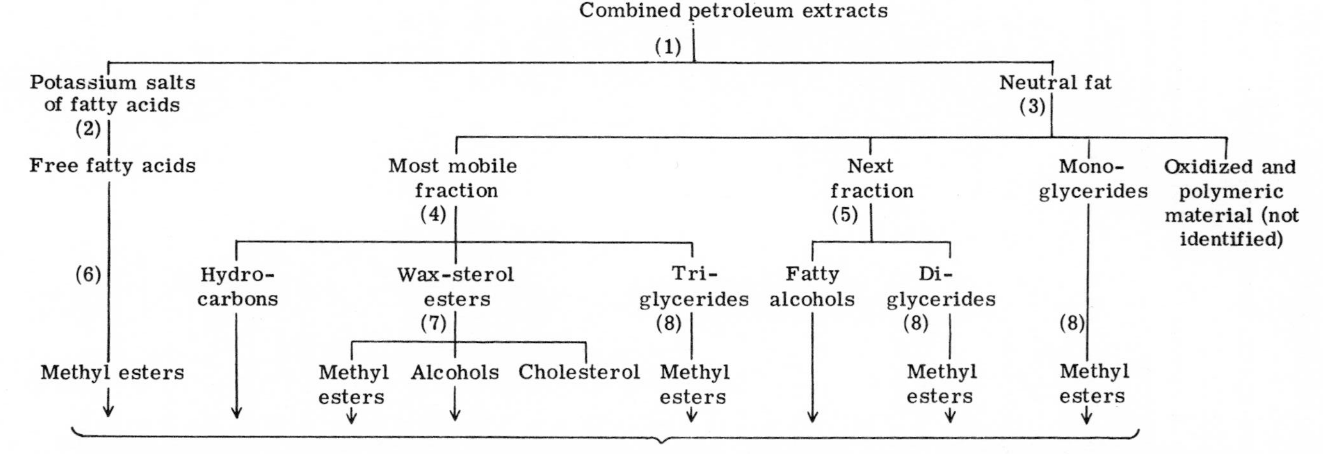

(1) In light petroleum; treat with 0.1-N potassium hydroxide in ethanol-water (50:50 vol./vol.).
(2) Acidify and then extract with light petroleum.
(3) Preparative thin-layer chromatography (Prep. TLC) with hexane-diethyl ether (60:40), followed by removal from plate.
(4) Prep. TLC with hexane-diethyl ether (95:5).
(5) Prep. TLC with continuous ascending development with hexane-diethyl ether (75:25) for 2.5 h in S-chamber (2b) open at upper end to atmosphere.
(6) Esterification followed by (4).
(7) Methanolysis, followed by prep. TLC with hexane-diethyl ether (90:10).
(8) Methanolysis followed by (4).

[a]Reprinted from (188, p. 20) by courtesy of Journal of the Society of Dyers and Colourists.

soils. In this connection, Hensley et al. (97) foretold in 1955 that "it is believed that the use of radioactive soils in laboratory evaluations will permit studies with soil types approaching natural soils more closely than have the soils used in conventional methods."

Several excellent reviews of radioactive soils and test methods for measuring adsorption of surfactants onto fabric, as well as radiochemical techniques used in detergency studies, have been made (3, 6, 138). The general requirements for developing a small radiochemical laboratory suitable for investigative and analytical applications of radioactivity to textile and related problems has been discussed by Bernstein (206). Murdock (207) described applications of radioactive tracers in textile work. Safety precautions from the standpoints of personnel protection, storage and handling, and disposal of radioactive wastes have been discussed by Harris (3). According to Harris (3) criteria for selection of isotopes to use in tracers include half-life, whether tag is the same or equivalent to the stable isotope for which it is substituted, type and strength of radiation, and use hazard.

^{14}Carbon has been used in labeling soils comprising carbon black (97, 138, 208), dibutyl stearamide (3), tripalmitin (209-211), stearic acid (212, 213), and fatty alcohols, acids, as well as hydrocarbons and algal protein triglycerides (138, 213, 214). Detergent ingredients labeled by ^{14}C have been soaps (215), sodium carbonate (216), and sodium carboxymethyl cellulose (117, 217, 218). The ^{14}C isotope is a soft-beta-ray emitter of 0.155 meV maximum energy and has 5700 years half-life.

Clay has been tagged both by neutron irradiation [kaolinite (219)] and with ^{45}Ca by cation exchange [montmorillonite (220)]. ^{45}Calcium has also been used to label water hardness (221-223) and to interact with surfactants absorbed onto graphon (224). This calcium isotope emits soft beta rays of 0.25 meV maximum energy and has 165 days half-life.

^{59}Iron was used to label iron oxide soiling ingredients (202, 225). Tritium found application as a tag for sodium linear alkylbenzene sulfonate (226) and for organic soiling components (214) in studies by Gordon and coworkers. Tritium is a very soft beta source of 0.018 meV.

^{32}Phosphorous was used to label zirconyl phosphate soils (227, 228), to tag bacteria (229), and to label phosphate builder salts (216, 221, 230). ^{32}Phosphorous emits beta particles of 1.7 meV maximum energy and its half-life is 14.3 days. Sulfate and sulfonate surfactants have been tagged with ^{35}S (216, 231, 232); this is a soft beta emitter (0.167 meV) and has a half-life of 87.1 days. The radioisotope Zr-Nb-95P was used for labeling carbon black by absorption (9). This isotope emits beta (0.39 and 1.0 meV) and gamma (0.73 and 0.92 meV) radiations, half-life being 65 days. Numerous forms of tagged compounds are commercially available (233).

Lambert et al. (234) described the preparation and use of radioactive soils. Attempting the tagging of the inactive synthetic multicomponent soil (mcs) (72), three types of radioactive soils were developed:

(1) Carbon black prepared from ^{14}C-labeled barium carbonate and incorporated into the mcs.

(2) Mixed fission products (mfp) were added to the mcs in water. Aged fission products were absorbed onto the colloidal constituents of the mcs, especially the particulate matter. Radiation was beta and gamma.

(3) Neutron irradiated mcs; ^{59}Fe was thought to be the major source of induced activity, the latter being transitory.

Tagged soiled fabric was generally prepared by pipetting a 2% aqueous soil slurry onto prewetted swatches clamped in a Schiefer abrader equipped with a frosted glass disk and run for 20 revolutions. Prewetting was by rinsing the swatch in 0.1% carbon tetrachloride solution of the seven fatty substances of the mcs.

Harris (3) described the preparation of N, N-di-n-butyl stearamide from stearic acid containing a carbonyl ^{14}C. The amide was applied to metal specimens after mixing into SAE No. 60 oil and dissolving in carbon tetrachloride.

Detectors for radioactivity measurements comprise Geiger-Muller tubes, gas flow counters, scintillation counters, and ionization chambers. Historically, the Geiger tube, also known as Geiger-Muller or G-M tube, has enjoyed the greatest amount of use. It is least expensive, and requires relatively inexpensive instrumentation. In recent years, the G-M tube has been largely displaced by gas flow counters, particularly the end-window type, and also by windowless varieties, operated in either the Geiger or proportional mode. This has certain advantages over conventional Geiger tubes.

Counting of radioactive tracer soil on fabric by end-window Geiger tubes has some disadvantages, as pointed out by Gordon (235): (1) The end window is impervious to the relatively weak beta emanations from tritium; (2) if the fabric on which the labeled compound is deposited is less than infinitely thick for the isotope, the small variations in thickness will result in large variations in observed counts rates; and (3) uneven distribution of the labeled compound on the fabric will result in variations in count rate even though the quantity on the fabric is the same.

DuPont workers (208, 236) used soiled films mounted drumhead fashion over the end of a gas-flow Geiger tube in order to continuously monitor the desorption of ^{14}C-tagged particulate carbon, fatty acids, fatty alcohols, hydrocarbons, and triglycerides from cellulose nylon, polyethylene, terephthalate, and tetrafluoroethylenehexafluoropropylene into aqueous solutions of cationic, anionic, and nonionic surfactants.

Hensley and Inks (220) used gas-flow end-window counters with 1 mg/cm^2 window in connection with development of a ^{45}Ca-tagged clay soil. It should be noted that BASF Wyandotte Corporation workers have successfully used this tagged clay soil routinely for the evaluation of detergent action for many years (52, 64, 106).

Whirlpool workers (221, 222) determined the radioactivity of dry fabric swatches by placing them into a special swatch stand (228) mounted under a gas-flow counting tube and using proportion counting equipment. In one report (221) interposing of an absorber of a density of approximately 68 mg/cm^3 between the fabric sample to be counted and the window of the gas-flow counter tube provided separation of relatively stronger ^{32}P (1.7 meV maximum energy) beta particles from total count of these particles plus weaker ^{45}Ca particles (0.25 meV) read without the absorption. In the other report (222) parallel experiments were necessary, as the energies of the two tags, ^{35}S (0.167 meV) for sulfated surfactant and ^{45}Ca, were too similar to allow similar separation.

In a recent series of seven papers, Gordon and co-workers (163, 214, 219, 226, 235, 237, 238) used liquid scintillation counting to study adsorption of ^{35}S-or tritium-tagged surfactants onto cotton fabric, as well as removal and redeposition of doubly tagged sebum-type soil with respect to cotton, nylon, and polyester fabrics in both Terg-O-Tometer and household washer tests. His 7-component soil (214), which is soluble in toluene-based liquid scintillation counting mixture and is described in Table 9.26, resembled to some extent the artificial sebum components of Spangler et al. (76), but composition was dictated in part by commercial availability of components.

Sontang et al. (325) developed a double-label technique to determine quantitatively the retention of fats on cotton in the presence and absence of clay. Six sebum-type fats were tagged with ^3H or ^{14}C. Counting was by liquid scintillation.

Gordon and Bastin (219) also recently applied the liquid scintillation approach to particulate-type soil, a clay tagged by neutron radiation. In this case, counting was done both on fabric swatches and on aliquots of wash solutions, using a dioxane-based mixture. More recently he described use of a triply labeled particulate soil (326), and his co-workers (327) reported on cold water detergency studies using radiolabeled soils.

It should be noted that relatively expensive equipment is needed for the approaches used by Gordon. He pioneered the liquid scintillation counting in absorption and detergency measurements. However, this scintillation technique can use only liquids, such as wash waters, and cannot be used for direct determination of radioactive tracer soils on fabrics. The latter is handled by extraction of the soil from cloth with a suitable solvent; in

TABLE 9.26

Composition of Synthetic Soil[a]

Component	Weight, %	Label
Hydrocarbon oil (medium viscosity lubricating oil)	25	^{3}H
Tristearin	10	^{3}H
Arachis oil	20	None
Stearic acid	15	^{14}C
Oleic acid	15	^{14}C
Cholesterol	7	^{14}C
Octadecanol	8	^{14}C

[a]Reprinted from (214, p. 290) by courtesy of Journal of the American Oil Chemists' Society.

case of synthetic sebum, the toluene-based liquid scintillation counting mixture serves well.

A literature summary of applications of radioactive tracer technology to detergency and related studies is given in Table 9.27.

Aside from detergency studies, Geiger sensing devices have been used to monitor the laundering of fabrics which become contaminated with isotopes during industrial usage of garments and other textile items.

Takada *et al.* (251) studied washing of radioactive isotopes from cotton cloth contaminated with ionic radioisotopes accompanying oily soils. *n*-Octadecane plus liquid paraffin as nonpolar soiling agents and octacedenol-1 plus tripalmitin as polar soiling agents were used to study removal of ^{32}P, ^{131}I, and ^{60}Co from cotton cloth. ^{32}P and ^{131}I were readily removed by washing with water and little advantage was obtained by addition of various detergents. In contrast, ^{60}Co was removed effectively by EDTA and detergents used above their critical micelle concentrations. Kang and Chung (252) removed radioactive uranium nitrate from cotton and woolen fabrics using EDTA-2 Na and synthetic detergents.

Bailey *et al.* (253) studied the effects of uranium contamination on laundry operations, using a 3-wire poppy alpha proportional counter as a detector for radioactivity. Clark (254) reported beta and alpha monitoring of laundry before and after washing.

TABLE 9.27

Detergency and Related Studies Using Radioactive Tracer Techniques

Type of Study	Radiotracer		Substrate	Washer	Detection	Literature source
	Chemical composition	Label				
Adsorption	Calcium bicarbonate	^{45}Ca	Cotton	Flask	G-M tube	Lambert '50 (223)
Detergency	Dibutyl stearamide	^{14}C	Steel	Beaker	G-M tube, mica window	Harris *et al*. '49 (239)
Detergency	Undecane	^{14}C	Metal	Beaker	G-M tube, mica window	Harris *et al*. '50 (240)
Detergency	Stearic acid	^{14}C	Steel	Beaker	Geiger tube	Hensley *et al*. '52 (212)
Adsorption	Alkyl aryl sulfonate Na palmitate	^{35}S ^{14}C	Cotton, wool	–	Geiger tube, mica window	Meader & F. '52 (215)
Detergency	Carbon black Fission products	^{14}C Mixed	Cotton	–	G-M tube, mica window	Lambert '54 (234)
Detergency	–	–	Painted surfaces	–	–	Shelburg *et al*. '54 (241)
Sanitation	Bacterial suspension	^{32}P	–	–	–	Armbruster & R. '55 (229)
Detergency	Carbon black	^{14}C	Cotton	Mini-washer	End window Geiger tube	Hensley *et al*. '55 (97)
Detergency	Carbon black	Zr-Nb-95P	Cotton	–	G-M tube	Phansalkar & V. '55 (9)
Detergency	Carbon black Glyceryl tristearate Algal protein	^{14}C ^{14}C ^{14}C	Cotton	Deter-Meter	Geiger tube	Ashcraft '56 (138)

Adsorption	NaDDBS Sodium Phosphates	^{35}S ^{32}P	Cotton	Beaker	Windowless counter	Boyd & B. '56 (216)
Soil removal	Tripalmitin	^{14}C	Cotton	Household washer	Geiger tube	Ehrenkranz '56 (209, 210)
Detergency	Zirconyl, phosphate	^{32}P	Cotton	Agitator washer	D-47 counter, "P" gas	Diamond & L. '57 (228)
Dairy cleaning	Milk	^{45}Ca	Metal	–	Windowless counter	Firsching & E. '57 (242)
Detergency	$FeCl_3$-ink-oils	^{59}Fe	Cotton, rayon, wool	–	G-M tube	Tolgyessy & R. '57 (225)
Detergency	Tripalmatin	^{14}C	Cotton, dacron	Household washer	Geiger tube	Ehrenkranz & J. '58 (211)
Detergency	Tristearin, tagged in position 1	^{14}C	Glass	TOT		Anderson et al. '59 (243)
Detergency	Zirconyl, phosphate	^{32}P	Cotton	TOT	D-47 counter, "P" Gas	Diamond & G. '59 (227)
Adsorption	NaCMC	^{14}C	Cotton, wool, orlon, acetate	Flash	End window Geiger counter	Hensley & Inks '58 (217)
Adsorption	NaDDBS	^{35}S	Cotton	–	–	Jayson '59 (232)
Detergency	TSPP	^{32}P	Cotton	–	–	Jayson '59 (230)
Soil removal	Fe_3O_4	^{59}Fe	Cotton	–	–	Sauerwein '60 (202)
Soil removal	Algal protein, stearic acid, tristearin, triolein	^{14}C	Nylon, Teflon, etc.	–	–	Harris & S. '61 (244)

Type of Study	Radiotracer Chemical composition	Label	Substrate	Washer	Detection	Literature source
Adsorption	Polyphosphate water hardness	^{32}P ^{45}Ca	Cotton	TOT	Gas flow counter tube	Rutkowski & M. '61 (221)
Adsorption	NaDDBS Water hardness	^{35}S ^{45}Ca	Cotton	Washing machine	Gas flow counter tube	Schwarz et al. '62 (222)
Detergency	Fatty compounds	^{14}C	Cotton, nylon, etc.		G-M end window counter	Wagg & B. '62 (245)
Detergency	Fatty compounds trialkyl phosphates	^{14}C ^{32}P	Cotton, PVC		G-M counter	Zeidler '62 (246)
Detergency	Tristearin	^{14}C	Stainless steel	Flow system	G-M tube	Bourne & J. '63 (154)
Detergency	Tagged clay	^{45}Ca	Cotton	Mini-washer	End window gas flow counter	Hensley '65 (64)
Soil removal	Tagged clay	^{45}Ca	Cotton	Mini-washer, TOT	End window gas flow counter	Schmolka & H. '65 (106)
Soil removal	Fatty soils	^{14}C	Nylon polyesters, copolymers, cellulose	Beaker + magnetic stirrer	End window Geiger counter	Fort et al. '66, '68 (213, 236)
Adsorption	NaCMC	^{14}C	Cotton, w/wo resin	Flask	End window G-M tube	Evans & E. '67 (218)
Soil removal	Fatty soils	^{14}C	Nylon, polyester, cellulose	Beaker + magnetic stirrer	End window Geiger counter	Grindstaff et al. '67 (208)

Detergency	Fatty soils	^{14}C	Metals, glass, polyethylene	Lab	–	Uhlich et al. '67 (247)
Diffusion	Fatty soils	^{14}C	Cotton, polyester	–	–	Yokoyama et al. '67 (248)
Adsorption	Surfactants	^{35}S tritium	Cotton	Household agitator	Liquid scintillation	Gordon '66 (163, 226) et al.
Detergency	Synthetic sebum, 7-component	^{14}C	Cotton, nylon, Dacron, PEC	TOT	Liquid scintillation	Gordon et al. '67 (214)
Detergency	Fatty soil	^{12}C	Cotton nylon, Dacron	TOT	Liquid scintillation	Gordon & B. '68 (219)
Biological clearability	Bacteria, biological compounds	^{32}P	Various	Ultrasonic	–	Gelezunas & B. '66 (249)
Detergency	Fe black	^{32}P	–	–	–	Markiewicz '66 (250)

Talboys and Spratt (255) contaminated fabric with ^{89}Sr, ^{91}Y, ^{141}Ce, ^{59}Fe, ^{32}P, and ^{131}I. They showed a photograph of the lead disk atop a wire cloth support between which was held the cloth while making activity counts. The support surrounded a mica end-window Geiger counter. The window of the tube was positioned $\frac{1}{4}''$ from the cloth sample being counted. A thin plastic sheet was attached to prevent the radioactive particles from dropping onto the tube window.

D. Neutron Activation Analysis

Netzel et al. (256) reported in 1964 a nondestructive neutron activation method for determining the amount of kaolinite clay soil washed from cotton fibers. The aluminum constituents of the kaolinite lattice structure were bombarded with neutrons to produce short-lived ^{28}Al. The amount of particulate clay soil present on cotton cloth before and after washing was determined by gamma scintillation counting of the ^{28}Al.

E. X-Ray Fluorimetry

Richards et al. (257) used X-ray fluorescence spectroscopy for determining the amount of soil on wool fabric treated with resin. Their soil was composed of oily and clay materials. The oily portion, which was a multicomponent mixture adapted from that used by McLendon and Richardson (125), contained dibromostearic acid, the bromine content measured by X-ray fluorescence to assess the amount of dibromostearic acid, and from that, the amount of oily soil present. The clay portion of the soil was bentonite-containing ferric oxide, and an iron count was taken to show the amount of clay present on the fabric. Correlations were all significant between X-ray fluorescence results and those obtained by conventional solvent extraction for oily soil followed by ashing of fabric for clay soil determination. The X-ray fluorescent data were obtained in 2 hours; extraction and ashing required 15-20 hours.

F. Microscopic Examination

Photomicrographs, both by light and by electron techniques, have been used by many workers to examine fabric for presence of dirt.

In 1937, Cunliffe (258) used photomicrographs of wool fibers removed from soiled fabric to show that the pigment particles were held on the fibers mainly by the edges of the scales. Some fibers showed a local concentration of particles, such fibers being derived from the outer ones of the fabric. On washing, the particles were gradually removed, leaving a fiber identical in appearance to an unsoiled one.

In 1953, Stevenson (259) presented approximately ten pages of photomicrographs to show soiling of textile fibers.

In 1959, Powe (81) depicted the nature of tenaciously bound soils on cotton. His electron micrographs, one of which is shown in Fig. 9.11, suggest that clay minerals 0.02-1 μ in diameter, rather than carbon particles, are the major particulate material causing soil buildup on cotton fibers.

Fort et al. (260) presented a microscopic study of naturally soiled textile fabrics. Both fatty organic and particulate inorganic soil were identified. Distribution of the soil appeared to be a function of treatment.

Meek (261) studied microscopically the behavior of water-soluble salt and sugar on wool fibers immersed in dry-cleaning baths. In another study (262), rolling up and emulsification of fatty and skin dirt were observed. Komeda et al. (328) studied the removal of natural sebum and oily soil from twisted yarn and fiber filaments in washing solutions.

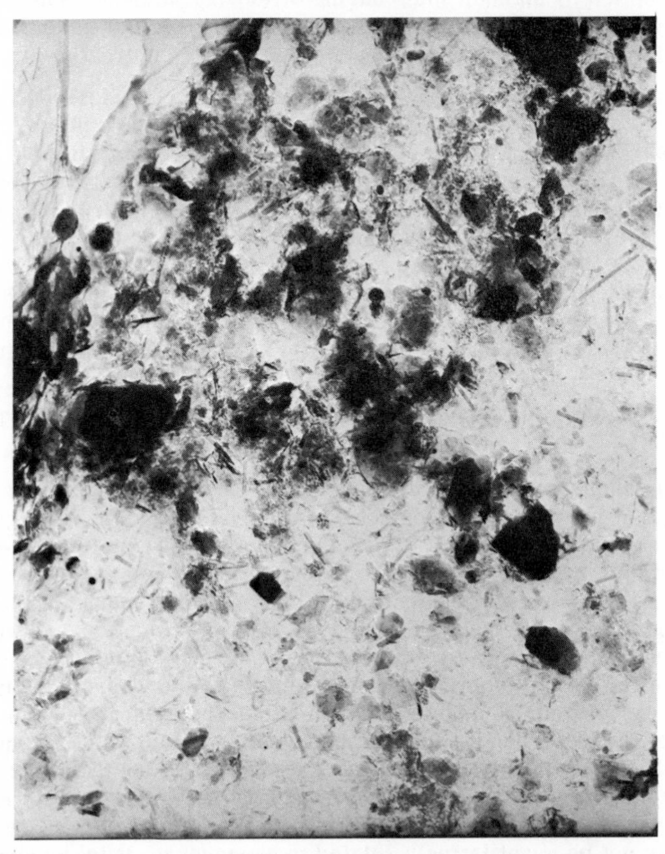

FIG. 9.11. Electron micrograph showing nature of tenaciously bound soil on cotton fiber (courtesy of W. C. Powe, Whirlpool Corporation Research Laboratory).

Nettelnstroth (263) used electron microscopy to examine WFK soiled cotton fabric Indications were that the pigment (86% kaolinite, 8% red iron oxide, 4% black iron oxide and 2% yellow iron oxide) particles become smaller and fewer with increasing brightness of fabric read by reflectance.

Several recent studies have used a scanning-type electron microscope to examine soiled fibers. Schimauchi and Mizushima (98) observed soil redeposition on polyester fibers in comparison with that on cotton fibers, redeposition being more rapid with the synthetic fibers. Komeda et al. (329) observed adhering states of natural soils on the collar and other parts of white shirts and underwear.

Arai and Maruta (264) similarly studied adherent natural soils on fiber surfaces. Fibers were from straight cotton underwear and 100% polyester fabrics soiled by an adult male during three-day wearing. The photomicrographs indicated that oily soil was present on fibers, spreading of the soiling being evident, particularly on the lipophilic polyester.

Brown et al. (33) used a Cambridge scanning electron microscope to examine staple and filament cloths before soiling, after soiling, and after one wash. Magnification was 55 × and 280 ×. Heavy deposits of fat were present on the staple fabric, being held between crisscrossed fibers, possibly due to wicking of oil in capillaries between fibers. No such deposits were found on the filament fabrics: instead, the oil was uniformly spread over the filaments themselves. Washing removed the filament soil, but residues persisted on the staple fabric, particularly at points of fiber contact. However, additional work using crimped and bulked filament polyester fabric disclosed deposits of fat between separated filaments.

IV. CORRELATION BETWEEN OPTICAL ASSESSMENT AND ACTUAL SOIL CONTENT OF FABRIC

That an understandable relationship exist between appearance and actual soil content of fabric is of considerable importance to the detergent chemist. Preferably, it should be a precise quantitative relationship of a relatively simple nature. Attempts to mathematically relate reflectance readings and actual soil contents have been made by means of the Kubelka-Munk (140) and soil additional density equations (73). In cases where this approach does not seem to apply, greater emphasis may be assigned to appearance, as the importance of this quality to the wearer of garments is well recognized, and has even been called the most important factor (74). On the other hand, accurate measures of actual soil content are basically indispensable in detergency studies. Gordon (219) has stated that "the radiotracer approach is particularly attractive in this regard because it is unequivocal. While it is true that the quantity of residual clay may not be quantitatively related to appearance, it is, nevertheless, important to know just how much of each component of a test soil is removed to understand the mechanism involved and also to arrive at a more realistic evaluation of detergents."

Bubl (322) found that extraction of sebum-type fats and ashing cotton fabric to assess for clay soil indicated the accumulation of mixed soils during 24 soil-wash cycles while reflectance measurements (Rd and b) did not do so consistently.

Kubelka-Munk Relationship

It is well known that the relationship between reflectance readings on soiled fabrics and their actual soil content is basically not a simple linear one, particularly at medium and heavy soil loading (11, 265).

In 1947, Utermohlen and Wallace (142) soiled cotton fabric in baths of carbon tetrachloride containing various levels of lampblack, carbon black, or magnetic iron oxide, as well as mineral and vegetable oils. It was assumed that the amount of soil picked up by the cloth was proportional to the content of soil in the soiling bath. This assumption was verified in the case of the iron oxide, which could be chemically analyzed. Graphical plots showed approximately straight line relationships between reflectance values on the three soiled fabrics and log of relative concentrations of the blacks in the soiling bath. The straight lines were analyzed mathematically, in accordance with the general formula $y = mx + b$, where the two linear variables x and y are, respectively, reflectance and log of relative black pigment content, m is the slope, and b is the intercept on the y axis.

Armstrong et al. (101) showed a relationship between Photovolt type 610 reflectometer reading and actual graphite soil present on the fabric. The relationship was almost linear at reflectance values greater than 35. Morrisroe and Newhall (141) reported results with a lampblack soil which agreed with those of Utermohlen and Wallace (142), indicating that the relationship between reflectance and quantity of soil is logarithmic.

In 1948, Bacon and Smith (140) described the application of the Kubelka-Munk concept (12) to the task of determining the actual soil content from reflectivity of soiled fabric. The equation of Kubelka-Munk (266) states that,

$$K/S = \frac{(1 - R)^2}{2R}$$

where K is the coefficient of reflectivity; S is the coefficient of light scattering; and R is the observed reflectivity for monochromatic light. It was held that the monochromatic light restriction would not invalidate use of the Kubelka-Munk equation in the case of nonchromatic soils such as lampblack. By use of K/S values, Bacon and Smith calculated the percent lampblack removed between any two values of reflectivity by using:

$$\frac{K/S \text{ for soiled fabric} - K/S \text{ for scoured fabric}}{K/S \text{ for soiled fabric} - K/S \text{ for unsoiled fabric}} \times 100 = \% \text{ black removed}$$

In 1949, Utermohlen and Ryan (267) reported that application of a Kubelka-Munk equation to washed cloth to determine absolute soil content was only of limited value when the soil is black iron oxide, basis ability to both measure reflectance and determine iron content. Sample calculations of K/S values are shown.

Reich et al. (268) have commented that semilog mathematical expressions of the general form $R = A \log G + B$, where R is the percent reflectance of soiled cloth, G is the relative pigment content, A and B being test condition constants, are basically empirical equations valid for specific soils over a limited range of intensities, as reflectances required for clean cloth and for highly soiled cloth would be infinite and negative, respectively.

Using the data of Utermohlen and Wallace (142), Reich et al. (268) showed that application of the Kubelka-Munk equation gave a linear relationship of K/S versus soil content only for low values of the latter, while a straight line over the entire range was given by plotting values for log $[(K/S)\text{ soiled} - (K/S)\text{ clean}]$ against log of soil content. The latter finding was interpreted to mean that $[(K/S)\text{ soiled} - (K/S)\text{ clean}]$ is proportional not to soil content but to some power n of soil content. This reasoning provided their generalized equation:

$$\log [(K/S)\text{ soiled} - (K/S)\text{ clean}] = n \log G + D$$

where G is soil content and D is log C. Further pursuit of their concepts led to indications that n varies according to both the degree of agglomeration of the particulates soil and the number of soil particles present: $n = \frac{2}{3}$ at a high degree of agglomeration, as during artificial wet soiling of fabric; $n = 1$ during washing of fabric, indicating removal of aggregates rather than deflocculation.

Strauss (269) studied the relationship between reflectance and surface iron content of cotton fabric. This content was determined on the basis of chemical analysis and surface area data on fabric. By means of both absorption and desorption, an empirical equation was derived which allowed calculation of iron oxide concentration on cotton fabric surface in grams per square centimeter directly from reflectance values. Kubelka-Munk concepts provided interpretation of the derived equation which was only valid at low soil concentrations.

Numerous other workers have studied the Kubelka-Munk concept quite carefully. Hart and Compton (149) looked for effects of physical factors related to soil particles, after obviating effects of yarn and fabric construction by using a chopped fiber technique. In order for the Kubelka-Munk equation to be valid, they found that soil particles must be uniformly distributed in the fabric; must be either randomly oriented if anisotropic or dimensionally istropic; and must be of either constant or known specific (light) absorbance.

Weatherburn and Bayley (270) described methods for measuring the comparative tendencies of yarns composed of various natural and man-made fibers to become soiled with vacuum cleaner dirt. After soiling, the yarns were chopped into 1-mm lengths, mixed thoroughly, and formed into flat rectangular pads. Subsequent reflectance readings were shown to be dependent on two factors: weight and particle size distribution of soil; and optical properties of the fibers. They found "effective soil content" (esc), defined as weight of soil retained multiplied by the specific light absorbance (149), to be a more meaningful criterion of soiling than either the weight of retained soil or the reflectance decrease. Relative esc was calculated from reflectance readings using the Kubelka-Munk equation.

Ashcraft (138) used both reflectivity measurements converted to soil removal values by means of the Kubelka-Munk equation, and radioactive isotope tracers to study detergency. The tracer technique allowed use of artificial, ^{14}C-tagged carbon black soil at low levels comparable to those found with many natural soils. Applying the concept of Reich et al. (268), Ashcraft found n values quite close to unity in water washes, indicating that the predominant washing mechanism is removal of aggregates of soil, while addition of a detergent gave a mean n value of 0.64, indicating deflocculation of aggregates.

Vold and Phansalker (170), using defatted Sterling NS carbon, found that neither the original Kubelka-Munk nor the Reich et al. modification (268) could be used to accurately convert reflectance data to weight of carbon when average size of soil agglomerates changed markedly with concentration of detergent. Wagg (271), working with deposition of ilmenite, showed that reflectance of cotton fabric varied with nature and solution concentration of surfactant, in presence and absence of sodium carboxymethyl cellulose.

Harris et al. (272) checked on the validity of the Kubelka-Munk equation by reading reflectance on cotton fabric soiled with graphite, the latter being quantitatively determined by a turbidimetric method after dissolving the cellulose away from the soiling pigment. A plot of reflectance versus the log of graphite content gave a linear relationship leading to the general equation $R = -A \log G + B$, where R is reflectance and G is concentration of graphite on the cotton fabric in milligrams per gram. In the working range, and especially for higher reflectances, they found the K/S value of colloidal graphite on cotton to be essentially a linear function of graphite concentration. This article includes statistical treatment of experimental data.

Since workers (140, 268, 272) showed that the Kubelka-Munk equation generally applies to soils such as graphite, carbon black, and lampblack, Loeb et al. (155) used three types of commercially available carbon soils and assumed that the Kubelka-Munk equation applied. Smith et al. (65) noted that the Kubelka-Munk calculations of soil removal are not valid with rug-beater soil.

In 1954 Bacon and Smith (273) reported studies of the relationship between soil content of lampblack-soiled fabric and reflectance. Linear plots between reflectance and log of fractional soil content were obtained.

Netzel et al. (256) reported that the validity of the applicability of the Kubelka-Munk equation relating reflectance to soil content on cotton fabric has been experimentally confirmed for soiled cloths having high reflectance, based on neutron activation analysis to determine actual amount of particulate kaolinite clay soil removed during a wash cycle.

Many workers have used the Kubelka-Munk equation to convert reflectance data into soil content values. Among these are Frishman et al. (274), Weatherburn and Bayley (270), workers at Southern regional Research Laboratories of the USDA and at Harris Research Laboratories (275-277), Davis (83), Martin and Davis (69), Nuessle et al. (278), O'Brien et al. (279), Trost (280), Ginn et al. (95), and Rutkowski (84). Numerous reviews of the activities of workers with the Kubelka-Munk equation are found in the literature (7, 272, 138).

Martin (281) in 1961 described a simplified form of Kubelka-Munk equation,

$$S = e^{-6.20 \Delta R} = \text{Fraction of soil remaining on fabric after washing where } \Delta R \text{ is change in reflectance upon washing.}$$

Schmolka and Hensley (106) reported use of the Martin modification of the Kubelka-Munk equation when performing detergency testing of nonionic surfactants with various types of commercially available and other soiled fabrics based on carbon black, vacuum cleaner, and clay soils. Darbey (282) presented several graphical procedures based on K/S values of tristimulus reflectance values. Although designed to aid in dyeing applications, these graphs may be helpful to the detergent chemists dealing with chromatic phenomena in soils and fabrics.

Allen (283) found that absorption of ultraviolet radiation seemed to obey the Kubelka-Munk law of reflectance, linear relations being obtained in plots of $(1 - R)^2/2R$ versus concentration of whitener, except at highest concentrations. This was used to support the conclusion that at least a practical limit to fluorescent whitening agent fluorescence exists. For example, calculations on the basis of the Kubelka-Munk equation indicates that an additional absorbance of 10% more light with consequent increase of 10% in fluorescence would require about 70% more whitener.

Wilson (284) reviewed use of the Kubelka-Munk equation, and attempted a confirmation of its application to single layers of fabric by using a green, nonsubstantive dye in the dry, particulate form. The dye was applied by rubbing in a wear tester. Reflectance measurements on the "soiled" fabric were made with an EEL reflectance spectrophotometer using diffuse light.

Two backgrounds were used: black, and sixteen thicknesses of soiled fabric. The dye was then stripped off the fabric and estimated colorimetrically. Fairly good linear relationship between dye content and soil present as calculated by the regular Kubelka-Munk equation was obtained.

Rees (73) reviewed activities with Kubelka-Munk-type equations in detergency applications, and concluded that none of the linear equations were valid for experimental data over any range of soiling level, however small. He derived the "soil additional density" approach, defined as:

$$\text{SAD} = \log \underline{R}_0/\underline{R}_s$$

where $\underline{R}_0$ is the reflectance factor of unsoiled fabric and $\underline{R}_s$ is the reflectance factor of soiled fabric.

His experimental work in using dry ferric oxinate particles disclosed three distinct relationships between reflectance and soil content, and between SAD and soil content for the three modes of soiling a certain cotton fabric: filtration, contact, and deposition. Further relationships were found with deposition of wet oxinate, and more relationships were expected to exist. From these observations, Rees decided that it was obvious that a unique relationship between a visual assessment of soiling and actual soil content to cover all modes of soiling simply cannot exist. Further, changes in fabric construction and in finish would be expected to lead to further, different relationships.

Rydzewska (74) used the SAD concept with a goniophotometric method for measuring degree of soiling of fabric with kaolin plus graphite and soot, as well as natural dust. Byrne et al. (188) applied the SAD equation and an analysis of "graying additional density" (GAD, a measure of visible dirtiness remaining after washing) to assessment of soiling and graying of nylon underskirts during wearing-washing cycles. The GAD of individual and hems of underskirts after 24 cycles was correlated to some extent with total fat content (%), $\underline{F}$, determined by solvent extraction:

$$\text{GAD} = 0.027\underline{F} + 0.009$$

And yellowness (%), $\underline{Y}$, of individual pieces of underskirt after 24 cycles was also moderately correlated with fat content by

$$\underline{Y} = 1.44\underline{F} + 4.8$$

Additional use information with the SAD concept is provided by Lord and Rees (285) in connection with soiling of garments as a function of static electrical effects.

Flath (286) surveyed different forms of textile soiling and lab assessment of degree of soiling. The SAD values were said to be more sensitive to degree of soiling of fabrics, while the Kubelka-Munk index expresses quantitative dirt content.

V. DETERGENT SOLUTION EVALUATION BY PHYSICOCHEMICAL METHODS

In 1929, Rhodes and Brainard (55) pointed out that the first methods suggested for measuring and comparing the detergent power of soap solutions were based not on direct measurements of washing action but rather on the determination of some physical or physicochemical property which was supposed to correlate with detergency. They reviewed these early test methods which comprised:

(1) surface tension of soap solutions against air or against oils;

(2) sudsing properties;

(3) emulsification of oils;

(4) protective colloidal action, such as gold number determination;

(5) amount of pigment soil which could be caused to pass with the detergent solution through a filter.

In 1941, Sarin and Uppal (287) determined the umber deflocculating and emulsifying power of soap nut powder in comparison to that of six other materials. However, they recognized that, "while—[these] measurements are indicative of the detergent value of the solution, they may not be the true index of its detersive efficiency. Actual washing tests are now considered to be the only satisfactorily final criterion of detergent efficiency."

In 1958 Vallee and Guillaumin (288) described a stretching method, using the Thiband apparatus, and comparing results with those obtained in Launder-Ometer tests. Detergency properties were evaluated irrespective of soils, fabrics, stirring, temperature, rinsing, bleaching agent, or fluorescent whitening agents. Nine synthetic detergents at various concentrations were studied. It is claimed for this method that it permits:

(1) measurement of surface tension;

(2) discernment between wetting agents and detergent products;

(3) measurement of detergent power, molecularity, and independent of persalts and fluorescent whitening agents;

(4) selection of the optimum molecular structures desired for functions of proprietary detergents;

(5) measuring action of additives on detergent power;

(6) measuring synergistic effects of mixtures of detergents;

(7) measurement of critical micelle concentration.

Sonntag and Strenge (289) used interferometric observations of the contact zone soil-polymer to determine "specific washing power," expressed as soil suspension. Polyamide, polyester, polyethylene, and Teflon films were soiled with oleic acid in cyclohexane. Surfactants comprised sodium oleate, sodium dodecyl sulfate, nonylphenol-ethyleneoxide, and cetyl trimethyl ammonium bromide. The process was regarded as one of a heterocoagulation. The end point in cleaning was the formation of a mechanically stable, approximately saturated absorption layer of surfactants, both at the oil-surfactant and at the polymer-surfactant interfaces.

The cloud point of a nonionic surfactant has been related to its optimum washing temperature (290). Contact angle measurements have been made in conjunction with detergency studies (330, 331).

VI. EVALUATION OF WASHED CLOTH BY PHYSICOCHEMICAL METHODS

Industrial fabrics such as cotton wiping towels are examined for rewetting properties after washing. A usual test is for the time required for water to "wick" up the dry fabric to vertical heights of 1 and 2 inches. The latter height must be traversed in less than 45 sec. Alternately, the towel may be tightly rolled up, retained by rubber bands, and gently placed atop water. It should sink in approximately 15 sec.

Petzel (291) used the Larose device (292) to measure water absorption of fabric and reported results as a function of repetitive laundering. The weight of solution taken up by a piece of immersed goods, or Berbig, has been used to assess detergent power (56). Wolfram and Nuessle (293) described a wick-rise absorbancy test which used a red dye in the water in order to facilitate observation of rate of capillary rise. For assessing washing ability, Sivadjian (332) disclosed a hygrophotographic method which measures the water absorbancy of cellulose fibers before and after washing. The picture obtained by contact with fabrics washed with synthetic detergents is more complete and blacker, indicating a higher moisture content and greater cleanliness than the picture of the same fabric washed with soap.

VII. EXPERIMENTAL DESIGN AND STATISTICAL METHODS

A highly rewarding way to conduct research is the design for experiments which provide more information than that apparent from the actual experimental data alone. One such design is that developed by Box and Wilson (294). The object of the approach is to develop a mathematical expression which states how a dependent variable such as soil removal is governed by changes in parameters of detergent concentration, temperature, time, etc. The experimental system also lends itself to locating optimum performance conditions and to statistical examination of dependability of results.

Ferris et al. (295) presented an application of the Box method in designing and analyzing the effects of five variables in the evaluation of a detergent. The variables were concentration, water hardness, soil load, wash time, and temperature.

Ehrenkranz and Jebe (210) described the basis of selection of a general split-plot design utilizing three replicates, imposed onto which was a partially confounded 3^2 factorial (296). Their study concerned developing a ^{14}C labeling method for measuring effectiveness of different home laundering conditions on soil removal.

Feuell and Wagg (297) applied factorial experimentation to the study of washing and suspending powers of a ternary system of nonionic surfactants-soda ash-sodium carboxymethyl cellulose. Details of the study, including effects of individual ingredients as well as analysis for errors and interactions, were presented. The reader is referred to standard texts on experimental design (296, 298).

Statistical methods are of interest to the detergent chemist because he soon finds out that reproducibility of reflectance measurements and assessments made for soil content of fabric seems to be definitely inferior to certain other measurements that he has had experience with, such as acidimetry-alkalimetry titrations.

Fundamental facts about the difference between precision and accuracy have been stressed by Lambert and Sanders (11) as well as by Weatherburn and Bayley (116). Precision is the degree of agreement or reproducibility among replicate determinations and its indicator is the standard deviation of the mean. Accuracy is a measure of agreement between the determined mean and a detergent's true performance, and hence reflects the validity of data.

General agreement is that validity can be gauged only by rather extensive field trials. Hence, lab data always tend to be questioned no matter how high reproducibility may be. Lambert and Sanders (11) cited an example of high precision but complete inaccuracy of lab swatch test in light of both practical towel tests and actual field trials on two different detergent formulations. It was suggested that improvements in precision and statistical treatments of data would not be particularly helpful unless validity is improved simultaneously.

In view of this, it seems likely that some factors of recent shifts away from heavily soiled, carbon black fabrics toward "realistic" soils have been basically in the interest of better validity for lab detergent tests.

Schwartz and Perry (2) noted that detergency results obtained with particle-containing soils, measured by reflectance, have a low degree of reproducibility. Even with much replication and averaging, they cited a variation of about $\pm$ 10-15% in the calculated detersive efficiency (DE):

$$DE = \frac{\text{Units of soil removed by test detergent}}{\text{Units of soil removed by standard detergent}} \times 100$$

where units of soil are proportional to reflectance readings.

The precision of the Woodhead et al. (299) method, although lengthy and laborious, gives a deviation in the calculated DE of only $\pm$3.4%.

Harris (3) reported on statistical presentation of detergency data. Numerous technical papers on detergency contained statistical treatment of data. Among these, Lambert (151) applied statistics to multicycle washing-soiling. Leonard (184) showed a trisigma control chart for wool grease determinations. Ashcraft (138) investigated both radiotracer and reflectance reproducibility. Bacon et al. (114) used statistical treatment quite extensively in connection with the development of a standard washing procedure needed to investigate the utility of artificially soiled fabrics. The data obtained required statistical treatment to establish the limits of reliability. The 95% confidence limits obtained with a soiled cotton fabric from Testfabrics, Inc., are $\pm$9% with 20 replicates and $\pm$12.5% with ten replicates within laboratories. The agreement between the cooperating laboratories was $\pm$12.5% with 20 replicates.

Armour workers (95, 105, 300) presented a series of three papers on statistical approach to detergency with respect to: (1) standardized Terg-O-Tometer detergency test method; (2) correlation of performance data with gas chromatography patterns of alkylbenzenes; and (3) relative effectiveness of several proprietary detergents as measured with or without Kubelka-Munk treatment of reflectance readings on three commercially available test fabrics. After standardization of techniques with respect to machines, soiled fabrics, wash times, and water hardness, and using two anionic heavy-duty detergents at 0.2% concentration, standard deviations of 0.74% soil removal for a water hardness of 50 ppm, and 1.03% soil removal for a water hardness of 135 or 300 ppm were obtained. In the second paper it was found that improved precision resulted from increase in detergent concentration to 0.4-0.5%; the standard deviation was 0.56% soil removal units, and precision at a 95% confidence level was $\pm$1.12% soil removal units. In the third paper, various calculation methods and statistical treatments of detergency data were discussed; quantities of soil removal were related to detergent type, and to hydrophilicity of the soil and cleaning media.

Trowbridge (333) reported on statistical analysis of detergency tests with a natural soil, while Komeda et al. (334) explored the contribution of the various parameters of washing on the basis of analysis of variance.

The reader is referred to standard texts on statistical handling of experimental data (296, 301).

VIII. FUTURE POSSIBILITIES

In 1963, Ginn (6) reviewed and discussed a number of concepts and approaches to evaluation of fabric detergents which appeared to have future potential. One was an instrumental approach to the microscopic examination of oily soil removal by Harker (302). A 35-mm camera and microscope attachment were used to photograph what was visually observed; the "rolling up" of soil as a function of time and relating this to fiber and soil polarities seemed to be of particular interest. Another concept was that of "residual work of laundering" developed by Kling and Lang (303). Ginn expressed the hope that differences would be found among detergent systems with respect to the residual work concept.

Other future means of evaluation of cleanliness of fabric probably exist in the rapidly merging interests of detergency, textile, and equipment technologies. Appropriate soils to use in detergency evaluations and colorimetric appearance of washed fabric have become deeply enmeshed with development in textiles—lipophilic fibers, fabric made with blends of fabrics, durable press, and soil release features.

Many of these aspects were discussed during a symposium entitled The Next Phase: Textiles and Detergents of the '70's, conducted by the American Oil Chemical Society on April 2, 1968, during their 1968 joint meeting with the American Association of Cereal Chemists in Washington, D.C. Fortress of Celenase presented a paper entitled, "The Impact of New Fibers and Fabrics and New Levels of Consumer Performance Requirements." Robert Hunter of Colgate-Palmolive spoke on, "The Effects of the Changing Textile Technology on Detergent Formulations." Donavan of General Electric described changes in hardware, particularly instrumentation of washing machines.

Further discussion along the same lines has been provided by Davis (304), under the title "Developments in Appliances Which Might Influence the Detergent and Textile Industries." It was pointed out how interrelation of researches in detergents, appliances, and textiles can result in a major development in one industry causing reactions by the other industries. It was suggested that a water softener built onto the washer might affect detergent composition. Some predictions about the technology of the future are presented.

ACKNOWLEDGMENT

The writer gratefully acknowledges the advice and assistance of James W. Hensley.

NOMENCLATURE

"a"	Green-red reflectometer scale
ACH	American Conditioning House, Inc.; 11-17 Melcher St.; Boston, Massachusetts
"b"	Blue-yellow reflectometer scale
B	Blue filter reflectance
CD	Color difference
DM	Deter-Meter, American Conditioning House, Inc.
D-25	Hunterlab Color Difference Meter[a]
D-40	Hunterlab D-40 Reflectometer[a]
Elrepho	Elrepho Photoelectric Reflection Photometer[b]
EMPA	Eidgenoessische-Material-Prulfungs-Anstalt (91)
FDS	Foster D. Snell (92)
FWA	Fluorescent whitening agent
G	Green filter reflectance
GE	General Electric[c] Schenectady, New York
GMP	Gardner Multi-Purpose reflectometer[d]
H.D.	Heavy duty
HMP	Hunter Multi-Purpose reflectometer. Manufactured by Gardner Laboratories[d]
L	Reflectance scale
$\underline{L}$	Lightness in NBS units (25)
LOM	Launder-Ometer, Atlas Electric Devices Co. 361 West Superior St. Chicago, Illinois
NaCMC	Sodium carboxymethyl cellulose
NaDDBS	Sodium dodecylbenzene sulfonate
NBS	National Bureau of Standards, Washington, D.C.
PEC	Polyester-cotton
PV, PV610	Photovolt reflectometer Photovolt Corporation 95 Madison Avenue New York, New York
Rd	Reflectance scale
$\Delta \underline{S}$	Chromaticity difference in NBS units
TFI	Testfabrics, Inc. (91)
TOT	Terg-O-Tometer, U.S. Testing Company (93)
TSPP	Tetrasodium pyrophosphate
UST	U.S. Testing Laboratories, Inc. (93)
w/o	Without
Zeiss	Reflectometer[b]

[a] Hunter Associates Laboratory, Inc., 9529 Lee Highway, Fairfax, Virginia.

[b] Carl Zeiss, Inc., 44 Fifth Avenue, New York, New York.

[c] General Electric, Schenectady, New York.

[d] Gardner Laboratories, Inc., P.O. Box 5728, Bethesda, Maryland.

REFERENCES

1. A. M. Schwartz, J. W. Perry, and J. Berch, Surface Active Agents and Detergents, Vol. II, Wiley-Interscience, New York, 1958.
2. A. M. Schwartz and J. W. Perry, Surface Active Agents, Vol. I, Wiley-Interscience, New York, 1949.
3. J. C. Harris, Detergency Evaluation and Testing, Wiley-Interscience, New York, 1954.
4. A. R. Martin and G. P. Fulton, Drycleaning, Technology and Theory, Wiley-Interscience, New York, 1954.
5. H. Stüpel, Synthetische Wasch- und Reinigungsmittel, 2nd ed., Konradin-Verlag Robert Kohlhammer G.m.b.H., Stuttgart, 1957.
6. M. E. Ginn, J. Amer. Oil Chem. Soc., 40, 662 (1963).
7. K. A. Schoenberg, Amer. Dyest. Rep., 50, 660 (1961).
8. R. B. Diaz, H. Paitchel, and J. A. Woodhead, ASTM Bulletin, 202, TP242 (1954).
9. A. K. Phansalkar and R. D. Vold, J. Phys. Chem., 59, 885 (1955).
10. J. J. Benischeck et al., Amer. Dyest. Rep., 49, 565 (1960).
11. J. M. Lambert and H. L. Sanders, Ind. Eng. Chem., 42, 1388 (1950).
12. D. B. Judd, Color in Business, Science, and Industry, Wiley-Interscience, New York, 1952.
13. F. W. Billmeyer, Jr., and M. Saltzman, Principles of Color Technology, Wiley-Interscience, New York, 1966.
14. H. Hemmendinger and J. M. Lambert, J. Amer. Oil Chem. Soc., 30, 163 (1953).
15. A. Kling and J. Kurz, Wascherei-Tech. u.-Chem., 14, 255 (1961).
16. E. Allen, Amer. Dyest. Rep., 46, 425 (1957).
17. R. T. Hunter, J. Amer. Oil Chem. Soc., 45, 362 (1968).
18. S. Goldwasser, Soap Chem. Spec., 38 (8), 47; (9), 62 (1962).
19. D. L. MacAdam, J. Opt. Soc. Amer., 24, 188 (1934).
20. R. R. Blakey, J. Soc. Dyers Colour., 84, 120 (1968).
21. R. M. Evans, J. Opt. Soc. Amer., 39, 774 (1949).
22. R. S. Hunter, S. Pulp Pap. Mfr., 30 (8), 64 (1967).
23. F. W. Billmeyer, Jr., Sci. Technol., No. 78, 26 (1968).
24. Anon., Chem. Eng. News, 44 (33), 50 (1966).
25. F. L. Diehl, Proc. Int. 4th Congr. Surface Activity, Brussels 1964, 3, 29 (1967).

26. W. Löcher, "Fachlehrbuch für das Wäscherei- u. Plättereigewerbe", Wäscherei- Fachverlag Glagow & Co., Baden-Baden, 1959.

27. W. Spring, Kolloid-Z., 4, 11 (1909).

28. A. A. Shukov and P. I. E. Shestakow, Chem.-Ztg., 35, 1027 (1911).

29. W. Stericker, Ind. Eng. Chem., 15, 244 (1923).

30. Procter and Gamble Co., Belgian 670, 531 (1966).

31. J. E. Kiefer (to Eastman Kodak Co.), U.S. Pat. 3,355,413 (1967).

32. O. Viertel, Melliand Textilber., 48, 1074 (1967).

33. C. B. Brown, S. H. Thompson, and G. Stewart, Text. Res. J., 38, 735 (1968).

34. J. D. Carter, Ind. Eng. Chem., 23, 1389 (1931).

35. H. K. Schilling, I. Rudnick, C. H. Allen, P. B. Mack, and J. C. Sherrill, J. Acoust. Soc. Amer., 21, 39 (1949).

36. W. K. Griesinger and J. A. Nevison, J. Amer. Oil Chem. Soc., 27, 96 (1950).

37. J. G. Jarrell and H. B. Trost, Soap Chem. Spec., 28 (7), 40 (1952).

38. Deering Milliken, Inc., Box 1926, Spartanburg, South Carolina.

39. R. Tsuzuki and N. Yabuuchi, Amer. Dyest. Rep., 57, 472 (1968).

40. J. M. Kennedy and E. E. Stout, ibid., 57, 2 (1968).

41. W. B. Reese, ibid., 47, 49 (1958).

42. F. J. Rizzo, ibid., 50, 211 (1961).

43. W. A. Coppock, ibid., 54, 343 (1965).

44. G. Anders, J. Soc. Dyers Colour., 84, 125 (1968).

45. A. Rauchle and W. Schramm, Textilveredl., 2, 719 (1967).

46. M. S. Furry, P. L. Bensing, and J. L. Kirkley, Amer. Dyest. Rep., 48 (8), 59 (1959).

47. K. McLaren, J. Soc. Dyers Colour., 83, 438 (1967).

48. L. F. Luechauer, Linen Supply News 48 (2), 26 (1965).

49. N. MacBeth, Jr. and W. B. Reese, Illum. Eng., 59 (6), 461 (1964).

50. Amer. Ass. Textile Chemists and Colorists Technical Manual, 1967 Ed., Res., Triangle Park, N. C. Tentative Test Method AATCC 121-1967T (1967).

51. 1968 Book of ASTM Stand., Part 21, D-1721-61, 359 (1968).

52. J. W. Hensley, private communications, 1968.

53. O. M. Morgan and J. G. Lankler. Anal. Chem., 14, 725 (1942).

54. R. M. Reinhardt, T. W. Fenner, J. D. Reid, M. S. Furry, and M. Walsh, Amer. Dyest. Rep., 50, 771 (1961).

55. F. H. Rhodes and S. W. Brainard, Ind., Eng. Chem., 21, 60 (1929).

56. J. P. Sisley and P. J. Wood, Amer. Dyest. Rep., 36, 457 (1947).

57. R. S. Hunter, J. Res. Nat. Bur. Stand., 25, 581 (1940).

58. H. H. Willard, L. L. Merritt, Jr., and J. A. Dean, Instrumental Methods of Analysis, 3rd. ed., Van Nostrand, Princeton, N.J., 1958, p. 26.

59. R. S. Hunter, "A History of Photoelectric Tristimulus Colorimetry", paper presented at 50th Anniversary of the Optical Society of America, Washington, D.C., March 18, 1966.

60. J. C. Harris and E. L. Brown, J. Amer. Oil Chem. Soc., 27, 564 (1950).

61. H. Machemer, Fette Seifen, 54, 324 (1952).

62. R. S. Hunter, et al., Amer. Dyest. Rep., 50, 812 (1961).

63. T. H. Trimmer, et al., ibid., 55, 997 (1966).

64. J. W. Hensley, J. Amer. Oil Chem. Soc., 42, 993 (1965).

65. W. H. Smith, M. Wentz, and A. R. Martin, ibid., 45, 83 (1968).

66. M. G. Kramer, H. Paitchel, and W. A. Tidridge, Amer. Soc. Test. Mater., Spec. Tech. Publ. 173 (1956).

67. A. M. Schwartz, J. Berch and H. Peper, ibid., Spec. Tech. Publ. 316 (1962).

68. T. H. Vaughn and H. R. Suter, J. Amer. Oil Chem. Soc., 27, 249 (1950).

69. A. R. Martin and R. C. Davis, Soap Chem. Spec., 36 (4), 49 (1960).

70. O. E. Ford, Text. Res. J., 38, 339 (1968).

71. J. C. Harris and E. L. Brown, J. Amer. Oil Chem. Soc., 27, 135 (1950).

72. H. L. Sanders, and J. M. Lambert, ibid., 27, 153 (1950).

73. W. H. Rees, J. Text. Inst., 53, T230 (1962).

74. D. Rydzewska, ibid., 54, P156 (1963).

75. R. C. Ferris and L. O. Leenerts, Soap Chem. Spec., 32 (7), 37 (1956).

76. W. G. Spangler, H. D. Cross, III, and B. R. Schaafsma, J. Amer. Oil Chem. Soc., 42, 723 (1965).

77. S. V. Vaeck and E. Maes, Tenside, 5, 4 (1968).

78. T. G. Jones and J. P. Parke, Proc. 3rd Int. Congr. Surface Activity, Cologne 1960, 4, 178 (1960).

79. A. S. Weatherburn and C. H. Bayley, Res., 11 (4), 141 (1958).

80. T. G. Jones, R. A. Hudson, and J. P. Parke, Proc. 3rd Int. Congr. Surface Activity, Cologne 1960, 4, 188 (1960).

81. W. C. Powe, Text. Res. J., 29, 879 (1959).

82. W. C. Powe and W. L. Marple, J. Amer. Oil Chem. Soc., 37, 136 (1960).

83. R. C. Davis, Soap Chem. Spec., 39 (8), 47 (1963).

84. B. J. Rutkowski, J. Amer. Oil Chem. Soc., 44, 103 (1967).

85. B. J. Rutkowski, ibid., 45, 266 (1968).

86. W. G. Spangler, R. C. Roga, and H. D. Cross, III, ibid., 44, 728 (1967).

87. J. H. Gustafson, Proc. Chem. Spec. Mfr. Ass., 53rd Mid-Year Meeting, 195 (1967).

88. K. J. Nieuwenhuis, Riv. Ital., Sostanze Grasse, 44 (1), 13 (1967).

89. G. Weder, Amer. Dyest. Rep., 51, 154 (1962).

90. H. Stüpel, Fette Seifen, 54, 143 (1952).

91. Testfabrics, Inc., 55 Vandam St., New York, New York 10013.

92. Foster D. Snell, Inc., 29 West 15th St., New York, New York 10011.

93. United States Testing Co., 1415 Park Ave., Hoboken, N.J. 07030.

94. A. Warren, Soap Chem. Spec., 39 (3), 50 (1963).

95. M. E. Ginn, G. A. Davis, and E. Jungermann, J. Amer. Oil Chem. Soc., 43, 317 (1966).

96. F. L. Diehl and J. B. Crowe, ibid., 31, 404 (1954).

97. J. W. Hensley, M. G. Kramer, R. D. Ring, and H. R. Suter, ibid., 32, 138 (1955).

98. S. Shimauchi and H. Mizushima, Amer. Dyest. Rep., 57, 462 (1968).

99. O. C. Bacon, ibid., 34, 556 (1945).

100. J. R. Clark and V. B. Holland, ibid., 36, 734 (1947).

101. L. J. Armstrong, et al., ibid., 37, 596 (1948).

102. H. J. Wollner and G. S. Freeman, ibid., 40, 693 (1951).

103. G. E. Barker and H. J. Ranauto, J. Amer. Oil Chem. Soc., 32, 249 (1955).

104. H. A. Ludeman, J. A. Balog and J. C. Sherrill, ibid., 35, 5 (1958).

105. W. M. Linfield, E. Jungermann and J. C. Sherrill, ibid., 39, 47 (1962).

106. I. R. Schmolka and J. W. Hensley, Soap Chem. Spec., 41 (8), 47 (1965).

107. A. M. Schwartz and C. A. Rader, J. Amer. Oil Chem. Soc., 42, 800 (1965).

108. W. Gerhardt, Tenside, 3, 141 (1966).

109. R. R. Pollard, Hydrocarbon Process., 45 (11), 197 (1966).

110. C. A. Rader and A. W. Schwartz, Det. Age, 3, 28 (1966).

111. B. M. Finger, G. A. Gillies, G. M. Hartwig, E. E. Ryder, Jr., and W. M. Sawyer, J. Amer. Oil Chem. Soc., 44, 525 (1967).

112. K. J. Nieuwenhuis, J. Polym. Sci., 12, 237 (1954).

113. C. H. Lindsey, Amer. Dyest. Rep., 43, 760 (1954).

114. O. C. Bacon et al., ibid., 45, 946 (1956).

115. I. J. Gruntfest and E. M. Young, J. Amer. Oil Chem. Soc., 26, 236 (1949).

116. A. S. Weatherburn and C. H. Bayley, ibid., 37, 20 (1960).

117. G. A. Johnston and R. A. C. Bretland, J. Appl. Chem. 17, 288 (1967).

118. A. R. Morris (to Amer. Cyanamide Co.), U.S. Pat. 3,335,086 (1967).

119. J. R. Trowbridge and J. Rubinfeld, Proc. 4th Int. Congr. Surface Activity, Brussels 1964, 221 (1967).

120. R. T. Hunter, C. R. Kurgan, and H. L. Marder, J. Amer. Oil Chem. Soc., 44, 494 (1967).

121. A. E. Vandegrift and B. J. Rutkowski, ibid., 44, 107 (1967).

122. G. Hoff, Tenside, 4, 381 (1967).

123. M. S. Furry, P. L. Bensing, R. K, Taube, N. D. Poole, and E. S. Ross, Amer. Dyest. Rep., 48 (8), 67 (1959).

124. M. S. Furry, P. L. Bensing, and M. L. Johnson, ibid., 50, 50 (1961).

125. V. McLendon and F. Richardson, ibid., 52, 112 (1963).

126. V. McLendon and F. Richardson, ibid., 54, 305 (1965).

127. R. E. Wagg, D. M. Meek, and A. L. D. Willan, Proc. 4th Int. Congr. Surface Activity, Brussels 1964, 3, 269 (1967).

128. G. M. Leigh, ibid., 3, 19 (1967).

129. W. B. Mihalik and H. D. Cross, III, J. Amer. Oil Chem. Soc., 42, 177 (1965).

130. L. Loeb, E. D. Morey, and R. O. Shuck, Amer. Dyest. Rep., 55, 703 (1966).

131. R. K. Taube and N. D. Poole, ibid., 56, 172 (1967).

132. N. D. Poole, E. S. Ross, and R. K. Taube, ibid., 51, 2 (1962).

133. R. Bernstein and H. Sosson, Proc. CSMA 41st Annual Meeting, 166 (1954).

134. A. M. Schwartz and J. Berch, Soap Chem. Spec., 39 (5), 78 (1963).

135. L. E. Weeks, J. L. Staubly, and W. A Millsaps, ASTM Spec. Tech. Publ. 177A (1964).

136. D. W. Stephens and C. B. Brown, ASTM Bull., 214, TP103, (1956).

137. O. Oldenroth, Fette, Seifen, Anstrichm., 59, 414 (1957).

138. E. B. Ashcraft, ASTM Spec. Tech. Bull., 268, 27 (1956).

139. J. V. Beninate, E. L. Kelly, G. L. Drake, Jr., and W. A. Reeves, Amer. Dyest. Rep., 55 (2), 25 (1966).

140. O. C. Bacon and J. E. Smith, Ind. Eng. Chem., 40, 2361 (1948).

141. J. J. Morrisroe and R. G. Newhall, ibid., 41, 423 (1949).

142. W. P. Utermohlen, Jr., and E. L. Wallace, Text. Res. J., 17, 670 (1947).

143. W. E. Thompson, J. Amer Oil Chem. Soc., 26, 509 (1949).

144. P. T. Vitale, ibid., 31, 341 (1954).

145. J. Ross, P. T. Vitale, and A. M. Schwartz, ibid., 32, 200 (1955).

146. O. Oldenroth, Tenside, 3, 9 (1966).

147. O. Oldenroth, Chemiefasern, 17, 726 (1967).

148. J. Powney and A. J. Feuell, Res., 2, 331 (1949).

149. W. J. Hart and J. Compton, Text. Res. J., 23, 418 (1953).

150. J. Compton and W. J. Hart, Ind. Eng. Chem., 43, 1564 (1951).

151. J. M. Lambert, Proc. CSMA 38th Annual Meeting, 103 (1951).

152. F. D. Snell, Chem. Eng. News, 27, 2256 (1949).

153. A. J. Stirton, R. G. Bistline, Jr., E. B. Leardi, and M. V. Nunez-Ponzoa, J. Amer. Oil Chem. Soc., 44, 99 (1967).

154. M. C. Bourne and W. G. Jennings, ibid., 40, 517 (I), 523 (II) (1963); 42, 546 (III) (1965).

155. L. Loeb, P. B. Sanford, and S. D. Cochran, ibid., 41, 120 (1964).

156. D. Nickerson, Amer. Dyest. Rep., 33, 252 (1944).

157. O. Oldenroth, Wascherei-Tech. u.-Chem., 17, 771 (1964).

158. K. J. Nieuwenhuis, J. Amer. Oil Chem. Soc., 45, 37 (1968).

159. P. S. Stensby, Soap Chem. Spec., 43 (7), 80 (1967).

160. A. Berger. private communication, 1968.

161. H. J. Selling and L. C. F. Friele, Appl. Sci. Res., B I 453 (1950).

162. J. C. Illman, G. M. Hartwig, and J. W. Roddewig, J. Amer. Oil Chem. Soc., 46, 70 (1969).

163. B. E. Gordon, W. T. Shebs, D. H. Lee, and R. U. Bonnar, ibid., 43, 525 (1966).

164. E. Walter, Fette Seifen, 53, 322 (1951).

165. R. Neu, ibid., 54, 636 (1952).

166. T. H. Vaughn, A. Vittone, Jr., and L. R. Bacon, Ind. Eng. Chem., 33, 1011 (1941).

167. T. H. Vaughn and C. E. Smith, J. Amer. Oil Chem. Soc., 25, 44, (1948).

168. T. H. Vaughn and C. E. Smith, ibid., 26, 733 (1949).

169. G. E. Barker and C. R. Kern, ibid., 27, 113 (1950).

170. R. D. Vold and A. K. Phansalkar, Recueil, 74, 41 (1955).

171. K. J. Nieuwenhuis, Proc. 2nd Int. Congr. Surface Activity, London 1957, 4, 150 and 156 (1957).

172. W. P. Evans and M. Camp, Proc. 4th Int. Congr. Surface Activity, Brussels 1964, 3, 259 (1967).

173. M. S. Furry and E. M. O'Brien, Amer. Dyest. Rep., 41, 763 (1952).

174. M. S. Furry and P. L. Bensing, ibid., 44, 786 (1955).

175. H. E. Wixon (to Colgate-Palmolive Co.), U.S. Pat. 3,360,470 (1967).

176. E. M. McCabe, Soap Chem. Spec., 30 (12), 44 (1954); 31 (1) 42 (1955).

177. S. V. Vaeck, Wascherei-Tech. u.-Chem., 17, 26 (1964).

178. I. Gailey, J. Soc. Dyers Colour., 83, 481 (1967).

179. H. Paitchel and R. S. Hunter, "An Instrument for Measuring the Whiteness and Cleanliness of Naturally Soiled Fabrics," paper presented at Amer. Oil Chem. Soc. 11/2/54 meeting in Chicago.

180. H. B. Mann and T. H. Morton, J. Soc. Dyers Colour., 75, 522 (1959).

181. E. S. Shanley, Amer. Dyest. Rep., 47, 445 (1958).

182. W. Fong, A. S. Yeiser, and H. P. Lundgren, Text. Res. J., 21, 540 (1951).

183. J. C. Dickinson and R. C. Palmer, J. Text. Inst., 42, T6 (1951).

184. E. A. Leonard, Amer. Dyest. Rep., 39, 813 (1950).

185. W. C. Powe, J. Amer. Oil Chem. Soc., 40, 290 (1963).

186. W. I. Lyness and D. E. O'Connor (to Procter and Gamble Co.)., U.S. Pat. 3,329,618 (1967).

187. H. Arai, I. Maruta, and T. Kariyone, J. Amer. Oil Chem. Soc., 43, 315 (1966).

188. G. A. Byrne, F. H. Holmes, and J. Lord, J. Soc Dyers Colour., 84, 20 (1968).

189. T. Tachibana, A. Yabe, and M. Tsubomura, J. Colloid Sci., 15, 278 (1960).

190. E. A. Kantor, Tr. Nanch-Issled Tekhnokhim Byt. Obshizh, No. 5, 98-106 (1965).

191. S. Tomiyama and M. Iimori, J. Amer. Oil Chem. Soc., 42, 449 (1965).

192. V. B. Holland and A. Petrea, Amer. Dyest. Rep., 32, 534 (1943).

193. W. H. Smith and A. R. Martin, "The Importance of Liquid-to-Cloth Ratio in Detergency," paper presented at Joint Meeting of AACC-AOCS, April 2, 1968, Washington, D.C.

194. H. Schott, Text. Res. J., 35, 612 (1965).

195. W. H. Smith, M. Wentz, and A. R. Martin, "Estimation of Machine and Detergent Efficiency in Drycleaning Systems," paper presented at the AOCS Meeting in Chicago, Oct. 1967.

196. W. H. Smith, M. Wentz, N. McCullough, and A. R. Martin, "Measuring Soil Dispersion Power of Detergents by Filtration," paper presented at the AOCS Meeting in Chicago, Oct. 16, 1967.

197. J. Tuzson and B. A. Short, Text. Res. J., 32, 111 (1962).

198. B. A. Short, ibid., 35, 474 (1965).

199. Coulter Electronics Industrial Div. of Coulter Electronics, Inc., 2601 Mannheim Road, Franklin Park, Illinois 60131.

200. R. S. Krieger, "A Particle Size Analysis of Laundry Water and How the Particle Size Distribution is Affected by Mechanical Filtration," paper presented at CSMA May 1968 Meeting.

201. E. Gotte, First World Congr. Surface Active Agents, Paris 1954, 1, 267 (1954).

202. K. Sauerwein, Proc. 3rd Int. Congr. Surface Activity, Cologne 1960, 3, 265 (1960).

203. R. E. Wagg and D. F. B. Kevan, J. Text. Inst., 53, T102 (1962).

204. N. A. Matlin, Amer. Dyest. Rep., 40, 44 (1951).

205. A. Schaeffer, Klepzigs Text.-Z., 44, 467 (1941).

206. I. A. Bernstein, Amer. Dyest. Rep., 47, 297 (1958).

207. R. E. Murdock, ibid., 51, 18 (1962).

208. T. H. Grindstaff, H. T. Patterson, and H. R. Billica, Text. Res. J., 37, 564 (1967).

209. F. Ehrenkranz, Soap Chem. Spec., 32 (3), 41 (1956).

210. F. Ehrenkranz and E. H. Jebe, Nucleonics, 14 (3), 96 (1956).

211. F. Ehrenkranz and E. H. Jebe, Soap Chem. Spec., 34 (4), 47 (1958).

212. J. W. Hensley, H. A. Skinner, and H. R. Suter, ASTM Spec. Tech. Publ. 115, 18 (1952).

213. T. Fort, Jr., H. R. Billica, and T. H. Grindstaff, J. Amer. Oil Chem. Soc., 45, 354 (1968).

214. B. E. Gordon, J. Roddewig, and W. T. Shebs, ibid., 44, 289 (1967).

215. A. L. Meader, Jr., and B. A. Fries, Ind. Eng. Chem., 44, 1636 (1952).

216. T. F. Boyd and R. Bernstein, J. Amer. Oil Chem. Soc., 33, 614 (1956).

217. J. W. Hensley and C. G. Inks, "Adsorption of Labeled Sodium Carboxymethyl Cellulose by Textile Fibers," paper presented at 134th Nat. Meeting of the ACS at Chicago, Illinois, Sept. 1958.

218. P. G. Evans and W. P. Evans, J. Appl. Chem., 17, 276 (1967).

219. B. E. Gordon and E. L. Bastin, J. Amer. Oil Chem. Soc., 45, 754 (1968).

220. J. W. Hensley and C. G. Inks, ASTM Spec. Tech. Publ. 268, 27 (1959).

221. B. J. Rutkowski and A. R. Martin, Text. Res. J., 31, 892 (1961).

222. W. J. Schwarz, A. R. Martin, and R. C. Davis, ibid., 32, 1 (1962).

223. J. M. Lambert, Ind. Eng. Chem., 42, 1394 (1950).

224. A. C. Zettlemoyer, J. D. Skewis, and J. J. Chessick, J. Amer. Oil Chem. Soc., 39, 280 (1962).

225. J. Tolgyessy and T. Robinson, Text., 12, 414 (1957).

226. B. E. Gordon, G. A. Gillies, W. T. Shebs, G. M. Hartwig, and G. R. Edwards, J. Amer. Oil Chem. Soc., 43, 232 (1966).

227. W. J. Diamond and J. E. Grove, Text. Res. J., 29, 863 (1959).

228. W. J. Diamond and H. Levin, ibid., 27, 787 (1957).

229. E. H. Armbruster and G. M. Ridenour, Soap Chem. Spec., 31 (7), 47 (1955).

230. G. G. Jayson, J. Appl. Chem., 9, 429 (1959).

231. W. J. Schwarz and G. W. Bain, Kolloid-Z., 177, 157 (1961).

232. G. G. Jayson, J. Appl. Chem. 9, 422 (1959).

233. J. L. Sommerville, The Isotope Index, 10th anniv. ed., Scientific Equipment Co., Indianapolis, Ind., 1967.

234. J. M. Lambert, J. H. Roecker, J. J. Pescatore, G. Segura, Jr., and S. Stigman, Nucleonics, 12 (2), 40 (1954).

235. B. E. Gordon, J. Amer. Oil Chem. Soc., 45, 367 (1968).

236. T. Fort, Jr., H. R. Billica, and T. H. Grindstaff, Text. Res. J., 36, 99 (1966).

237. B. E. Gordon, W. T. Shebs, and R. U. Bonner, J. Amer. Oil Chem. Soc., 44, 711 (1967).

238. W. T. Shebs and B. E. Gordon, ibid., 45, 377 (1968).

239. J. C. Harris, R. E. Kamp, and W. H. Yanko, ASTM Bull. 158, 49 (1949).

240. J. C. Harris, R. E. Kamp, and W. H. Yanko, ASTM Bull. 170, 82 (1950).

241. W. E. Shelberg, J. L. Machin, and R. K. Fuller, Ind. Eng. Chem., 46, 2572 (1954).

242. F. H. Firsching and H. E. Everson, J. Amer. Oil Chem. Soc., 34, 547 (1957).

243. R. M. Anderson, J. J. Satanek, and J. C. Harris, ibid., 36, 286 (1959).

244. J. C. Harris and J. Satanek, ibid., 38, 244 (1961).

245. R. E. Wagg and C. J. Britt, J. Text. Inst., 53, T205 (1962).

246. H. Zeidler, Int. Koloristenkongr., Budapest, 4, 141 (1962).

247. H. Uhlich, K. Schumann, and W. Nowak, Tenside, 4, 133 (1967).

248. S. Yokoyama, S. Shimauchi, and H. Mizushima, Sen-i Gakkaishi, 23 (9) 449 (1967).

249. V. L. Gelezunas and A. J. Bryce, AEC Accession No. 41218, Rep. No. AED-CONF-66-077-1 (1966).

250. W. Markiewicz, Krajove Symp. Zastosow., Izotop Tech. 3rd, Stettin, Poland, 1966 (Sect. 148).

251. K. Takada, Y. Wadachi, Y. Yamaoka, S. Noguchi, and M. Muramatsu. Radioisotopes (Tokyo), 14 (6), 487 (1965).

252. M. S. Kang and H. P. Chung, Kisul Yon'guso Pogo, 2, 79 (1963).

253. J. C. Bailey, S. A. Kingsbury, and J. R. Knight, Rep. No. K-1344, "The Effects of Uranium Contamination on Laundry Operations," Oak Ridge Gaseous Diffusion Plant operated by Union Carbide Nuclear Co. (1957).

254. W. A. Clark, U.S., AEC, LRL-120, Livermore Res. Lab., Calif. Res. & Dev. Co., Livermore, California (May 1954).

255. A. P. Talboys and E. C. Spratt, U.S., AEC, NYO-4990, Johns Hopkins Univ., Baltimore, Maryland (March 1, 1954).

256. D. A. Netzel, C. W. Stanley, and D. W. Rathburn, J. Amer. Oil Chem. Soc., 41, 678 (1964).

257. S. Richards, M. A. Morris, and T. H. Arkley, Text. Res. J., 38, 105 (1968).

258. P. W. Cunliffe, J. Text. Inst., 28, T341 (1957).

259. D. G. Stevenson, ibid., 44, T12 (1953).

260. T. Fort, Jr., H. R. Billica, and C. K. Sloan, Text. Res. J., 36, 7 (1966).

261. D. M. Meek, J. Text. Inst., 58, 58 (1967).

262. D. M. Meek, ibid., 57, T337 (1966).

263. K. Nettelnstroth, Tenside, 4, 8 (1967).

264. H. Arai and I. Maruta, J. Amer. Oil Chem. Soc., 45 448 (1968).

265. F. D. Snell, C. T. Snell, and I. Reich, ibid., 27, 62 (1950).

266. P. Kubelka and F. Munk, Z. Tech. Phys., 12, 593, (1931).

267. W. P. Utermohlen, Jr., and M. E. Ryan, Ind. Eng. Chem., 41, 2881 (1949).

268. I. Reich, F. D. Snell, and L. Osipow, ibid., 45, 137 (1953).

269. W. Strauss, Kolloid-Z., 150, 134 (1957).

270. A. S. Weatherburn and C. H. Bayley, Text. Res. J., 25, 549 (1955).

271. R. E. Wagg, J. Text. Inst., 43, T325 (1952).

272. J. C. Harris, M. R. Sullivan, and L. E. Weeks, Ind. Eng. Chem., 46, 1942 (1954).

273. O. C. Bacon and J. E. Smith, Amer. Dyest. Rep., 43, 619 (1954).

274. D. Frishman, et al., ibid., 43, 751 (1954).

275. L. W. Mazzeno, Jr., R. M. H. Kullman, R. M. Reinhardt, H. B. Moore, and J. D. Reid, ibid., 47, 299 (1958).

276. J. V. Beninate, E. L. Kelly, and G. L. Drake, Jr., ibid., 52 (20), 26 (1963).

277. J. Berch, H. Peper, and G. L. Drake, Jr., Text. Res. J., 34, 29 (1964).

278. A. C. Nuessle, L. M. Nageley, and E. O. J. Heiges, ibid., 33, 146 (1963).

279. S. J. O'Brien, et al., Amer. Dyest. Rep., 52, 377 (1963).

280. H. B. Trost, J. Amer. Oil Chem. Soc., 40, 669 (1963).

281. A. R. Martin, "A New Relationship between Reflectance Data and Soil Removal," paper given at AOCS Meeting, Chicago, Oct. 31, 1961.

282. A. Darbey, Amer. Dyest. Rep., 46, 465 (1957).

283. E. Allen, Soap Chem. Spec., 35 (7), 51 (1959).

284. D. Wilson, J. Text. Inst., 53, T1 (1962).

285. J. Lord and W. H. Rees, ibid., 51, T419 (1960).

286. H. J. Flath, Deut. Textiltech. 18, No. 1/2, 97 (1968).

287. J. L. Sarin and M. Y. Uppal, Ind. Eng. Chem., 33, 666 (1941).

288. J. P. Vallee and R. Guillaumin, Proc. 3rd Int. Congr. Surface Activity, Cologne 1960, 3, 151 (1960).

289. H. Sonntag and K. Strenge, Tenside, 4, 129 (1967).

290. "Handling Articles Containing Dacron in the Laundry Industry," Technical Information Bulletin D-200, March 1967, Textile Fibers Dep., Technical Service Section, E. I. DuPont de Nemours & Co., Wilmington, Del.

291. F. E. Petzel, Amer. Dyest. Rep., 46, 569 (1957).

292. P. Larose, ibid., 31, 105 (1942).

293. R. E. Wolfram and A. C. Nuessle, ibid., 42, 753 (1953).

294. G. E. P. Box and K. B. Wilson, J. Roy. Statist. Soc., B13, 1 (1951).

295. R. C. Ferris, M. Patapoff, and J. F. Pietz, Proc. CSMA 44th Annual Meeting, 130 (1957).

296. W. G. Cochran and G. Cox, Experimental Designs, Wiley-Interscience, New York, 1950.

297. A. J. Feuell and R. E. Wagg, Res. Suppl. 2-7, 334 (1949).

298. K. A. Brownlee, Industrial Experimentation, 4th ed., Chemical Publishing, New York, 1952.

299. J. A. Woodhead, P. T. Vitale, and A. J. Frantz, J. Am. Oil Chem. Soc., 21, 333 (1944).

300. E. Jungermann, G. A. Davis, E. C. Beck, and W. M. Linfield, ibid., 39, 50 (1962).

301. W. J. Yonden, Statistical Methods for Chemists, Wiley-Interscience, New York, 1951.

302. R. P. Harker, J. Tex. Ints., 50, T189 (1959).

303. W. Kling and H. Lange, J. Amer. Oil Chem. Soc., 37, 30 (1960).

304. R. C. Davis, ibid., 45, 409 (1968).

305. Anon., Materials Research and Standards, 10 (8), 25 (1970).

306. A. C. Nuessle and E. O. J. Heiges, Text. Chem. Color., 2 (13), 225 (1970).

307. I. Kashiwa, T. Hirabayashi, Y. Kawasaki, T. Tsunoda, and Y. Oba, Yukagaku, 19 (2), 76 (1970).

308. M. M. Win, Diss. Abstr. Int. B., 30 (6), 2672 (1969).

309. T. Tsunoda, Y. Oba, and I. Kashiwa, Kogyo Kagaku Zasshi, 72 (8), 1904 (1969).

310. I. Kashiwa, H. Kuwamura, Y. Kawasaki, H. Nishizawa, T. Tsunoda, and Y. Oba, Yukagaku, 19 (3), 158 (1970).

311. S. Shimauchi, R. Tsuzuki, and H. Mizushima, Amer. Dyestuff Reptr., 58 (16), 15 (1969).

312. O. Viertel and O. Oldenroth, Chemiefasern Text.-Anwendungstech, 20 (3), 216 (1970).

313. L. Labhard and M. A. Morris, Textile Res. J., 40 (5), 445 (1970).

314. H. E. Bille, A. Eckell, and G. A. Schmidt, Text. Chem. and Color., 1 (27), 600 (1969).

315. O. Oldenroth, Zeitschrift fur die gesamte Textilindustrie, 71 (9), 597 (1969).

316. G. A. Byrne and J. C. Arthur, Jr., Text. Res. J., 41 (3), 271 (1971).

317. S. H. Hierons, Canadian Tex. J., 86 (2), 30 (1969).

318. M. S. Sontag, M. E. Purchase, and B. F. Smith, Amer. Dyestuff Reptr. 58 (26), 19 (1969).

319. R. Knaggs, Linen Supply News, 53 (6), 58 (1970).

320. L. D. Johnson, Tex. Res. J., 40, 650 (1970).

321. T. Tsunoda, Y. Oba, and I. Kashiwa, "Study of Detergency," Paper presented at Am. Chem. Soc. meeting, September 14-18, 1970, in Chicago, Ill.

322. J. L. Bubl, Tex. Res. J., 40, 637 (1970).

323. K. Takada, Y. Wadachi, and M. Muramatsu, Radioisotopes, 19 (5), 253 (1970).

324. T. H. Grindstaff, H. T. Patterson, and H. R. Billica, Proc. Chem. Specialties Mfg Ass'n., 55th Mid-year Meeting, 110 (1969).

325. M. S. Sontag, M. E. Purchase, and B. F. Smith, Tex. Res. J., 40 529 (1970).

326. B. E. Gordon and W. T. Shebs, J. Amer. Oil Chem. Soc., 46 (10), 537 (1969).

327. J. C. Illman, B. M. Finger, W. T. Shebs, and T. B. Albin, J. Amer. Oil Chem. Soc., 47 (10), 379 (1970).

328. Y. Komeda and J. Mino, Yukagaku, 19 (6), 410 (1970).

329. Y. Komeda, J. Mino, and K. Yoshizaki, Yukagaku, 19 (3), 163 (1970).

330. J. C. Stewart and C. S. Whewell, Tex. Res. J., 30, 903 (1960).

331. Y. Komeda and J. Mino, Yukagaku, 19 (6), 420 (1970).

332. J. Sivadjian, Tenside, 6 (5), 266 (1969).

333. J. R. Trowbridge, J. Amer. Oil Chem. Soc., 47 (4), 112 (1970).

334. Y. Komeda, J. Mino, T. Imamura, and F. Tokiwa, Tex. Res. J., 40 (8), 733 (1970).

Chapter 10

TEST EQUIPMENT

W. G. Spangler
Colgate Palmolive Company
Piscataway, N. J.

I. INTRODUCTION

The twentieth century has created a new look in the washing procedure of the average housewife, particularly in the United States. The application of mechanical energy to the washing machine created many new situations. The programmed washer for the home, the appearance of Laundromats in the community, and the modernized commercial laundry are all part of a new pattern. In addition, the chemical industry has introduced many new types of synthetic detergents to compete with and overtake the use of soap. The textile industry has also provided a variety of synthetics which compete with the natural fibers. The application of special treatments to both the natural and synthetic fibers for acquiring specific functional properties is now a possibility. Lastly, the "soapers" are no longer content to merely remove dirt, they try to satisfy the customer with the right feel, a good smell, a whiter white, a more sparkling color, a safe bleach, and germproofing.

These changes have increased the stress and importance of an evaluation section. Previously the formulator took the leading role, but now a great deal of the success of a product rides on the predictions of the evaluation section and the formulator receives guidance from their reports of acceptance or rejection. Products must be tested in real or simulated conditions. The instrumental or visual measurements must include an evaluation of the overall parameters and be meaningful to the marketing division.

The first part of this chapter is devoted to a discussion of the washing machines (miniaturized, prototype, or full-scale) used throughout the industry for evaluating the washing procedure in its entirety. The remainder deals with the instrumental and visual procedures for assessing the results accumulated in the first part. The discussion is limited mainly to the cleaning of fabric substrate.

II. LAUNDRY TEST MACHINERY–LABORATORY SCALE

A. History

The history of "cleanliness" somewhat parallels the efforts of man to obtain shelter, food, and clothing. In the beginning garments were generally washed by beating them with sticks or stones. This was usually done near a natural body of water and in the absence of a detersive agent. Soap was eventually discovered but was not commonly used until the early part of the nineteenth century. Concurrently, crude washing devices made their appearance. These early machines were friction devices. The kicker, dasher, cradle, dolly, and others were designed to achieve mechanically what the Egyptians did by hand, namely to beat the dirt out of clothes (1).

B. The Launder-Ometer

The American Association of Textile Chemists and Colorists was founded in 1921 with the purpose of promoting the increase of knowledge of the application of dyes and chemicals in the textile industry (2). Their interest in testing the fastness of dyed textile material to washing and laundering led to the development of the Launder-Ometer (3). This was a standard machine for conducting laboratory washing tests. An agreement was entered into between the Association and the Atlas Electric Devices Co., whereby the latter was to have exclusive rights to manufacture and sell the standard washing machine (4, 5).

The Launder-Ometer consists of a mounted reservoir with a 15-gal. capacity (Fig. 10.1). Inside the tank is a rotor which has a capacity for holding 20 metal containers. Five types of stainless-steel specimen containers are available for use in the rotor, three of which are normally only used in the high-temperature model. The rotor is driven at a uniform rate of 42 rpm. This machine is operated by a control panel which is usually placed over the preheating loading table. Metal balls are put in the containers when in use to produce a desired abrasive action. Thus the machine permits the testing of 20 samples under controlled temperature and mechanical agitation. Launder-Ometer procedures are described in the *Technical Manual* of the AATCC and by Harris (6).

Several years after the introduction of the Launder-Ometer, soap was challenged as a detergent by the production of alcohol sulfates and synthetic

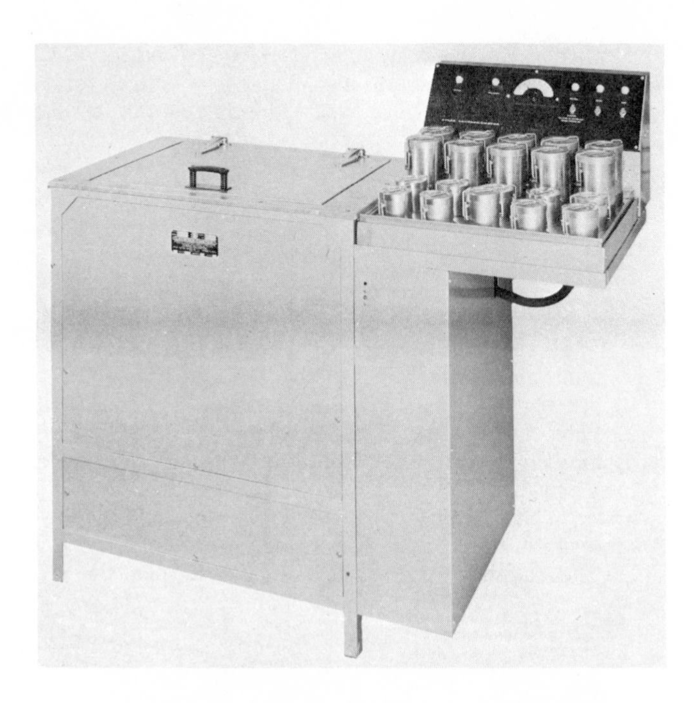

FIG. 10.1. Launder-Ometer (courtesy Atlas Electric Devices Co.).

detergents from petroleum. It became desirable to compare these new detergents with each other and with the traditional soaps. The accepted method for making comparisons until that time was to wash in the Launder-Ometer some type of soiled fabric made from carbon and "grease." During 1930-1955 many papers were published on testing the performance of the new detergents. These writings are too numerous to list but categorically they dealt with soiling procedures, new test equipment, new philosophies on testing, reproducibility of data, etc. Many experimenters were of the opinion that laboratory washing equipment should employ reciprocating mechanisms if the results were to be meaningful. A simple laboratory device (Fig. 10.2) was developed in the Research Department of the Colgate-Palmolive Company using this principle (7).

This device was called a "one-lung" reciprocator and the angle through which the paddle rotated, as well as the speed of rotation, could be regulated. Originally, tests were run with a Dewar flask for temperature control. General Aniline and Film Corp. also described a miniature oscillating agitator type of washer developed in their laboratories during the same period (8). In May 1949, Baker Instrument Co. announced the Terg-O-Tometer as a device for simulation of home washers of the agitator type on a laboratory scale (9). Publicity regarding the introduction of the machine was given by McCutcheon (10). It was developed by Baker Instrument Co., Orange, N.J., in cooperation with the U.S. Testing Co., Hoboken, N.J., the marketer of the device.

C. The Terg-O-Tometer

The Terg-O-Tometer [Fig. 10.3—also see (11)] consists of a thermostatically controlled constant temperature bath in which four separate 2-liter

FIG. 10.2. "One-Lung" Reciprocator (courtesy Colgate-Palmolive Co.).

FIG. 10.3. Terg-O-Tometer (courtesy U.S. Testing Co.).

beakers are mounted, each fitted with a motor-driven agitator of scaled proportions. The temperature of the bath can be regulated from 77 to 212°F. The agitator speed may be regulated in the range 40-200 cpm. The control panel is located in the lower right-hand portion of the instrument and a tachometer in the upper right location reads in cycles per minute. The machine is readily modified to increase its versatility. A modification (Fig. 10.4) designed for foam measurements has been made by Hartwig (12). The author has used an external cooling unit and inserted a large resistance into the bath so that the temperature can be varied from 40 to 212°F. This permits testing at European conditions of washing at the boil and also making comparisons with products especially designed for cold-water washing. A washing procedure using the Terg-O-Tometer is also described by Harris (6).

D. Role of Mechanical Input

The relationship of mechanical action to detergent composition is very important in evaluating a washing procedure. In the past, mechanical

FIG. 10.4. Modified "Terge" for foam evaluation (courtesy Shell Chemical Co.).

action was the sole means of removing soil. Soap and synthetic detergents were gradually used in increasing amounts in order to minimize the work that must be applied to the system. Today, other adjuvants are added to the "pot," such as bleaches, softeners, brighteners, foam stabilizers, solvents, enzymes, redeposition agents, etc., which contribute to the overall effect. Their action is also related to the work put into the system. Most of all, fabric damage helps to govern the amount of mechanical energy applied. Bacon and Smith (13) studied the relationship between detergent, mechanical force, and time, using the Launder-Ometer as the washing machine. They varied the mechanical force by changing the number of steel balls in the jar and by varying the speed of rotation. Harris (14) established a quantitative relationship between the Launder-Ometer and the Terg-O-Tometer. Tuzson and Short (15) studied the power input in a Terg-O-Tometer and found the net power input increases with increasing water bath volume and increasing cloth-to-water ratio. It would seem obvious that all testing must include power input as an important parameter to consider before making final conclusions. This applies to miniaturized equipment and also differences in full-size equipment.

E. The Deter-Meter

The Deter-Meter (16) is a precision laboratory instrument for detergent studies that features a close control of mechanical input. The energy input is the vertical component of a sinusoidal wave motion. The length of the intercept, frequency, and amplitude can be precisely controlled. A disk of soiled fabric, approximately 4 in. in diameter, is placed in a metal cylinder (ends fabricated with screening) that is designed to oscillate along its vertical axis. The oscillation may be adjusted. The cylinder, with the fabric, is immersed in a 4-liter beaker of detergent solution and operated at a constant temperature. The data permit a mathematical analysis of the work input to the fabric load-detergent system. At the present time this instrument is undergoing design modifications.

F. Dexter Twin Washer

The Dexter Twin Washer (17) is often used for making foam comparisons in the laboratory (18). It has the advantage of making a quick comparison between standard and experimental sample using equipment that approaches full size with respect to wash load, agitator size, and tub capacity.

III. HOME LAUNDRY EQUIPMENT

A. Necessity for Use of Home Equipment

The housewife has the option of sending her wash to the laundry, going to the coin-op, or doing it at home. The fact that in 1962 more than 50 million automatic washing machines were owned by American families is evidence that a large part of the laundry is still being done in the home. This may be due to the improvements in washing machines, cost, convenience, or because so much of the laundry of today has to be custom washed because of the many new fabrics and treatments being used. Miniaturized apparatus is essential for screening purposes, but it is necessary to use home equipment for making final decisions. Washers differ in load size, water capacity, power applied, agitator types, speed, rinsing procedures, draining methods, temperature control, water extraction, etc., and therefore it is natural to expect the same formulation to behave differently in the various types of machines tested. In addition to differences in washer design, other variables such as fabric colors, composition, and treatment will influence the results. Finally, in addition to soil removal, other effects such as brightening, conditioning, "odorizing," repellency, wrinkling, etc., are being tested in order to make a complete product performance report. Hence the washer manufacturer must use representative products for evaluating his machines; the detergent manufacturer must use representative washers to evaluate his products. All of this does not negate the use of scaled-down equipment because it is still necessary for quick screening with small samples, but it must be placed in proper perspective.

B. Description of Home-Type Laundry Machine

Figure 10.5 is a view of one of the evaluating areas of the Colgate-Palmolive Research Center. Here are found representative machines of the leading foreign and domestic manufacturers. They include wringer and spinner washers, semiautomatic and automatic washers, and hand-operated equipment. Some of the machines are the front-loading, tumbling type, but the majority are top loaders. Most of the top loaders still use

FIG. 10.5. A research center evaluation area (courtesy Colgate-Palmolive Co.).

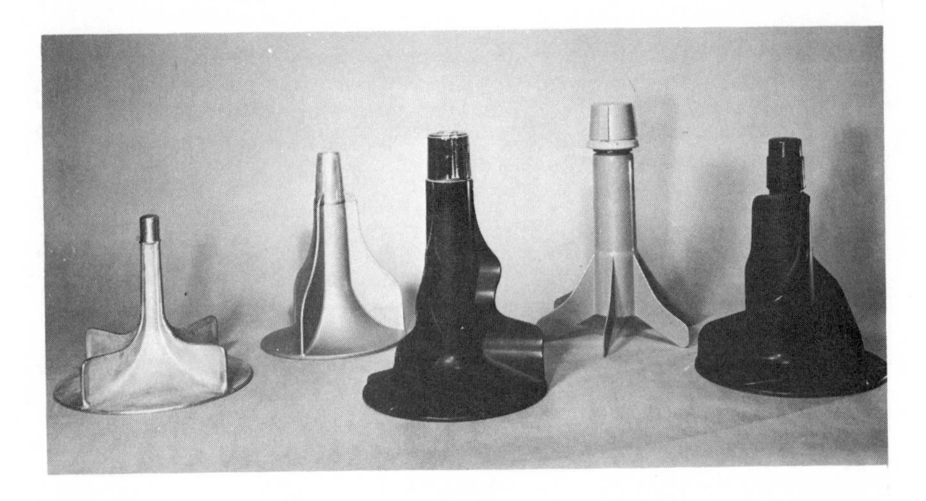

FIG. 10.6. Agitator types found in home laundry equipment.

modifications of the agitator designed by Howard Snyder in 1922 (1). Several of the agitators are shown in Fig. 10.6. They create currents of detergent solution within the tub which in turn flex the cloth, which helps to remove the soil and keep it in suspension.

The machines are filled in one of three ways: manually, timed, or metered. The hot and cold water used in the fill or rinse is the tap temperature. Mixtures of these are made by proportioning valves or heat-regulating valves. Some of the European models have heating elements in the tub. In the wringer washer, filling, timing the wash cycle, and wringing are usually manual procedures. The spinner washer provides a spinning tub in place of the wringer. Automatic washers usually fill, wash, rinse, and stop without attention. The main difference among them (other than the plane of rotation or agitator design) is the selection of washing conditions and options like dispensers of various kinds. A fully programmed machine is operated by pushing one button for a given type of wash, and the time, temperature, and speed are automatically set for that type of wash. Other automatics allow you more freedom in selecting time, temperature, or speed by pushing more buttons for a particular type of wash load. The rinsing procedure varies among washers. The removal of the wash water, replacing it with fresh water, agitating, and then draining is called a deep rinse. Introducing water during spinning is a spray rinse. Introducing water during washing is called an overflow rinse. The operation and performance of washing machines is more fully discussed in (1, 19).

IV. COMMERCIAL LAUNDRY MACHINERY

Commercial laundries generally use a rotating type of washer called a wash wheel. It is similar in some respects to the front loader used in the home. Vaughn et al. (20) made a miniature wash wheel for studying foam performance. Harris (6) suggests a procedure using a 25-lb pony wash wheel. Commercial machines are generally classified as open-cylinder, Pullman, or Y-pocket, these terms relating to the type of horizontal partitioning employed. Figure 10.7 is a picture of a recently developed Troy 400-lb washer-extractor-conditioner (21). This machine replaces the

FIG. 10.7. Troy side-loading 400-lb washer-extractor-conditioner (courtesy Ametek-Laundry and Drycleaning Equipment Div., East Moline, Illinois).

conventional units: washer, extractor, conditioning tumbler. It has fully automated controls and time injections of supplies. In the wash wheel, the effect produced is directly proportional to the amount of lift the load receives, and the speed of operation is dependent on the height of the ribs and the type of partitioning (22).

Commercial laundries use a lot of water. It has been estimated that the average laundry used 5 1/2 gal of water to process 1 lb of work, 75-80% of this being rinse water. "The washing process is a series of dilution baths, wherein the rate of dilution of the soil, which has been removed from the clothes by chemical and physical means, is dependent on the ratio of the water absorbed by the clothes to the total water in the wash wheel. This, in turn, is governed by the amount of free solution dumped after each operation" (23). Freed from the chemist's worries over wetting, removing, and suspending dirt, absorption and desorption, a mathematician sees a washing process as a dilution progression with removed dirt and detergents as the material to be diluted (24). The American Institute of Laundering (A.I.L.) is the leading research organization in the laundry industry (24).

V. INSTRUMENTATION FOR MEASURING LAUNDERING RESULTS

So far the discussion has dealt with equipment used for testing detergency and related processes. The remainder of the chapter is devoted to equipment commonly used for measuring the changes in the appearance of the substrate after it has been washed. The eye enables us to distinguish relative size and shape of objects, the texture, uniformity, color, transparency, gloss, etc. The instruments discussed will be used most often in measuring the changes in color.

A. Color Measurements

1. Introduction

Color measurements are made frequently in the analytical laboratory. The use of a colorimetric procedure for determining the concentration of a colored ion is as common as weighing, titrating, gas analysis, and other common analytical procedures. In all cases experimental design, treatment of data, proper sample preparation, etc., are involved in the experiment. The average analyst, however, makes transmittance measurements more often than reflectance measurements. He is more knowledgeable of the laws governing transmittance (Lambert-Beer) than he is of the Kubelka-Munk theory, which deals with the properties of light scattering of colorants. His results are usually expressed in terms of concentration (physical terms such as grams per liter) or absorbance rather than "color terms." Color is sometimes expressed in simple terms like percent reflectance but more often its physical measurements are transposed into psychophysical terms and the data obtained represent coordinates which plot the location of the color and show its position in color space.

"Instruments are aids to the eye. Their function is to express color differences or the color of materials in terms of what the eye perceives" (25). Figure 10.8 is a simplified drawing showing some of the instruments that contribute to the measurement of color. The eye is the reference instrument since visual examination is still the most common means of making comparisons and most data are related to its perception. Excellent descriptions of the eye and how it functions can be found in (26, 27, 28).

In order for the eye or an instrument to measure color it must have three essential parts, namely, a light source, a sample to be evaluated,

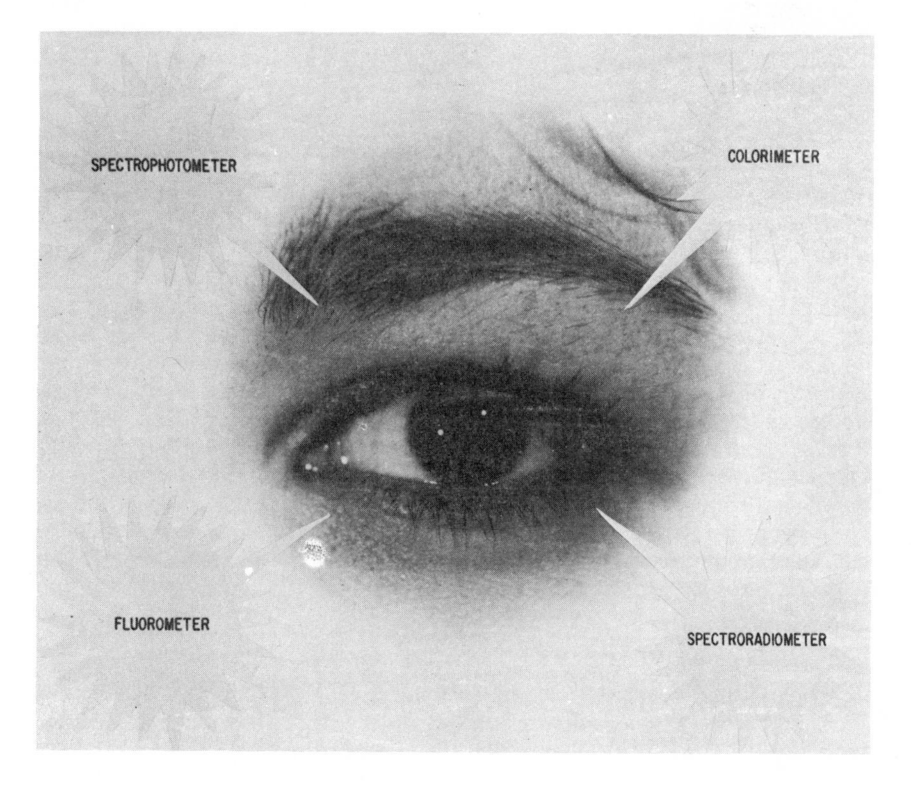

FIG. 10.8. The eye and related instruments.

and a detector. The eye is extremely complicated but it is a well-designed instrument. At the risk of being facetious, one might say it has an automatic windshield wiper, constant lubrication, self-adjusting lens, is thermostatically controlled, adapted for day and night vision, and has other features too numerous to mention. For a given light source it does an excellent job of comparing surface appearances. But, alas, it suffers from fatigue and loss of memory, and it changes with age. Having an instrumental substitute to come to its assistance and share the burden is indicative of scientific advance. There are those who predict that an instrument may someday outperform the human eye, but a conservative thinker is not inclined to agree. As a consequence, during the last few years many color-measuring instruments and their accessories have made their appearance. To review all of these that are used in the evaluation of detergent products is nigh impossible, but to illustrate a few is the goal. The desire is not for endorsement or omission but for enlightenment. At the end of the chapter a partial list of manufacturers of color-measuring instruments, gleaned from magazine ads, brochures, textbooks, "yellow pages," etc. will be given. In the near future, Subcommittee 24 of the Inter-Society Color Council will publish a catalog of color-measuring instruments.

2. Spectrophotometers

The spectrophotometer is a very useful instrument for the research or control laboratory. It is designed to measure either the spectral transmittance or reflectance of objects. The radiant flux leaving the object is compared with the incident light at each wavelength. Sometimes the analyst is satisfied with a simple transmittance statistic; sometimes he can solve the problem with a spectral curve of the sample; at other times he may wish to integrate the area under the curve with a tristimulus integrator. Two of the most familiar spectrophotometers to the instrumentalist are the Beckman DU® and the General Electric Recording. The former is an all-purpose, prismatic-type instrument used chiefly for measurement of absorption of radiant energy in the visible and ultraviolet portions of the spectrum. The latter represents a recording type and is often regarded as the referee instrument for color measurement. An excellent guide for selecting a spectrophotometer is found in the literature (26).

One of the first commercially available spectrophotometers was the Beckman Model DU. It was known as the standard of the world in spectrophotometry for nearly a quarter of a century. The vast number of references to it in the literature would appear to substantiate the fact. Its chief use is probably measuring the absorbance or transmittance of liquid, solid, or gas samples in the uv or visible range. The Beckman Model DU-2 is an improved version of the original DU (Fig. 10.9) (29, 30, 31). This is a single-beam, null balancing instrument. Alternate light

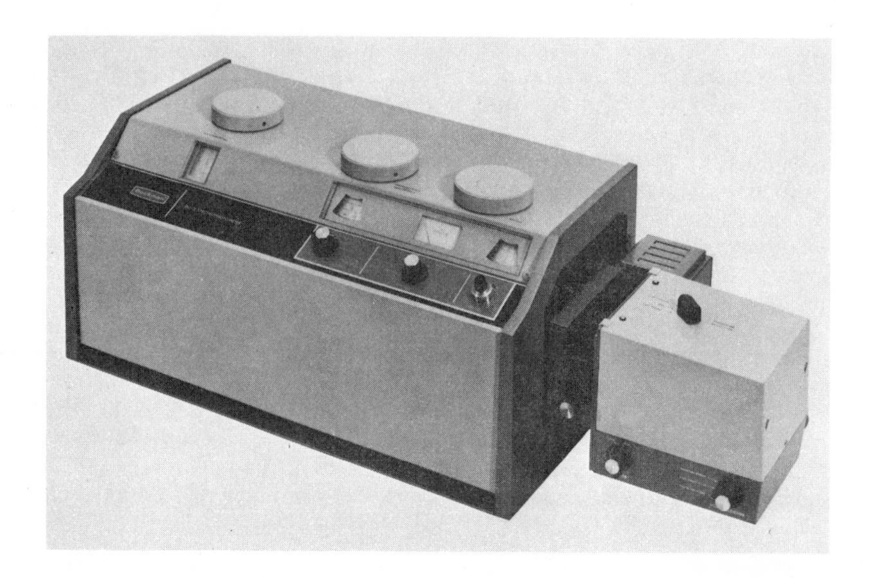

FIG. 10.9. Beckman DU-2 Spectrophotometer (courtesy Beckman Instruments, Inc.).

sources (tungsten, hydrogen, deuterium) provide an expanded wavelength of 190-1000 nm. A quartz or fused silica prism is used as the diffracting medium. Blue- and red-sensitive photocells serve as detectors. Additional sensitivity is achieved in the ac-line operated instrument by using a multiplier phototube. The amount of light transmitted or absorbed by a sample at a discrete wavelength, compared to a known reference, provides a quantitative measurement of the sample. The interrelationship of concentration, absorptivity, and path length is defined by Bouguer-Beer's law. Accessories are available that can increase its range of usefulness. The Unicam SP-500, the Hilger UVspek, and the Russian SF-4 are similar instruments (32).

The General Electric Recording Spectrophotometer (GERS) was invented by Professor Hardy of M.I.T. (33). It was first manufactured commercially in 1935. This instrument stimulated interest in color measurements because of its recording features, the advantage of avoiding metamers, and its reliability. In 1957 a round robin study of color measurements on the GERS was carried out by 15 participating laboratories. "The results clearly show, as in the repeatability and reproducibility studies, that the G.E. Spectrophotometer, when suitably calibrated and operated, can still be considered the referee instrument for accurate color measurements" (34).

In principle, the GERS is a double-beam spectrophotometer with a double monochromator. Each part of the visible spectrum is reflected from a sample and a standard, and means are provided for recording the reflectance of the sample along with that of the standard on the same graph sheet (Fig. 10.10). The double monochromator produces a spectral band of width determined by the size of the slits, and one which is virtually free of unwanted energy from other spectral regions. As the spectrum is traversed, the mechanical widths of the three slits are continuously altered so as to transmit a constant bandwith of 10 nm. After emerging from the final slit, the light is split into two beams which are polarized at right angles to each other, and by means of the rotating polarizing filter and separate double lenses it is conducted into a spherical enclosure wherein the now separated light beams are reflected from the sample and standard. The inside of the sphere is white and hence the reflected light is diffused and a small portion is conducted to the phototube by means of a plastic rod. The reflecting conditions represent a 6° diffuse geometry (31, 35). Supplementary devices are manufactured by Davidson and Hemmendinger (36).

The preceding spectrophotometers isolated narrow bands of radiant energy by using slits and prisms. Narrow bands can also be isolated by filters and point-by-point measurements can be made. These instruments

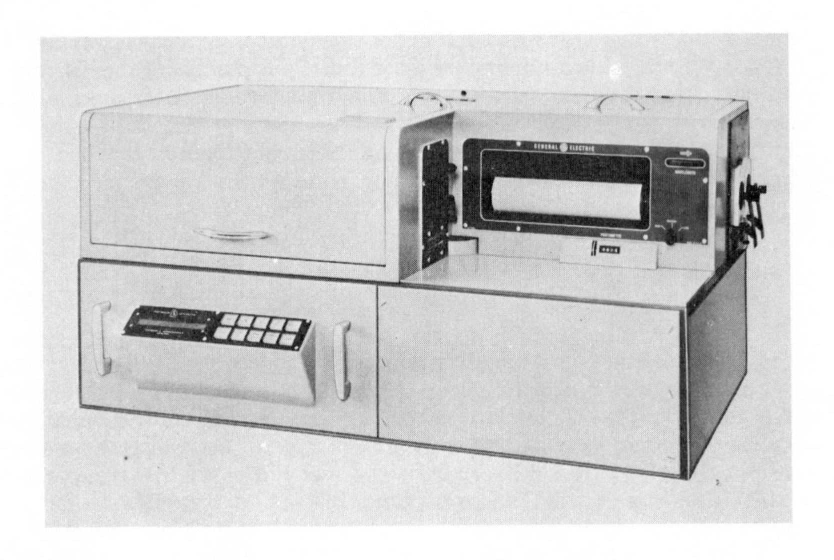

FIG. 10.10. G. E. Recording Spectrophotometer (courtesy General Electric Co.).

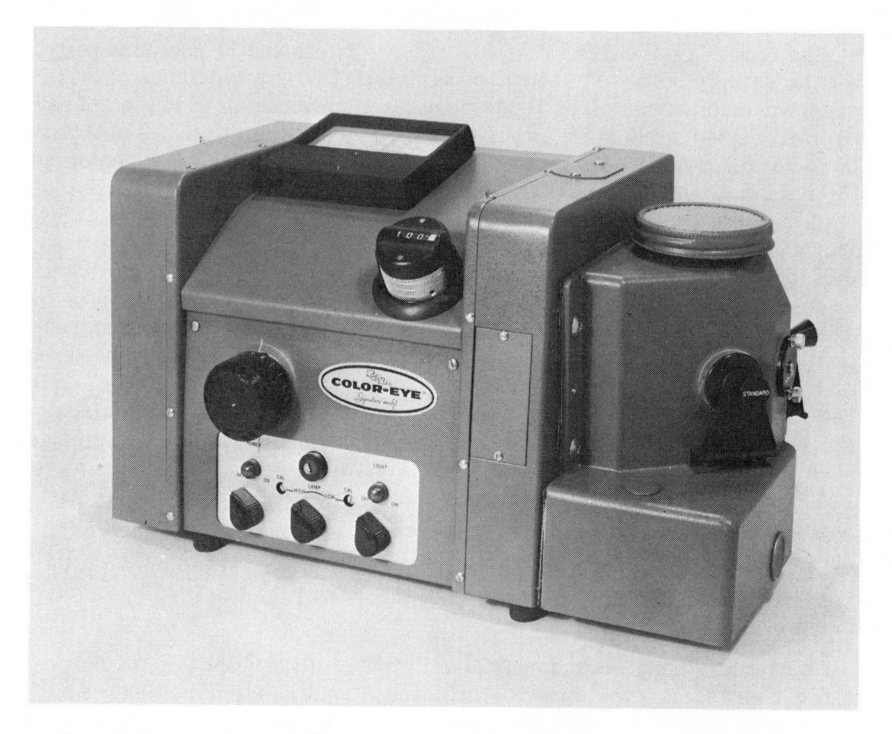

FIG. 10.11. Color-Eye (courtesy Instrument Development Laboratories).

are sometimes referred to as abridged spectrophotometers. Instrument Development Laboratories has developed a combination colorimeter and abridged spectrophotometer which is used mainly for measuring reflectance. It is called the Color-Eye (Fig. 10.11). This instrument gives the option of making measurements with three tristimulus filters or 16 narrow band interference filters which cover the visible spectrum at 20-nm intervals. The sample is illuminated before light enters the optical system. This is the reverse of the GERS and permits the Color-Eye to be used in the measurement of fluorescent samples (37, 38). The basic instrument can be modified to accommodate a large sample area viewing and external illuminator attachment. This is designed especially for highly textured surfaces and products which utilize fluorescent colorants or optical brighteners (39). The Color-Eye instrument is designed to supply available computers with the essential and accurate input information.

The analysis of fluorescent materials requires a different kind of spectrophotometer. "A true fluorescent spectrometer is an instrument in which two monochromators are used, one to supply the excitation radiation and the other to analyze the fluorescence emission" (32). The Hitachi Perkin-Elmer Fluorescence Spectrophotometer (MPF-2A) is an example of this type of instrument (Fig. 10.12) (40, 41). It can be equipped with a solid sample holder for scanning samples of fabric which contain fluorescent brighteners. The light source is a xenon lamp which is operated by a stabilized solid-state power supply. During operation, a small portion of radiation from the light source is reflected to the reference photomultiplier from a quartz plate. The resulting signal is fed to the recorder for ratio recording. The balance of the beam is passed through the excitation monochromator system to be dispersed by the grating. The dispersed monochromatic beam excites the sample. The fluorescence is analyzed by the emission monochromator and the corresponding signal recorded. The monochromators operate at 220-650 nm. This instrument features continuously adjustable bilateral entrance and exit slits on both monochromators, cutoff filters to eliminate second-order radiations, a beam splitter, and ratio recording. Other fluorescent spectrometers are the Aminco-Bowman, Baird-Atomic, Turner, Farrand, and Zeiss. Some instruments such as the Beckman DU and DK, the Cary, the Unicam SP, and the Zeiss PMQ 11 can be adapted to measure fluorescence (32). Geigy Corp. has measured spectral radiance by reversing the optics of the Beckman DK-2 and the GERS, as well as by using a large-sphere Color-Eye (42).

3. Colorimeters

Color-measuring instruments could well be divided into two broad categories, namely the spectrophotometers and the colorimeters. As previously stated, the spectrophotometer is the spectrum-scanning instrument and its response may be integrated with respect to the tristimulus values of the human eye and the light source that is being used. The colorimeter is usually a photoelectric tristimulus instrument which tries to match the source-filter-photodetector combination with the tristimulus functions of the standard observer. "Spectrophotometers give complete

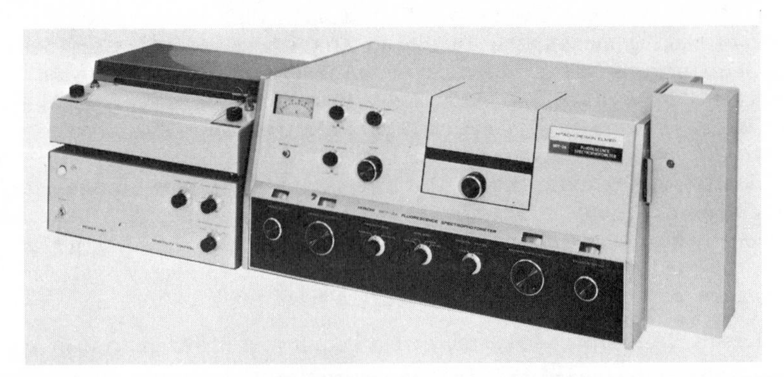

FIG. 10.12. Hitachi Perkin-Elmer Fluorescence Spectrophotometer MPF-2A (courtesy Perkin-Elmer Corp.).

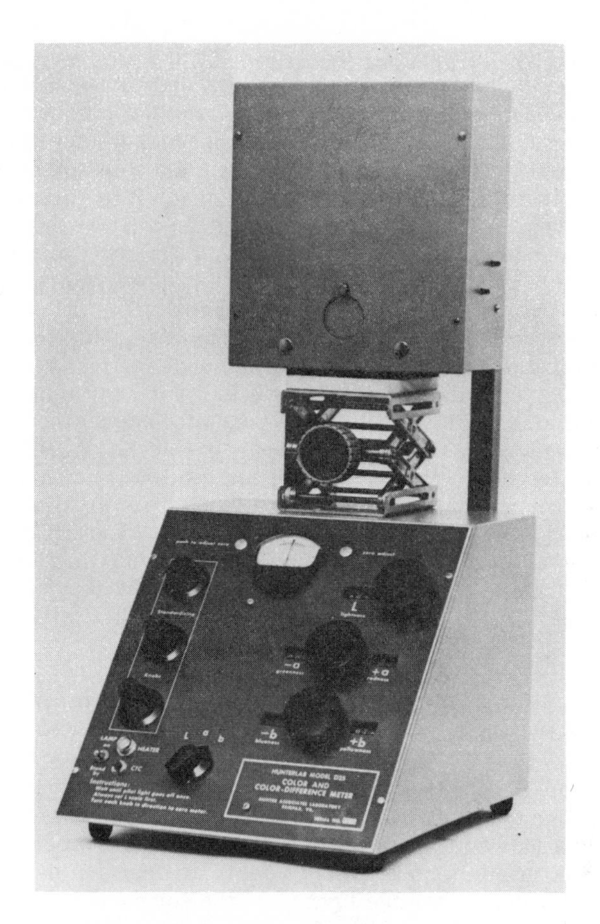

FIG. 10.13. Hunterlab Color Difference Meter D-25 (courtesy Hunter Associates Laboratory, Inc.)

information about the color of a specimen and give it in absolute terms. They are primarily research instruments, and are most valuable in situations where the detailed information they produce is essential, as, for example, color matching. Colorimeters, on the other hand, yield limited information, and yield it only in relative terms. Their most appropriate use is to compare the color of two similar samples. They are eminently suited for production control operations" (43).

Richard S. Hunter developed the method of photoelectric tristimulus colorimetry and designed the Multipurpose Reflectometer and Color Difference Meter which use this method of colorimetry (44, 45, 46). The Hunterlab Model D-25 (Fig. 10.13) is a color difference meter with three photodetectors, each with a separate filter (split filter for X response) and each receiving some of the light from the specimen. Signals from the photodetectors are measured by analog circuits that give rectangular coordinates for surface colors in close correspondence to their position in uniform color space. The instrument is divided into two parts, an exposure unit and a measurement unit. This arrangement gives versatility to sampling procedures. A 6 V instrument lamp is used as the light source, and the light-mixing device between specimen and detectors is a rectangular light pipe with prismatic diffusing face. The phototubes are in an aluminum block and are maintained at constant temperature. Lamplight and photodetector corrections are built into the various filters in order to match tristimulus functions. The instrument is intended to measure the color of surfaces as they appear to the normal observer under daylight, 45°, 0° observing conditions. With the "hitching post" technique, the high precision of the instrument makes it useful for many problems. Accessories such as the circumferential illumination unit for decreasing differences due to sample orientation, ultraviolet absorbing filter for optical brightener studies, or a built-in computer for color differences increase its usefulness (47). Some other manufacturers of colorimeters are Gardner Laboratory, Manufacturing Engineering and Equipment Co., and Carl Zeiss.

4. Fluorometers; Nephelometers

Some instruments compare fabric fluorescence by using filters to select the exciting wavelength. An example of a filter fluorimeter (also called fluorometer) is shown in Fig. 10.14. It is manufactured by the Engineering Equipment Co., licensed under patents of the American Cyanamid Co., Inc. It is called the G. M. (Galvanek-Morrison) Fluorimeter (Model EE-100B) and is a modification of the original instrument which was used for the analysis of uranium minerals. Two 4 W uv bulbs act as a light source. A cutoff filter removes visible light from the source before it excites the sample and another filter screens the reflected uv light before it strikes the photomultiplier tube. These instruments are generally used for quality control because they measure "gross fluorescence." When brighteners have similar excitation and fluorescence patterns, their buildup on fabric and their destruction can be relatively determined by this type of measurement (48, 49). Jarell-Ash, Photovolt, Beckman, and Turner make similar instruments. The amount of soil removed from clothing by washing is usually determined by observing the color change visually or by measuring the reflected light with a colorimeter. However, sometimes particulate removal is measured by the scattered light caused by the suspended matter in the wash water. Nephelometry is the science that measures light reflected by finely divided particles in suspension and it can be a valuable tool for determining material balance in some soil removal operations. The importance of this type of analysis is the fact that the amount of material in suspension can be determined without filtering, washing, drying,

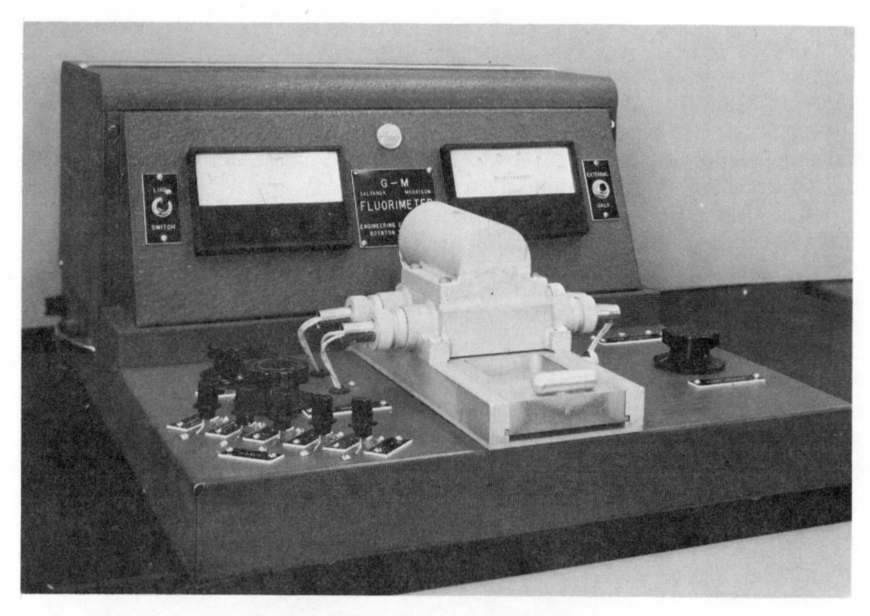

FIG. 10.14. G. M. Fluorimeter (courtesy Engineering Equipment Co.).

or weighing. On more concentrated suspensions, turbidity measurements are usually made. Figure 10.15 is a picture of the Coleman Nepho-Colorimeter. The sample is placed in a cuvette which in turn is placed in a cuvette wall. A collimated beam of light from a tungsten source is directed into the sample. Scattered light is measured by two photocells mounted at opposite sides of the cuvette well with their faces parallel to the exciting beam. Coleman has developed a series of standard suspensions (inorganic salt in a highly viscous styrene polymer) which enables one to assign a Nephelos number to any sample suspension. Nephelos measurements can be related easily to sample concentration (50).

B. Light Sources

In the recent history of man, many kinds of artificial light sources have become available: the candle, kerosene lamp, gas mantle, filament lamp, arc light, fluorescent lamp, etc. Add to this the variations of daylight and the knowledge that color is dependent on the nature of the light source, and one soon realizes the need for some type of standardization. Surely, a painter who daubs a portrait under the attic skylight deserves a similar gallery lighting for its display. In more mundane situations, the evaluator should be able to correlate his laboratory comparisons with the outdoor clothes line, the dimly lit cellar, or the newly furnished laundry room with its fluorescent fixtures.

"Lighting standardization is also complex because we are dealing simultaneously with objective physical data on one hand, and with their application on the other, which involves human perception and leads into subjective areas of color technology, i.e., physiological and psychophysical aspects, about which we really know very little" (51). In the United States the expert color matcher usually preferred north sky light. When it came to using an artificial source, the majority of color matchers selected filtered tungsten light whose color temperature was above 7000°K. Problem Committee 13 of the Inter-Society Color Council determined that the minimum value of preferred north sky light was approximately 7500°K. Most present-day instruments use light source C at 6770°K for colorimetry. This makes it difficult to correlate instrumental and visual comparisons. Unfortunately, most of the standardization work covered only the visible spectrum (400-700 nm), and did not include the near ultraviolet so essential in exciting optical brighteners. Figure 10.16 is a picture of the

FIG. 10.15. Nepho-Colorimeter (courtesy of Coleman Instruments, division of the Perkin-Elmer Corp.).

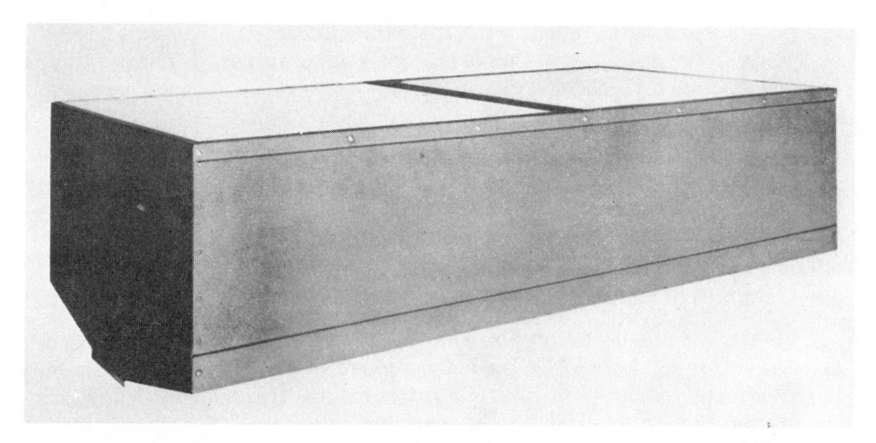

FIG. 10.16. Macbeth Examolite (courtesy Macbeth Corp.).

Macbeth Examolite with the addition of ultraviolet. This light gives three lighting options: incandescent, north sky daylight, and ultraviolet. Viewing first under north sky and then incandescent light will exaggerate the visual differences in metameric pairs. Viewing with north sky alone and then with north sky plus ultraviolet will show the effects of optical brightener (52).

Over the years there has been a number of color systems used by those working in the field of color. In 1931 the Commission Internationale de l'Eclairage (C.I.E.) adopted the visibility curve of the normal eye, standard light sources, tristimulus values of the standard observer, and the chromaticity diagram. The three standard light sources were A, B, and C. Source A is representative of a gas-filled incandescent lamp operating at a color temperature of 2854°K. Source B is representative of noon sunlight and operates at a temperature of 4870°K. Source C is representative of average daylight and operates at a color temperature of about 6770°K.

The new spectroradiometric daylight data being collected worldwide have included the ultraviolet component (53-56). These studies of the spectral power distribution of typical daylight brought about certain recommendations that have significant effects on both the practice of color measurement and the interpretation of the results. The C.I.E. Committee E-1.31 in 1966 adopted recommendations for a series of daylight illuminants designated by the letter D, with a subscript denoting their correlated color temperature. The recommendation specifies illuminant D_{6500} representing daylight with that correlated color temperature and notes that for general use in colorimetry, illuminant A and D_{6500} should suffice. Most of the body of color measurement results are based on illuminant C. Since its continued use will be discouraged when the new recommendations are put into practice, much of the instrumental data on file will have to be recalculated. At present no artificial source is recommended to represent standard illuminant D_{6500} (57).

The British have taken the lead by adopting for general use a special standardized fluorescent light source which is inexpensive, convenient, and easily available (British Standard 950, Part I, 1967, "Artificial Daylight for Colour Matching and Colour Appraisal"). The tubes are newly developed with a substantially thicker coating of five phosphors which balance the light and reduce the amount of mercury-emission lines that are transmitted through the radiating surface (58).

In closing, it seems fitting to mention the work reported by the Rhode Island Section of the AATCC as being typical of the efforts of those engaged in color measurements to improve instrumental techniques. They tackled the problem of "Instrumental Evaluation of the Color of Fluorescent Textiles in Terms of Average Observer Response." Their conclusion was that the use of a properly designed instrument and techniques similar to those used in nonfluorescent applications will provide a satisfactory means for the control of the application of fluorescent colorants. It is to be hoped that data of this type may eventually lead to a standardization of illuminating and viewing conditions for instrumental and visual comparisons.

VI. THE DRUM-TYPE EUROPEAN HOME WASHING MACHINE

(Editor's Note: It is assumed that the reader of this volume is familiar with American domestic laundry appliances or can easily familiarize himself with such equipment through references cited in this chapter. European domestic laundry processes and appliances differ substantially from American ones. To acquaint the reader with European domestic laundry appliances, we have provided the following section through the courtesy of Dr. Ing. O. Viertel, Waeschereiforschung Krefeld, West Germany.)

A. Operating Mode

The European drum-type washing machine (Fig. 10.17) is characterized by a horizontally mounted, cylindrical drum which alternately rotates to the left and right. The laundry put into it is thereby rolled around, tumbled, loosened, and penetrated by the wash and rinse waters. The revolving drum is surrounded by a stationary tank which serves as water tank and which accommodates all fixtures required for operation.

A number of conditions must be met to attain the maximum of laundry performance for the washing and rinsing processes.

A maximum of laundry performance is attained when the demand for good laundering effect as well as the demand for gentle laundry treatment is met.

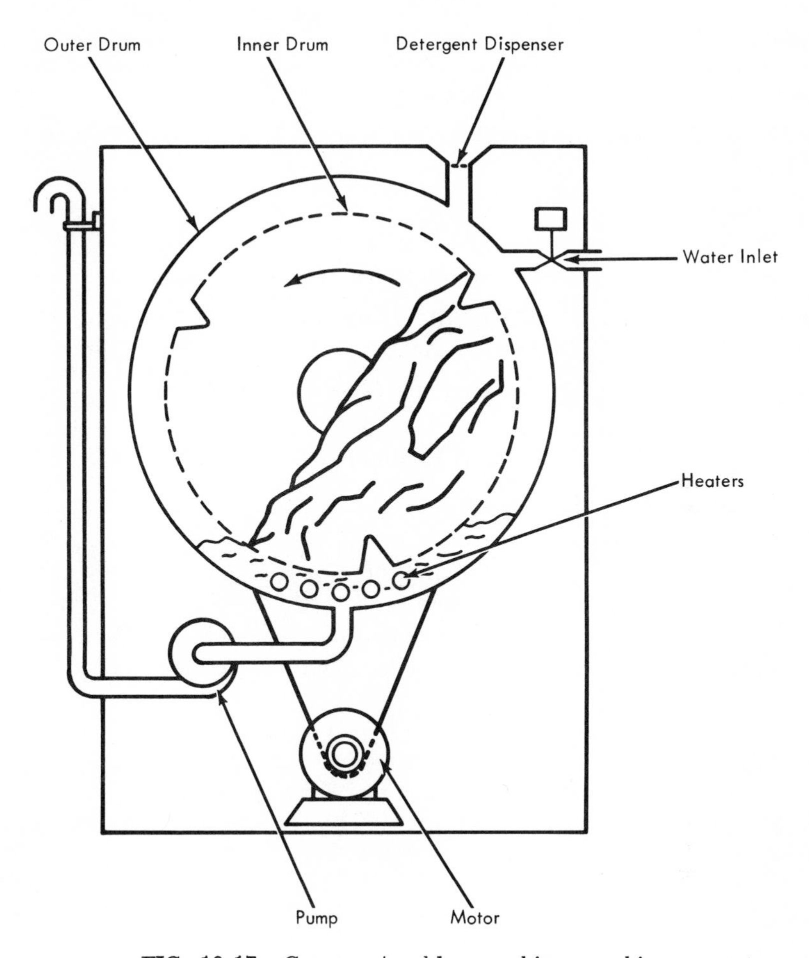

FIG. 10.17. German tumbler washing machine.

The following factors have an influence on the mechanics:

centrifugal acceleration

$$\underline{B} = \frac{dn^2}{1800000} (\underline{G})$$

(relative to the drum diameter), where $\underline{d}$ is the drum diameter (millimeters), $\underline{n}$ is the drum speed (rev/minute), and $\underline{G}$ is the acceleration due to gravity;

filling ratio

$$\underline{z} = \underline{M}/\underline{V} \text{ (kg/l)} \quad (2)$$

where $\underline{M}$ is the laundry weight (kilograms) and $\underline{V}$ is the drum volume (liters);

load ratio

$$\underline{f} = \underline{M}/\underline{F} \text{ (kg/l)} \quad (3)$$

where $\underline{M}$ is the laundry weight (kilograms) and $\underline{F}$ is the amount of water (liters);

running time of the drum during one laundering process:

$$[\underline{Lz} \text{ (min)}] \quad (4)$$

reversing rhythm of the drum

$$\frac{\text{Revolving Time}}{\text{Idle Time}} = \underline{R}_r \left(\frac{\text{sec}}{\text{sec}}\right) \quad (5)$$

These factors must be varied, depending on type of laundry and degree of soiling, to select the most favorable conditions as shown in Table 10.1.

TABLE 10.1

Mechanical Factors in the Washing Process

	B, g	z, kg/liter	f, kg/liter Wash	f, kg/liter Rinse	Lz, min	R_r, sec/sec
Normally soiled boiling laundry, such as cottons	0.7	1/12	1/5	1/7	40	12/3
Normally soiled boiling laundry, such as prints	0.7	1/14	1/5	1/7	30	12/3
Normally soiled delicate laundry, such as woolens	0.2	1/40	1/30	1/30	5	3/12

The laundering process in the drum-type washing machine is influenced, besides the above-mentioned factors, by the temperature and the amount of detergent, both of which must be determined by load type and degree of soiling. In practice a number of standard programs are provided.

One distinguishes between boiling laundry (95°C), hot laundry (60°C), and delicate laundry (30°C) and uses standard programs which take into consideration the degree of soiling and the degree of the laundry's ability to withstand mechanical treatment. Three standard programs are shown diagrammatically as examples in Figs. 10.18, 10.19, and 10.20. They show, plotted over the laundering and rinsing operations, the detergent quantity per kilogram dry laundry; the water ratio; the temperature; and the running time of the drum, that is, active running time.

So-called "fully automatic" machines also include a spinning operation (centrifuging) in the laundering and rinsing operations. It is employed for the acceleration of the rinsing operation between the rinsing phases, and it serves to remove the water from the laundry charge after the conclusion of the laundering and rinsing operations.

B. Construction

1. Drum Construction

The drums of the machines available on the European market differ little from one another. A certain influence is exerted by ribs and perforations. It is a basic rule that the number and the height of the ribs must be greater, the smaller the chosen centrifugal acceleration. Since a drum is always designed so that it produces maximum laundry motion, making possible short laundering periods, there results at optimum centrifugal acceleration (0.7-0.8 G) 3 ribs of a rib height approximately 6-8% of the drum diameter.

The drum perforations influence the way in which the laundry is exposed to water and the time required for water drainage. However, an open hole area of 4-6% of the drum diameter is sufficient. The optimum hole diameter amounts to approximately 6-7 mm; the hole edges are drawn outward slightly in funnel fashion, to avoid damage to the laundry.

The drum diameter plays an important role as regards the laundry motion, particularly in view of the size and bulkiness of individual laundry pieces. Because of space limitations, there is a tendency to select smaller and smaller drum diameters, but this increases the danger of reducing the laundering results considerably for large, bulky pieces of laundry. This results especially in such pieces of laundry being washed unevenly. The result for smaller and flexible pieces of laundry can, however, be quite satisfactory, provided the machine is not overloaded. In any event, drum diameters of 400-450 mm should be the minimum.

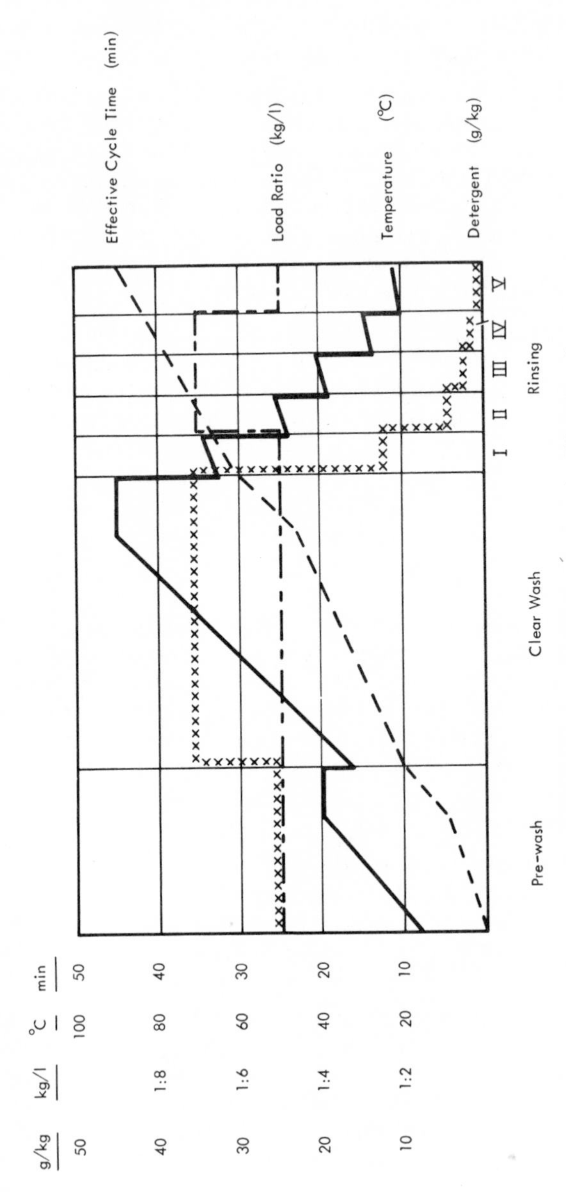

FIG. 10.18. Wash program for boiling wash.

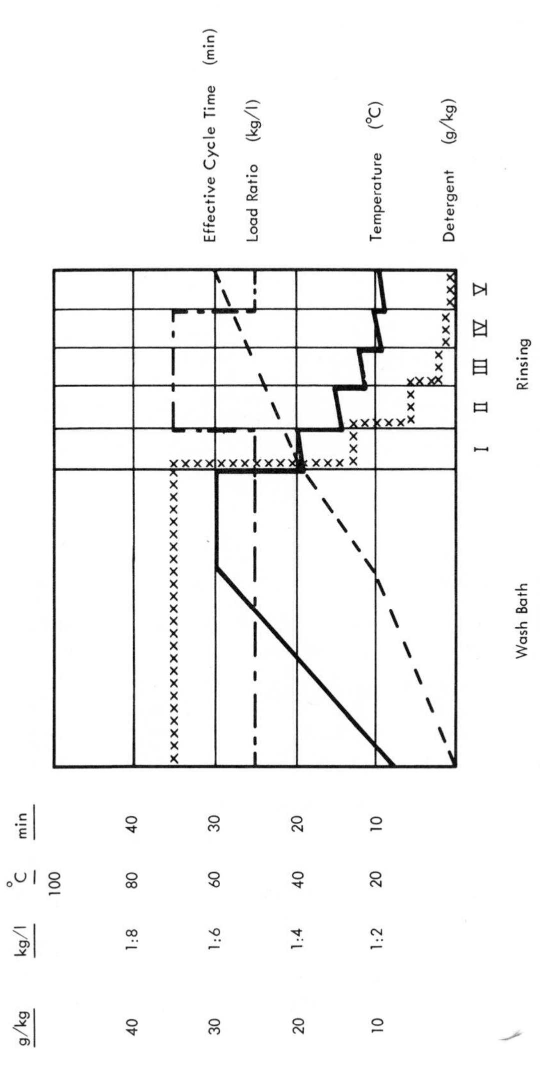

FIG. 10.19. Wash program for hot wash.

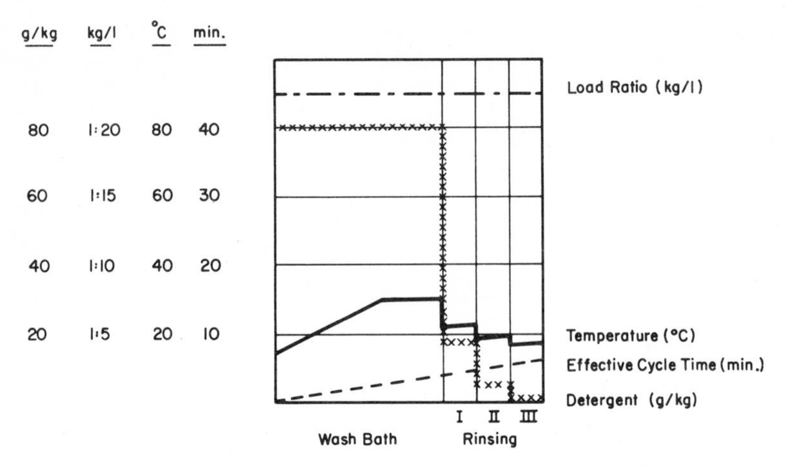

FIG. 10.20. Wash program for delicate wash.

For the same reasons, the drum depth should not fall below a minimum of 300 mm.

2. Type of Loading

Depending on the type of loading, one distinguishes between front-loading and side-loading drum-type washing machines.

Front-loading drums need no closure attached to the drum. The external door of the washer, arched inward in the manner of a pan, prevents the laundry from falling out. Since the external door is contacted by the wash water, it must be completely sealed. Opening of the door, which might perhaps be required for correcting the laundry charge, is possible only when the water level is low.

A disadvantage of small machines is the relatively low and inconvenient location of the loading door. It is for this reason that the side-loading arrangement frequently is employed. The machine opens at its upper side. The drum periphery has a loading hole which is closed by a removable lid, by flip-open doors, or by sliding doors which are reached through a cover of the outer tank. Neither the cover of the outer tank nor the closures of the drum need be watertight, except when the drum is used for spinning as well as laundering. In that case, the outer cover must be tight.

3. Drum Drive

An electric motor is used for driving the drum at the optimum laundering speed via a speed reducer. To avoid entanglement, and with that, poor laundering, a reversing apparatus is used which reverses the drive motor according to a certain reversal rhythm. An expansion of this method consists in modifying the idle and running times of the motor, thereby shortening the effective running time of the motor while the overall laundering time remains the same. This is important especially when the heatup times become long, due to poor heat input values, so that unnecessary stresses are not imposed on the laundry. This also is important when sensitive laundry must stay under the effect of the chemicals for longer periods of time than the time for the mechanical treatment.

Electronic speed regulation of the drive motor represents a current development. This system permits almost infinitely variable speed regulation; however, price is still a hurdle to the incorporation of this regulating device into production washers. A price reduction can be expected.

C. Supplying Operating Materials and Draining Waste Water

Operating materials are defined as water, detergents, and energy.

The water supply comes from the cold-water main and is adequate when approximately 15 liters/min water supply to the machine is available. A minimum water pressure of approximately 1.5 atm absolute is required for the satisfactory function of the electrically actuated inlet valves. This pressure is also required to flush the detergents into the laundry tank from the detergent reservoir at the proper moment in automatically controlled machines. The amount of detergent depends on type of laundry, degree of laundry soiling, and properties of the water (hardness, etc.) and should amount to approximately 30-60 g/kg dry laundry.

The water consumption depends on the above factors and may amount to 25-120 liters/kg dry laundry. The high figures occur with delicate laundry (such as woolens), where for reasons of low mechanical action, high water ratios must be applied.

The energy supply for the drive mechanism and for heating is taken exclusively from the public electric power network which furnishes single-phase ac (50 cycles, 220 V). In some new buildings, three-phase ac (50 cycles, 220/380 V) is supplied. The power consumption is limited and amounts to only 3 kW single-phase, often less. This means long heatup times (approximately 30-50 min for boiling laundry). Where three-phase ac is available, input values of 6 kW and more can be used, and the heatup times are correspondingly shorter.

The energy consumption depends on the type of laundry and the degree of laundry soiling and varies 0.5-1.5 kWH/kg dry laundry.

The removal of the waste water is usually accomplished by centrifugal pumps whose task it is to pump the waste water of the machine into the drainage sink located approximately 50-70 cm above the bottom of the laundry tank. A curved hose section hung over the edge of the sink serves as a drain. A drain valve is unnecessary.

D. Spinning in the Drum-Type Washer

Spinning the laundry for the purpose of extracting water requires considerable structural expenditures, depending on the speed employed. Up to a centrifugal acceleration of approximately 30 G, a rigid mounting of the drum in the machine housing is still used. The occurring imbalances thus cause a vibration of the entire machine. Eccentric rollers prevent the machine from moving from its location. When the accelerations are higher, the machine must either be anchored in its location or the laundering drum and the tank must be suspended in the machine frame from springs so as to swing free. The natural frequency is traveled through as quickly as possible during the acceleration to spinning speed. Vibration attenuators limit the vibration amplitudes. At extreme imbalances, limit switches are actuated which stop the drive motor. The distribution of the laundry in the drum can be influenced but little. A charge consisting of similar pieces of laundry and containing no large pieces such as bed sheets, etc., will distribute more evenly than one consisting of large pieces. A slow increase in centrifugal acceleration in the range from 0.8-1.5 G also favors laundry distribution within the rotating drum.

The extraction action of spinning with horizontal drums naturally is not as good as spinning with vertical drum mounting. At centrifugal accelerations of 20-200 G and spinning times of about 5 min, residual moisture contents of 140-70% (cotton) are obtained. Separate centrifuges achieve residual moisture contents of 50-40% at 600-1000 G (Fig. 10.21).

This is also one of the reasons why drum-type washing machines are being combined with a spinner accommodated in the same machine housing. The extraction results achieved thereby represent a noticeable shortening of the drying time and, when drying is done in tumble dryers, a reduction in the cost for power is realized. However, the larger space requirement and greater operating inconvenience must be taken into account.

E. Drying in the Drum-Type Washing Machine

Another step forward in the treatment of laundry is taken with drum-type washing machines which, besides spinning, also take over the drying of the laundry. Besides an additional heating system, these machines require a blower for heated air and for the transport of moisture. A blower can be omitted if heat is transferred by radiation and the evaporated moisture is removed by condensation. When blowers are used, the exhaust must be conducted into the open or through a condenser, which costs more. Because

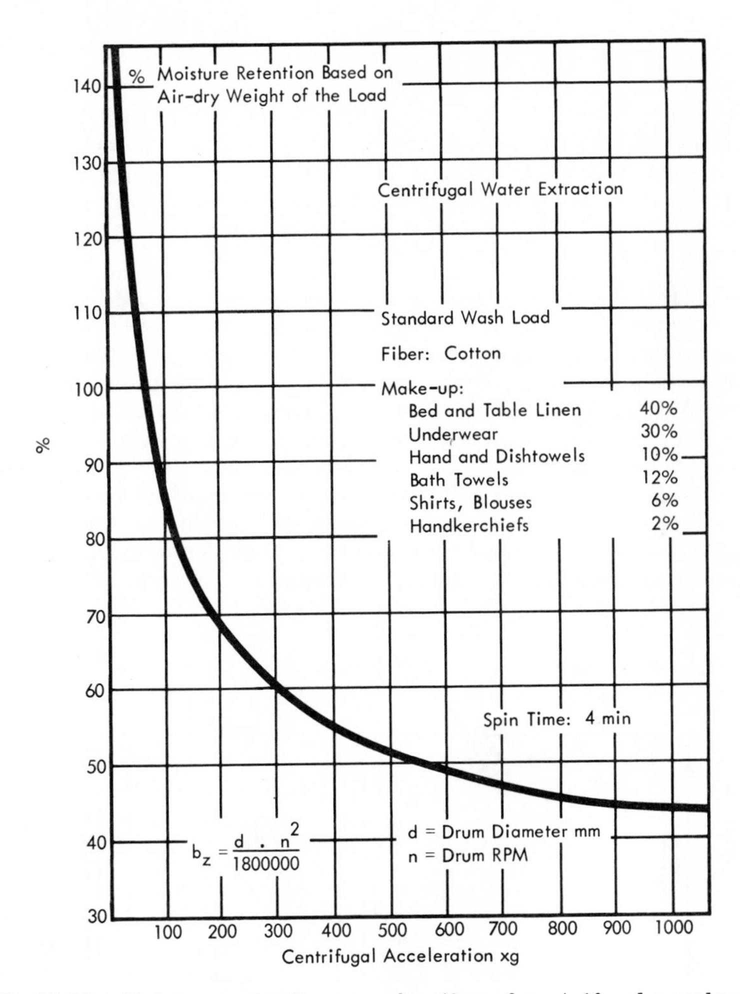

FIG. 10.21. Moisture retention as a function of centrifugal acceleration.

of the better extraction effect of separate centrifuges, washer combinations are also being used as dryers (Fig. 10.22). However, this requires reloading the laundry from the laundering drum to the spinning drum. The laundry is divided into two sections which are dried successively. This is not the case when laundering, spinning, and drying are done in one drum (Fig. 10.23). The charge ratio can amount to only about 1:30 kg/liter, namely the optimum capacity of the drum-type dryer. This means an increase in the consumption of operating materials for the laundering and also for the drying process, because more moisture must be evaporated. To compensate for this, operation is much simpler.

F. Control of the Drum-Type Washing Machine

It is almost a matter of course that the laundering process (possibly also spinning and drying) in a home washing machine is automatically controlled. This requires that this process be run off according to a certain program. The program is fixed in accordance with the type of laundry and the degree of laundry soiling; and determines the amount and temperature of the water, the amount of detergent, the intensity of the mechanical laundry treatment, and the treatment time.

The control apparatus consists of a time-dependent multiple switch (timer) which, driven by a synchronous motor, closes and opens at the desired moments the circuits which control the machine. Water quantity and temperature are limited by separate switch contacts (possibly step contacts), a level regulator, and a thermostat. The level control, temperature control, drive motor control are locked against each other in order to prevent, for example, the heat being turned on and the drum rotating without water in it. Moreover, the timer motor is disengaged as long as the predetermined temperature is not attained. With this control may simultaneously be associated the reversing rhythm of the drive motor in order to reduce the mechanical working of the laundry. This is

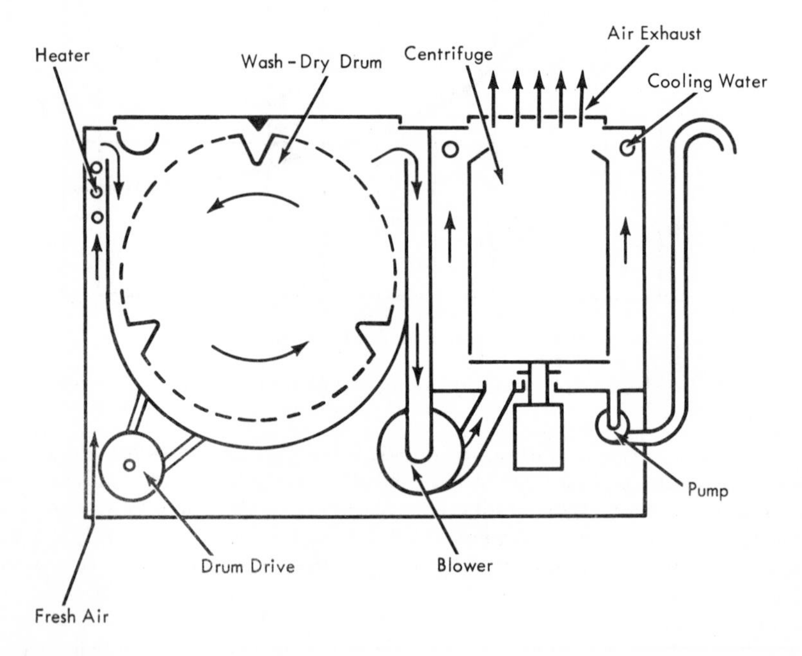

FIG. 10.22. Combination of tumbler washing machine, centrifuge (spinner), and tumbler dryer with condensation.

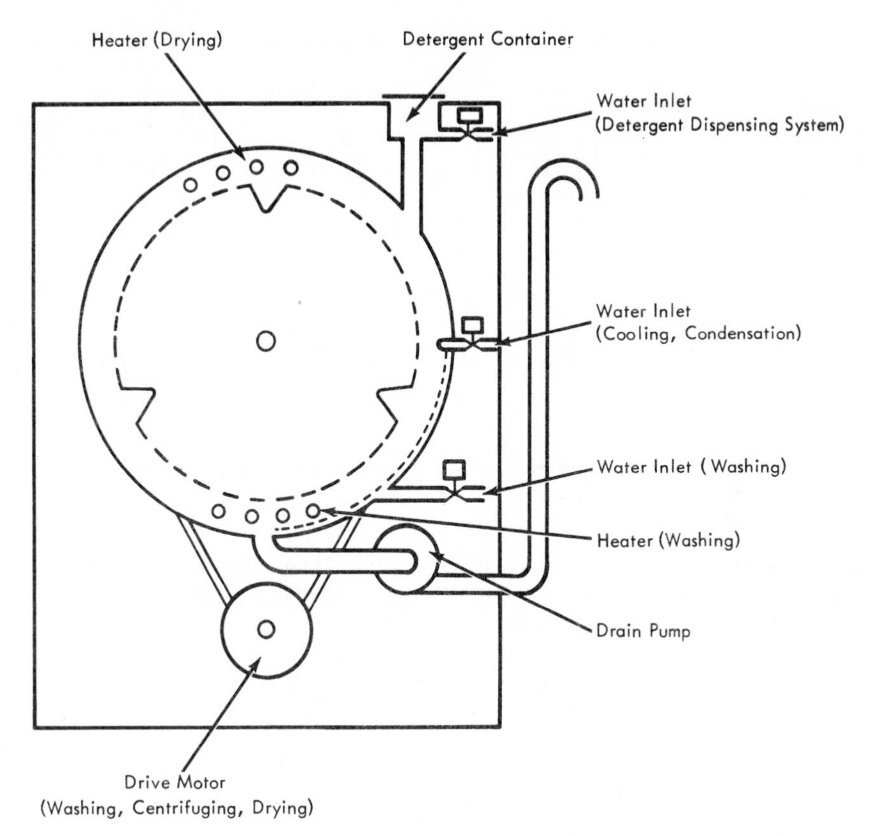

FIG. 10.23. Tumbler washing machine with centrifuging and drying means.

particularly recommended for machines with low current input values and, hence, long heatup periods.

The various switch contacts of the timer are actuated by cam disks, punched cards, or strip cards turned or advanced by the timer motor. Control systems with cam disks are laid out for one or several fixed programs. If there are several programs available, the desired program is preselected by a preselector switch (Fig. 10.24). The variations obtainable therewith are adequate for all practical application. The simple operation of this automatic machine, which consists in preselecting the program and in starting it, makes it especially suited for use in home washing machines.

Isolated cases in which punched card or strip card controls are used are found also, making it possible to run virtually any desired program

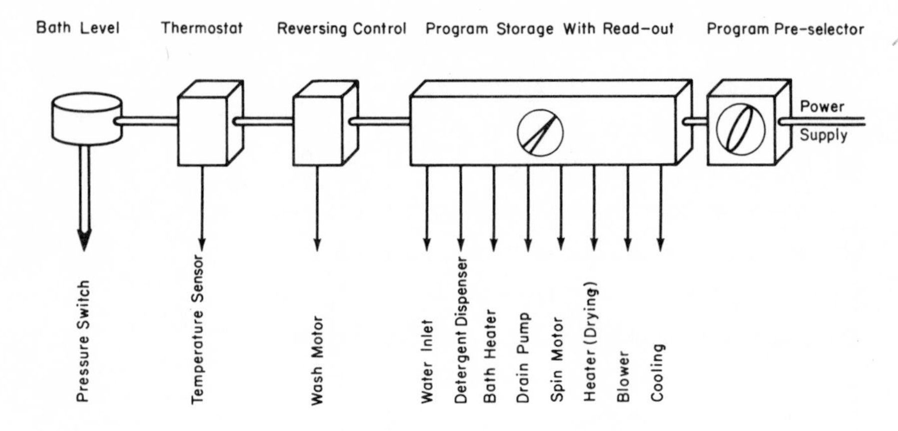

FIG. 10.24. Block schematic of the program control of an automatic washer with drying cycle.

with one control instrument by using various cards. Difficulties arising from handling and storing the cards have so far been a hurdle to the general introduction of these control systems.

VII. PARTIAL LIST OF INSTRUMENT MANUFACTURERS

1. American Instrument Co., Inc., Silver Springs, Md.
2. Baird-Atomic Inc., Cambridge, Mass.
3. Bausch and Lomb, Rochester, N.Y.
4. Beckman Instruments, Inc., Fullerton, Calif.
5. Cary Instruments, Monrovia, Calif.
6. Carl Zeiss Inc., New York, N.Y.
7. Coleman Instruments, Maywood, Ill.
8. Engineering Equipment Co., Boynton Beach, Fla.
9. Farrand Optical Co., Inc., Mt. Vernon, N.Y.
10. General Electric Co., West Lynn, Mass.
11. Gilford Instruments Labs Inc., Oberlin, Ohio
12. Hilger and Watts Ltd., London, England
13. G. K. Turner Assoc., Palo Alto, Calif.
14. Gardner Laboratory, Inc., Bethesda, Md.

15. Hunter Associates Laboratory, Inc., Fairfax, Va.

16. Instrument Development Laboratories, Attleboro, Mass.

17. Jarrell-Ash Corp., Waltham, Mass.

18. Klett Manufacturing Co., Inc., New York, N.Y.

19. Manufacturers Engineering & Equip. Corp., Warrington, Pa.

20. National Instruments Lab., Inc., Rockville, Md.

21. Perkin-Elmer Corp., Norway, Conn.

22. Phoenix Precision Inst. Co., Philadelphia, Pa.

23. Photovolt Corp., New York, N.Y.

24. Shimadzu Seisakusho Ltd., Tokyo, Japan

25. Unicam Instruments Ltd., Cambridge, England

REFERENCES

1. Maytag Encyclopedia of Home Laundry, Popular Library, New York, 1965, p. 20.

2. Technical Manual, Amer. Ass. Text. Chem. Color., 41, p. A-5 (1965).

3. W. D. Appel, W. C. Smith, and H. Christison, Amer. Dyest. Rep., 17, 679 (1928).

4. Technical Manual, Amer. Ass. Text. Chem. Color., p. 76 (1930).

5. Atlas Electric Devices Co., 4114 Ravenswood Ave., Chicago, Ill.

6. J. C. Harris, Detergency Evaluation and Testing, Wiley-Interscience, New York, 1954, p. 94.

7. A. J. Frantz and P. T. Vitale, private communication, 1967.

8. H. L. Sanders and J. M. Lambert, J. Amer. Oil Chem. Soc., 27, 153 (1950).

9. Instrument Notes, ASTM Bulletin 158, p. 43 (1949).

10. J. W. McCutcheon, Soap Sanit. Chem., 25, 83 (1949).

11. U. S. Testing Co., Inc., 1415 Park Ave., Hoboken, N.J.

12. G. M. Hartwig, 57th Annual AOCS Meeting, Philadelphia, Oct. 1966.

13. O. C. Bacon and J. E. Smith, Ind. Eng. Chem., 40, 2361 (1948).

14. J. C. Harris, J. Amer. Oil Chem. Soc., 29, 110 (1952).

15. J. Tuzson and B. Short, Text. Res. J., 30, 989 (1960).

16. American Conditioning House, Inc., 11 Melcher St., Boston, Mass.

17. Philco-Ford Corp., Tioga and C Streets, Philadelphia, Pa.

18. E. A. Knaggs, J. A. Yeager, M. Nussbaum, and E. J. Buerk, 58th Annual AOCS Meeting, New Orleans, May 1967.

19. A. C. Ludwig, Housekeeping for Young Homemakers, Bantam Books, New York, 1967.

20. T. H. Vaughn, H. R. Suter, and M. R. Kramer, J. Amer. Oil Chem. Soc., 30, 1 (1953).

21. Ametek-Laundry and Drycleaning Equipment Div., East Moline, Ill.

22. Gen. Wash Wheel Inform., (Gen. Notes #32), p. 17, 18, Am. Inst. of Laundering, Joliet, Ill.

23. H. Cohen and G. E. Linton, Chemistry and Textiles for the Laundry Industry, Textile Book Publishers, Inc., New York, 1961, pp. 190 and 278.

24. E. Allen, Power Laundry and Cleaning News, p. 961 (1964).

25. M. Saltzman, Color Eng., 1 (4) 12 (1963).

26. D. B. Judd and G. Wyszecki, Color in Business, Science, and Industry, Wiley, New York, 1963.

27. R. M. Evans, An Introduction to Color, Wiley, New York, 1948.

28. G. Wyszecki and W. S. Stiles, Color Science, Wiley, New York, 1967.

29. Brochure 7052-A, Beckman DU-2 Spectrophotometer, Beckman Instruments Inc., Fullerton, Calif.

30. Beckman Laboratory Instrumentation Catalogue, 2500-A, Beckman Instruments Inc., Fullerton, Calif.

31. M. G. Mellon, Analytical Absorption Spectroscopy, Wiley, New York, 1950.

32. H. H. Willard, L. L. Merritt, Jr., and J. A. Dean, Instrumental Methods of Analysis, 4th ed., Van Nostrand, Princeton, N.J., 1965.

33. A. C. Hardy, J. Opt. Soc. Amer., 25, 305 (1935).

34. F. W. Billmeyer, Jr., ibid., 55, 707 (1965).

35. Brochure HBK, General Electric Co., Instrum. Dep., West Lynn, Mass.

36. Bulletin 132-0966, Davidson and Hemmendinger, Easton, Pa. (now a part of Kollmorgen Corporation, Attleboro, Mass.)

37. E. L. Lewis, Color Eng., 5 (Sept/Oct. 1967).

38. Bulletin HB 520, Instrument Development Laboratories, Attleboro, Mass.

39. R. I. Section AATCC, Amer. Dyest. Rep., 55, 997 (1966).

40. Brochure, Hitachi (Model MPF-2A) Fluorescent Spectrophotometer, Hitachi Ltd., Tokyo, Japan.

41. S. D. Kullbom to W. G. Spangler, private communication, 1968.

42. R. C. Allison, Color Eng., 5, 41 (July/Aug. 1967).

43. F. W. Billmeyer, Jr., ibid., 1, 10 (1963).

44. R. S. Hunter, Can. Text. J. (April 1965).

45. R. S. Hunter, J. Opt. Soc. Amer., 50, 44 (1960).

46. R. S. Hunter, ibid., 48, 985 (1958).

47. Bulletin Model D-25, Hunter Associates Laboratory, Inc., Fairfax, Va.

48. E. Allen, Amer. Dyest. Rep., 46, 425 (1964).

49. Brochure, G. M. Fluorimeter, Engineering Equipment Co., Boynton Beach, Fla.

50. Catalogue B-298, Coleman Instrument Corp., Maywood, Ill.

51. N. Macbeth, Jr., and W. Reese, Illum. Eng., 49, 461 (1964).

52. Bulletins No. 1, 2, and 3., Macbeth Daylighting Corp., Newburgh, N.Y. (now a part of Kollmorgen Corporation, Attleboro, Mass.).

53. H. R. Condit and F. Grum, J. Opt. Soc. Amer., 54, 937 (1964).

54. S. T. Henderson and E. Hodgkiss, Brit. J. Appl. Phys. (March 1963).

55. D. B. Judd, D. L. MacAdam, and G. Wyszecki, J. Opt. Soc. Amer., 54, 1031 (1964).

56. A. H. Taylor and G. P. Kerr, ibid., 31, 3 (1941).

57. F. W. Billmeyer, Jr., Color Eng., 6, 34 (1968).

58. F. T. Simon, ibid., 5, p. 12 (July/Aug. 1967).